T0136988

Computer Communications and Networks

The **Computer Communications and Networks** series is a range of textbooks, monographs and handbooks. It sets out to provide students, researchers, and non-specialists alike with a sure grounding in current knowledge, together with comprehensible access to the latest developments in computer communications and networking.

Emphasis is placed on clear and explanatory styles that support a tutorial approach, so that even the most complex of topics is presented in a lucid and intelligible manner.

More information about this series at http://www.springer.com/series/4198

Muthu Ramachandran · Zaigham Mahmood
Editors

Software Engineering in the Era of Cloud Computing

 Springer

Editors
Muthu Ramachandran
School of Built Environment, Engineering,
and Computing
Leeds Beckett University
Leeds, UK

Zaigham Mahmood
Debesis Education
Derby, UK

Northampton University
Northampton, UK

Shijiazhuang Tiedao University
Hebei, China

ISSN 1617-7975   ISSN 2197-8433   (electronic)
Computer Communications and Networks
ISBN 978-3-030-33626-4   ISBN 978-3-030-33624-0   (eBook)
https://doi.org/10.1007/978-3-030-33624-0

To
My mother Guruvammal; my wife Vasuki;
and my daughters Virupa and Uma
—Muthu Ramachandran

To
My sisters Khalida Khanam and Irfana
Mahmood; and brothers Masood Zaigham,
Tahir Mahmood and Zahid Mahmood
—Zaigham Mahmood

Foreword

Software engineering has played a major role in the design, development, and management of all software-intensive systems for more than fifty years. Currently, service-oriented systems technologies and application environments such as Cloud Computing, Internet of Things, Fog and Edge Computing, Smart Home, Smart Cities, and Big Data are seamlessly integrated with the emergence of advancements in communication technologies. Therefore, this is a crucial moment adopting established software engineering principles and practices to service-based applications. In addition, it is beneficial to forge data science advancement to repositories of software engineering data such as PROMISE and other publicly available bug and failure data, thus creating a new era of *Software Engineering Analytics*. In this context, one of the main aims of this book is on the application of systematic approaches to the design, development, and implementation of cloud-based complex software systems and services that will allow parallelism, fast processing, integrated cloud-IoT-big data services, and real-time connectivity.

This authoritative text/reference describes the state of the art in Software Engineering in the Era of Cloud Computing (also known as cloud software engineering). A particular focus is on integrated solutions, which take into account the requirements engineering and domain modelling for cloud computing-based applications, software design methods for scalability, as well as flexibility, sustainability, and operability for distributed and cloud environments. Additionally, this book provides a discussion on software engineering analytics (a new area of interest in software engineering), software engineering best practices, as well as approaches on cloud-based testing, and software process improvement as a service. In particular, this reference text provides the following:

- Discusses the latest developments, tools, technologies, and trends in software engineering methods and techniques in the era of cloud computing.
- Reviews the relevant theoretical frameworks, practical approaches, and methodologies for cloud software engineering.

- Examines the key components of cloud software engineering processes and methods, namely cloud requirements engineering, cloud software design, cloud software development, cloud software testing, cloud software engineering analytics, and cloud software process improvement.
- Presents detailed contributions from renowned experts in the field of software engineering for distributed computing.
- Offers guidance on best practices and suggests directions for further research in distributed computing.

This illuminating volume is ideal for researchers, lecturers, and students wishing to enhance their knowledge of technologies, methods, and frameworks relevant to cloud software engineering for distributed environments. Software engineers and project managers interested in adopting the latest approaches will also find this book helpful as a practical reference. There are 13 chapters in this book and are organized in three parts:

Part I is on Cloud Requirements Engineering and Domain Modelling dedicated to providing a framework for service requirements, domain modelling approaches, and approaches to software security engineering for cloud computing.

Part II is on Cloud Design and Software Engineering Analytics with Machine Learning Approaches that presents chapters on design approaches to the design and development of cloud services and approaches on software engineering analytics with cloud-based machine learning tools.

Part III is on Cloud Testing and Software Process Improvement as a Service dedicated to providing chapters on cloud test as a service paradigm, Software Process Improvement as a Service (SPIaaS), thus providing automated testing and metrics to software industries.

In the light of the significant and fast emerging challenges that software engineers and service-oriented computing specialists face today, the editors of this book have done an outstanding job in selecting the contents of this book. In this context, I am confident that this book will provide an appreciated contribution to both software engineering, service computing, and cloud computing communities. It has the potential to become one of the main reference points for the years to come.

October 2019   Prof. Rajkumar Buyya
                                          The University of Melbourne, Melbourne, Australia

Preface

Overview

Software engineering (SE) is the application of engineering principles and technological knowledge for the development of software applications in a systematic manner. There are numerous approaches to SE, however, with the emergence of newer technologies and development platforms, and with the ever-increasing demands from the consumers for more sophisticated software applications, software engineers are now better able to build software that is much more complex, distributed, and scalable than before. Whereas, the traditional methods to building software are still valid if the user requirements are clear and well understood, newer approaches such as rapid application development (RAD), prototyping, and service-oriented software engineering (SOSE) are becoming much more attractive for various reasons including that large-scale highly complex, scalable, and distributed applications can be reasonably rapidly developed, embedding the highly desirable characteristics of functional independence, reuse, and maintainability, etc.

With the emergence of service computing vision and the cloud computing paradigm, software engineering has now moved into a new era. Although these are two different paradigms, there is much synergy between them in the sense that whereas service computing covers the whole life cycle of software applications development and provision, the cloud vision helps with the delivery and deployment of software as, e.g. the Software-as-a-Service (SaaS) and Software-as-a-Platform (SaaP).

Software engineers can combine the service and cloud computing paradigms in a SE framework to resolve some of the SE challenges, e.g. to manage the runtime quality-of-service of loosely coupled applications components (called services). Although cloud paradigm has its share of challenges, e.g. with respect to confidentiality, integrity, and security due to its multi-tenant environment, these are being sorted out with the passing of time.

With this background, although the above technologies are well developed, there still is an urgent need for even better integrated solutions to software engineering and provision, taking into account the consumer requirements of scalability at all levels, thorough flexibility and sustainability, around the clock availability, secure multi-tenancy, and operability for large-scale distributed computing applications, especially for business users. In this respect, the current text aims to extend the existing body of knowledge in the field of SE in the era of cloud computing.

This book aims to capture the state-of-the-art on the current advances in the said subject area. Majority of the contributions in this book focus on: requirements elicitation for software engineering, applications design, cloud testing, SE process improvement, and software provision. Thirty-three researchers and practitioners of international repute have presented latest research developments, methodologies, current trends, state-of-the-art reports, and suggestions for further understanding, development, and enhancement of subject area of cloud software engineering, especially for distributed computing environments.

Objectives

The aim of this volume is to present and discuss the state-of-the-art in terms of methodologies, trends, and future directions for *Software Engineering in the Era of Cloud Computing* (also known as cloud software engineering). The objectives include:

- Capturing the state-of-the-art research and practice relating to cloud software engineering and software engineering analytics with the use of data science, machine learning, and relevant processes.
- Discussing developments, tools, technologies, and trends in the subject area of cloud software engineering and software engineering analytics.
- Analysing the relevant theoretical frameworks, practical approaches, and methodologies for cloud software engineering and software engineering analytics.
- In general, advancing the understanding of the emerging new methodologies relevant to cloud software engineering and software engineering analytics.

Organization

There are 13 chapters in this book. These are organized into three parts, as follows:

- **Part I: *Cloud Requirements Engineering and Domain Modelling***. This section has a focus on approaches, research, and practices towards requirements elicitation. There are six chapters in this part. The Chap. 1 on Requirements Engineering Framework for Service and Cloud Computing (REF-SCC) discusses the use of BPMN as a method of requirement engineering in cloud business operations. The chapter also presents the requirements engineering framework for service and cloud computing (BPMN-REF-SCC). Chapter 2 presents an effective requirement engineering approach for cloud applications that examine different deployment approaches for cloud-based applications. Chapter 3 has a focus on approaches to requirements engineering for large-scale big data applications. The Chap. 4 discusses mechanisms for domain modelling and migrating from monoliths to cloud-based microservices using a large-scale banking industry case study. Chapter 5 probes further into cloud-enabled domain-based software development and the Chap. 6 in this section provides a systematic literature review of security challenges in software engineering for the cloud.

- **Part II: *Cloud Design and Software Engineering Analytics with Machine Learning Approaches*.** This part of this book comprises three chapters that focus on software design approaches with reference to cloud computing and software engineering analytics, which combine data science modelling and machine learning techniques. The Chap. 7 presents a novel software engineering framework for software defect management using machine learning techniques utilizing Microsoft Azure. The Chap. 8 illustrates an approach for sentiment analysis of twitter data; it uses machine learning techniques. The Chap. 9 contribution in this section suggests a framework for connection handlers to illustrate design pattern for recovery from connection crashes.
- **Part III: *Cloud Testing and Software Process Improvement as a Service*.** There are four chapters in this section that focus on Cloud Testing as a Service (CTaaS) and Software Process Improvement as a Service (SPIaaS). The Chap. 10 provides an analysis of approaches and techniques, considering a modern perspective on cloud testing ecosystems. The Chap. 11 contribution in this part addresses an approach Towards Green Software Testing in Agile and DevOps Using Cloud Virtualization for Environmental Protection. The Chap. 12 presents a novel technique for Machine Learning as a Service for Software Process Improvement (SPIaaS) which autonomically collects SPI data and performs analytics for process improvement. The Chap. 13 contribution of this book presents a set of methods on comparison of data mining techniques in the cloud for the software engineering perspective.

Target Audiences

The current volume is a reference text aimed at supporting a number of potential audiences, including the following:

- *Cloud Software Engineers, Cloud Service Providers and Consumers, Software Engineers*, and *Project Managers* who wish to adopt the newer approaches to ensure the accurate and complete system specifications.

- *Students and lecturers* who have an interest in further enhancing the knowledge of technologies, mechanisms, and frameworks relevant to cloud software engineering.
- *Researchers* in this field who require up to date knowledge of the current practices, mechanisms, and frameworks relevant to cloud software engineering.

Leeds, UK   Muthu Ramachandran
Northampton, UK/Hebei, China   Zaigham Mahmood

Acknowledgements

The editors acknowledge the help and support of the following colleagues during the review, development, and editing phases of this text:

- Dr. S. Parthasarathy, Thiagarajar College of Engineering, Tamil Nadu, India
- Dr. Pethuru Raj, IBM Cloud Center of Excellence, Bangalore, India
- Prof. Andrea Zisman, Open University
- Prof. Bashar Nuseibeh, Open University
- Prof. T. R. G. Nair, Rajarajeswari College of Engineering, India
- Prof. Zhengxu Zhao, Shijiazhuang Tiedao University, Hebei, China
- Dr. Alfredo Cuzzocrea, University of Trieste, Trieste, Italy
- Dr. Emre Erturk, Eastern Institute of Technology, New Zealand.

We would also like to thank the contributors to this book: all authors and co-authors, from academia as well as industry from around the world, who collectively submitted thirteen chapters. Without their efforts in developing quality contributions, conforming to the guidelines and meeting often the strict deadlines, this text would not have been possible.

Leeds, UK   Muthu Ramachandran
Northampton, UK/Hebei, China   Zaigham Mahmood
August 2019

Other Books by the Editors

By Muthu Ramachandran

Strategic Engineering for Cloud Computing and Big Data Analytics

This reference text demonstrates the use of a wide range of strategic engineering concepts, theories, and applied case studies to improve the safety, security, and sustainability of complex and large-scale engineering and computer systems. It first details the concepts of system design, life cycle, impact assessment, and security to show how these ideas can be brought to bear on the modelling, analysis, and design of information systems with a focused view on cloud computing systems and big data analytics. ISBN: 978-3-319-52490-0.

Requirements Engineering for Service and Cloud Computing

This text aims to present and discuss the state-of-the-art in terms of methodologies, trends, and future directions for requirements engineering for the service and cloud computing paradigm. Majority of the contributions in this book focus on requirements elicitation; requirements specifications; requirements classification; and requirements validation and evaluation. ISBN: 978-3-319-51309-6.

Enterprise Security

This reference text on Enterprise Security is a collection of selected best papers presented at the ES 2015 International workshop. Enterprise security an important area since all types of organizations require secure and robust environments, platforms and services to work with people, data, and computing applications. This book provides selected papers of the Second International Workshop on Enterprise Security held in Vancouver, Canada, 30 November–3 December 2016 in conjunction with CloudCom 2015. The 11 papers were selected from 24 submissions and provided comprehensive research into various areas of enterprise security such as protection of data, privacy and rights, data ownership, trust, unauthorized access and big data ownership, studies and analysis to reduce risks imposed by data leakage, hacking, and challenges of cloud forensics. ISBN: 978-3-319-54379-6.

By Zaigham Mahmood

The Internet of Things in the Industrial Sector: Security and Device Connectivity, Smart Environments, and Industry 4.0
This reference text has a focus on the development and deployment of the industrial Internet of things (IIoT) paradigm, discussing frameworks, methodologies, benefits, and inherent limitations of connected smart environments, as well as providing case studies of employing the IoT vision in the industrial domain. ISBN: 978-3-030-24891-8.

Security, Privacy, and Trust in the IoT Environment
This book has a focus on security and privacy in the Internet of things environments. It also discusses the aspects of user trust with respect to device connectivity. Main topics covered include: principles, underlying technologies, security issues, mechanisms for trust and authentication as well as success indicators, performance metrics, and future directions. ISBN: 978-3-030-18074-4.

Guide to Ambient Intelligence in the IoT Environment: Principles, Technologies, and Applications
This reference text discusses the AmI element of the IoT paradigm and reviews the current developments, underlying technologies, and case scenarios relating to AmI-based IoT environments. This book presents cutting-edge research, frameworks, and methodologies on device connectivity, communication protocols, and other aspects relating to the AmI-IoT vision. ISBN: 978-3-030-04172-4.

Fog Computing: Concepts, Frameworks, and Technologies
This reference text describes the state-of-the-art of Fog and Edge computing with a particular focus on development approaches, architectural mechanisms, related technologies, and measurement metrics for building smart adaptable environments. The coverage also includes topics such as device connectivity, security, interoperability, and communication methods. ISBN: 978-3-319-94889-8.

Smart Cities: Development and Governance Frameworks
This text/reference investigates the state-of-the-art in approaches to building, monitoring, managing, and governing smart city environments. A particular focus is placed on the distributed computing environments within the infrastructure of smart cities and smarter living, including issues of device connectivity, communication, security, and interoperability. ISBN: 978-3-319-76668-3.

Data Science and Big Data Computing: Frameworks and Methodologies
This reference text has a focus on data science and provides practical guidance on big data analytics. Expert perspectives are provided by an authoritative collection of 36 researchers and practitioners, discussing latest developments and emerging trends; presenting frameworks and innovative methodologies; and suggesting best practices for efficient and effective data analytics. ISBN: 978-3-319-31859-2.

Connected Environments for the Internet of Things: Challenges and Solutions
This comprehensive reference presents a broad-ranging overview of device connectivity in distributed computing environments, supporting the vision of IoT. Expert perspectives are provided, covering issues of communication, security, privacy, interoperability, networking, access control, and authentication. Corporate analysis is also offered via several case studies. ISBN: 978-3-319-70102-8.

Connectivity Frameworks for Smart Devices: The Internet of Things from a Distributed Computing Perspective
This is an authoritative reference that focuses on the latest developments in the Internet of things. It presents state-of-the-art on the current advances in the connectivity of diverse devices; and focuses on the communication, security, privacy, access control, and authentication aspects of the device connectivity in distributed environments. ISBN: 978-3-319-33122-5.

Cloud Computing: Methods and Practical Approaches
The benefits associated with cloud computing are enormous; yet, the dynamic, virtualized, and multi-tenant nature of the cloud environment presents many challenges. To help tackle these, this volume provides illuminating viewpoints and case studies to present current research and best practices on approaches and technologies for the emerging cloud paradigm. ISBN: 978-1-4471-5106-7.

Cloud Computing: Challenges, Limitations, and R&D Solutions
This reference text reviews the challenging issues that present barriers to greater implementation of the cloud computing paradigm, together with the latest research into developing potential solutions. This book presents case studies, and analysis of the implications of the cloud paradigm, from a diverse selection of researchers and practitioners of international repute. ISBN: 978-3-319-10529-1.

Continued Rise of the Cloud: Advances and Trends in Cloud Computing
This reference volume presents the latest research and trends in cloud-related technologies, infrastructure, and architecture. Contributed by expert researchers and practitioners in the field, this book presents discussions on current advances and practical approaches including guidance and case studies on the provision of cloud-based services and frameworks. ISBN: 978-1-4471-6451-7.

Software Engineering Frameworks for the Cloud Computing Paradigm
This is an authoritative reference that presents the latest research on software development approaches suitable for distributed computing environments. Contributed by researchers and practitioners of international repute, this book offers practical guidance on enterprise-wide software deployment in the cloud environment. Case studies are also presented. ISBN: 978-1-4471-5030-5.

Cloud Computing for Enterprise Architectures
This reference text, aimed at system architects and business managers, examines the cloud paradigm from the perspective of enterprise architectures. It introduces fundamental concepts, discusses principles, and explores frameworks for the

adoption of cloud computing. This book explores the inherent challenges and presents future directions for further research. ISBN: 978-1-4471-2235-7.

Cloud Computing: Concepts, Technology, and Architecture
This is a textbook (in English but also translated in Chinese and Korean) highly recommended for adoption for university-level courses in distributed computing. It offers a detailed explanation of cloud computing concepts, architectures, frameworks, models, mechanisms, and technologies—highly suitable for both newcomers and experts. ISBN: 978-0133387520.

Software Project Management for Distributed Computing: Life-Cycle Methods for Developing Scalable and Reliable Tools
This unique volume explores cutting-edge management approaches to developing complex software that is efficient, scalable, sustainable, and suitable for distributed environments. Emphasis is on the use of the latest software technologies and frameworks for life-cycle methods, including design, implementation, and testing stages of software development. ISBN: 978-3-319-54324-6.

Requirements Engineering for Service and Cloud Computing
This text aims to present and discuss the state-of-the-art in terms of methodologies, trends, and future directions for requirements engineering for the service and cloud computing paradigm. Majority of the contributions in this book focus on requirements elicitation; requirements specifications; requirements classification; and requirements validation and evaluation. ISBN: 978-3-319-51309-6.

User-Centric E-Government: Challenges and Opportunities
This text presents a citizen-focused approach to the development and implementation of electronic government. The focus is twofold: discussion on challenges of service availability, e-service operability on diverse smart devices; as well as on opportunities for the provision of open, responsive and transparent functioning of world governments. ISBN: 978-3-319-59441-5.

Cloud Computing Technologies for Connected Government
This text reports the latest research on electronic government for enhancing the transparency of public institutions. It covers a broad scope of topics including citizen empowerment, collaborative public services, communication through social media, cost benefits of the Cloud paradigm, electronic voting systems, identity management, and legal issues. ISBN: 978-1466-686298.

Human Factors in Software Development and Design
This reference text brings together high-quality research on the influence and impact of ordinary people on the software industry. With the goal of improving the quality and usability of computer technologies, topics include global software development, multi-agent systems, public administration Platforms, socio-economic factors, and user-centric design. ISBN: 978-1466-664852.

IT in the Public Sphere: Applications in Administration, Government, Politics, and Planning
This reference text evaluates current research and best practices in the adoption of e-government technologies in developed and developing countries, enabling governments to keep in touch with citizens and corporations in modern societies. Topics covered include citizen participation, digital technologies, globalization, strategic management, and urban development. ISBN: 978-1466-647190.

Emerging Mobile and Web 2.0 Technologies for Connected E-Government
This reference highlights the emerging mobile and communication technologies, including social media, deployed by governments for use by citizens. It presents a reference source for researchers, practitioners, students, and managers interested in the application of recent technological innovations to develop an open, transparent and more effective e-government environment. ISBN: 978-1466-660823.

E-Government Implementation and Practice in Developing Countries
This volume presents research on current undertakings by developing countries towards the design, development, and implementation of e-government policies. It proposes frameworks and strategies for the benefits of project managers, government officials, researchers, and practitioners involved in the development and implementation of e-government planning. ISBN: 978-1466-640900.

Developing E-Government Projects: Frameworks and Methodologies
This text presents frameworks and methodologies for strategies for the design, implementation of e-government projects. It illustrates the best practices for successful adoption of e-government and thus becomes essential for policy makers, practitioners, and researchers for the successful deployment of e-government planning and projects. ISBN: 978-1466-642454.

Contents

About the Editors

Dr. Muthu Ramachandran is a Principal Lecturer in the School of Built Environment, Engineering, and Computing at Leeds Beckett University in the UK. Previously, he spent nearly eight years in industrial research (Philips Research Labs and Volantis Systems Ltd, Surrey, UK) where he worked on software architecture, reuse, and testing. His first career started as a research scientist where he worked on real-time systems development projects. Muthu is an author/editor of several books including: *Software Components: Guidelines and Applications* (Nova Publishers 2008) and *Software Security Engineering: Design and Applications* (Nova Publishers 2011). He has also widely authored and published 9 books, over 100 journal articles, over 50 book chapters, and over 200 conferences papers on various advanced topics in software engineering, software security, cloud computing, and education. Muthu has led numerous conferences as chair and as keynote speakers on global safety, security and sustainability, emerging services, IoT, big data, and software engineering. Muthu is a member of various professional organizations and computer societies, e.g. IEEE, ACM, BCS (as Fellow), and HEA (as Senior Fellow). He has also been an invited keynote speaker on several international conferences. Muthu's research projects have included all aspects of software engineering, SPI for SMEs (known as Prism model), emergency and disaster management systems, software components and architectures, good practice guidelines on software developments, software security engineering, and service and cloud computing. Projects details can be accessed at the following sites:

- http://www.leedsbeckett.ac.uk/staff/dr-muthu-ramachandran/
- https://www.scopus.com/authid/detail.uri?authorId=8676632200
- https://scholar.google.co.uk/citations?user=KDXE-G8AAAAJ&hl=en
- https://www.amazon.co.uk/l/B001JP7SAK?_encoding=UTF8&redirectedFrom KindleDbs=true&rfkd=1&shoppingPortalEnabled=true
- https://www.linkedin.com/in/muthuuk/?originalSubdomain=uk
- https://twitter.com/muthuuk
- https://github.com/Muthuuk

Muthu can be reached at m.ramachandran@leedsbeckett.ac.uk and se4cloud-computing@gmail.com.

Prof. Dr. Zaigham Mahmood is a published author/editor of twenty-eight books on subjects including electronic government, cloud computing, data science, big data, fog computing, Internet of things, Internet of vehicles, industrial IoT, smart cities, ambient intelligence, project management, and software engineering, including: *Cloud Computing: Concepts, Technology & Architecture* which is also published in Korean and Chinese languages. Additionally, he is developing two new books to appear later in the year. He has also published more than 100 articles and book chapters and organized numerous conference tracks and workshops. Professor Mahmood is the Editor-in-Chief of *Journal of E-Government Studies and Best Practices* as well as Series Editor-in-Chief of the IGI book series on *E-Government and Digital Divide*. He is a Senior Technology Consultant at Debesis Education UK and Professor at the Shijiazhuang Tiedao University in Hebei, China. He further holds positions as Foreign Professor at NUST and IIU in Islamabad Pakistan. He has also served as a Reader (Associated Professor) at the University of Derby, UK, and Professor Extraordinaire at the North-West University, South Africa. Professor Mahmood is a certified cloud computing instructor and a regular speaker at international conferences devoted to cloud computing, distributed computing, and e-government. Professor Mahmood's book publications can be viewed at: https://www.amazon.co.uk/Zaigham-Mahmood/e/B00B29OIK6.

Part I
Cloud Requirements Engineering and Domain Modelling

Chapter 1
Requirements Engineering Framework for Service and Cloud Computing (REF-SCC)

Krishan Chand and Muthu Ramachandran

Abstract Requirements engineering (RE) is the most difficult and important stage of any business process or project development. This research endeavors to find out the characteristics and aspects of requirements engineering enforced by cloud computing. Business Process Modeling Notation (BPMN) has made an impact in the respect to capture the process and to make the changes accordingly for improvement in business operations. This chapter defines how BPMN can be used as a method of requirements engineering in cloud business operations. Furthermore, this chapter presents the requirements engineering framework for service and cloud computing (BPMN-REF-SCC) and will also discuss the reference architecture for service and cloud computing. Finally, the research delivers a case of financial cloud business, which has developed 15 hypotheses for the validation and evaluation through simulation.

Keywords Cloud computing · Business process modeling (BPM) and Business Process Modeling Notation (BPMN) · Requirements engineering framework (REF) · Service and cloud computing (SCC)

1.1 Introduction

While dealing with the cloud computing, the main problem is no one knows where the data has been saved and who can access the data; hence, software processes become more complex which is directly impacting the requirements engineering processes. Traditional software providers were not worried about the issues such as monitoring, evaluating performance, scalability, customization, and other concerns,

K. Chand (✉) · M. Ramachandran
School of Built Environment, Engineering, and Computing, Leeds Beckett University, Leeds
LS6 3QS, UK
e-mail: K.Chand7596@Student.Leedsbeckett.ac.uk

M. Ramachandran
e-mail: M.Ramachandran@leedsbeckett.ac.uk

© Springer Nature Switzerland AG 2020
M. Ramachandran and Z. Mahmood (eds.), *Software Engineering in the Era of Cloud Computing*, Computer Communications and Networks,
https://doi.org/10.1007/978-3-030-33624-0_1

though cloud providers need to address these non-functional application concerns which are quite essential for the success of cloud computing services.

For many years, researchers have been working in the field of cloud computing and requirements engineering. Some of them have also worked on the software process improvement areas but not discussed any specific design or framework for cloud computing requirements engineering process. Software engineering in cloud environment includes some major challenges such as software composition, query-oriented vs application programming interface (API)-oriented programming, availability of source code, execution model, and application management. In the respect to take advantage and to make cloud computing more useful, these challenges need to be addressed in the different software engineering processes and methodologies [1].

Some of the researchers have already tried prevailing tools, languages, and other methodologies in the cloud computing environment while considering requirement engineering methodologies which are typically focused on the object-oriented outcomes and service-oriented tools. The main problem in the cloud computing is the lack of standard, which can help to encounter the main objectives which are wrapping a different characteristic of cloud computing [2].

According to the research of Todoran et al., approaches and methodologies have been proposed. However, there was no practical suggestion made on the elicitation process to utilize by the cloud providers. In addition, another research of Respschlaeger et al. illustrated a framework which includes the evaluation benchmarks for the adoption of the cloud services. Furthermore, a framework has been deliberated by Schrödl and Wind for the validation of the traditional requirements engineering process with regard to implementation in the cloud. It has been also concluded that none of the collective models is appropriate to justify the requirements engineering under the cloud environment [3–5].

One of the researches of Guha, cloud computing, is the most modernized buildup for the IT business. However, there are a lot of challenges in the software engineering of a cloud computing platform which are not been examined yet. Moreover, no recommended cloud computing platform has been introduced. Furthermore, the cloud provider should also include the stakeholders in each stage or cycle of the extended programming methodologies of cloud environment [6].

BPMN helps researchers to create ideas and provide a platform to do the best research and development. Moreover, best practices and graphical presentation make it easily understandable. BPMN has different sub-processes and stages, which identifies the problem to make the best possible changes to improve the performance of a business process. Therefore, the research will use the BPMN as the requirements engineering process to collect the requirements for service and cloud business. This chapter is a detailed explanation of how BPMN can be important in the process of requirements engineering method. It can be seen in the previous research that investing less time and money in the requirements engineering process leads to project or product failure. BPMN plays a big role at the time of incorporating requirement engineering process to save a project to be failed.

This chapter will explain the importance of BPMN as a requirements engineering method for service and cloud computing in this research. Section 1.2 details briefly about the business process modeling and also explains the processes and sub-processes of BPMN and how these processes can be used as requirements engineering method. Section 1.3 provides a detailed explanation of the BPMN requirements engineering life cycle for service and cloud computing. Section 1.4 shows how BPMN works with other entities of the organization for service and cloud computing.With the help of above all sections, Sect. 1.5 introduces a unique requirements engineering framework for service and cloud computing (REF-SCC). Section 1.6 details a reference architecture diagram for REF-SCC. Finally, Sect. 1.7 depicts a requirements engineering framework for service and cloud computing (BPMN-REF-SCC) through experimental validation of a real cloud application of a credit card.

1.1.1 Business Process Modeling

The aim and objectives of an organization are accomplished by carrying out business maneuvers in a precise way, and this specific way is known as business processes. Business processes can be identified to fulfill the customer need and as per the detailed activities carried out by an enterprise to produce a product. Business processes are the main components of an organization and have a direct impact on the business performance and quality of the product for the success of the organization. From management to improvement of the processes, so many methodologies have been introduced to address the multiple characteristics of business processes.

As per the definition given by the Harrington, "a business process consists of a group of logically related tasks that use the resources of an organization to provide defined results in support of the organization's objectives" [7]. Resources of the organizations and the task-related are the key elements to achieving the business objective. Effective utilization of the resources and task structure is very significant for the time cost and the quality of the product and for the organization as well.

Business process modeling (BPM) is the process of collection of tools and methods to get an in-depth understanding of a business process to manage and improve the performance of an organization. Business process modeling is the activity of demonstrating the internal procedures of the business to find out the current situation in order to improve in the future. There are different graphical models available for the business process management like flowcharts and Unified Modeling Language (UML) diagrams.

Due to its existence importance and descriptive nature of the process, the characteristics representation for the activities such as business process improvement, business process re-engineering and process standardization, business process modeling is the first stage to success the organizational targets or objectives [8].

The communication of ideas is very important for business and stakeholders. Numerous techniques are available for the communication purpose such as documentary description and graphical representation. Graphical techniques used charts, diagrams, pictures, etc., for communication and exploration. As it is relating to pictorial art, it provides a spontaneous understanding of the ideas or concepts. The concept of addressing the problems related to business management operations in the graphic flora is known as business process modeling. Business process modeling helps the stakeholders and business operations to design and understand the business process and subsequently follow the analysis and improvement process until implementation [9].

Modeling and simulation are processes to reduce the complexity of the real-world business process. The main aim of the business process modeling and simulation is to review the complexity of a process directed to make it with fewer efforts, accordingly to ease the complexity of the business process and to make it simple and understanding. However, the main objective of the process modeler is to make the process understanding, to reduce the complexity in the practical world, and to design the complex models [10, 11].

Every single element or aspect of the business process needs to evaluate as these are used as a tool to control and advance the process. Numerous methods are used for the business process evaluation in the field of computer science. The main emphasis of computer science is to provide the support to carry out the business operations, database storage, computational methods and their other corresponding methods for graphical communications. Different stakeholders are involved in the organization in the numerous levels who evaluate the performance of the business process. According to the research of Lodhi, Koppen, and Sakke (2013), the different stakeholders and their involvement in the business process have been discussed. Executives provide the abstract-level evaluation such as profit and loss, and these figures are described in the graphical form and the textual descriptions. Managers are involved in the evaluation of low-level processes with more details of the business activities and the resources and also make some future projections [12].

Usually, the performance is evaluated in the quantifiable amounts which provide help to designate the quality of the process. Different techniques and methods are used for the business process evaluation and the processes, and its elements are evaluated in the aspect of time, cost, and quality. In the respect to get accurate and to get the real advantage of the evaluation process, it is necessary to involve all the participants of the organizations. In the evaluation process, it is important to have the full picture of the processes from the abstract level to a low level and it is also important to evaluate the overall impact of making any changes in the process.

1.1.2 Traditional RE Method

Requirements engineering includes the set of activities to discover, validate, elicit, analyze, document, and maintain the group of requirements for the desired process or system [13].

The main objective of requirements engineering is to discover the requirements of a business or product, which can provide quality and can be implemented into a business effectively. Requirements engineering is a crucial task that can impact on the current business activities. Requirements engineering is used as the most powerful tool for gathering the requirements of a business process with due respect to analyzing and documenting the requirement of a process [14].

Figure 1.1 shows a landscape of requirements engineering (RE) techniques and process. The RE process consists of main elicitation, modeling, verification, and validation activities. This paper mainly focuses on modeling requirements with BPMN process diagrams which allows us to elicit, develop requirements models, and validate the models with BPMN simulations. Therefore, it forms an effective RE tool for eliciting cloud requirements and can also build UML design and generate services.

Fig. 1.1 Traditional RE method

The first and important process of RE is elicitation, which consists of sub-processes such as process modeling, document analyses, interviews, observations, and brainstorming. After that, requirements specification is the next process, where requirements need to specify according to business goals and requirements.

After specification of the business requirements, all the business stakeholders and managements decide if there are any changes required in current specified requirements. Modeling and verification are the next processes followed by validation process to complete the requirements engineering process. Validation is the final process, where direct meeting of all management and staff validates the process.

The next section describes the BPMN process, and its sub-processes and BPMN can be used as RE method.

1.2 BPMN as Requirements Engineering Method

This section is the detailed explanation of how BPMN can be important in the process of requirements engineering method. It can be seen in the previous research that investing less time and money in requirements engineering process leads to project or product failure. BPMN plays a big role here to save a project to be failed. This section will explain the importance of BPMN in this research.

Figure 1.2 shows the different process stages of the business process modeling technique. The process starts with the assessment to identify the problem, which leads the process to design and simulate and execution process to get the results and make improvement in current business operations. And the final task is to validate and test the process as the need to neglect an uncertain task. All these different processes will be explained in the next subsections.

1.2.1 Assessment

Figure 1.3 shows the first step of the business process to achieve the maximum of the current state of the business. Interviewing people working within the organization will provide the problems associated with the current process. While interviewing, the observation method can also help to investigate the loopholes. Additionally, another way is the feedback from customers can tell the story of problems with due respect to the customer.

- Interview people working within the organization to achieve the current state.
- Observation is the best way to find out the problems and to make the decision to resolve.
- Get feedback from the customers about their experiences.

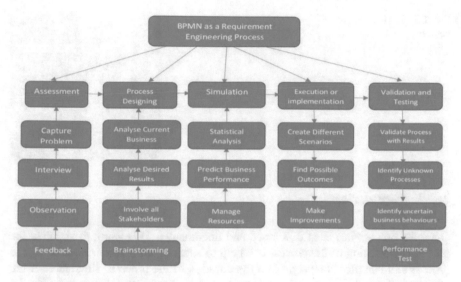

Fig. 1.2 BPMN as a requirements engineering method

Fig. 1.3 Business process
assessment

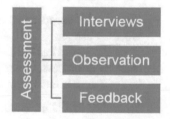

1.2.2   Process Design

Figure 1.4 describes the next stage; once the process is captured and the problem is
identified in the process, then the new business process can be developed or
designed. Keep in mind the current business process, and affirm the changes in the
process will get the desired results. Again, engaging all the major participants can
help in considering the different ideas and suggestions for process improvement.
Moreover, do not respite on one result, and brainstorming method can be used to
get multiple solutions to consider the best.

- Investigate the existing problem in the process before designing and making
  changes.
- Evaluate the results of changes in the process.
- Involve all major participants to get the exact current state of the business.
- Do not rely on one solution, and brainstorming is the process to get a different
  solution to one problem.

Fig. 1.4 Service design
process

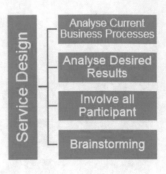

1.2.3 Simulation

Once the process has been developed and documented, it is worth to simulate the
process. Simulation of the process can help to categorize the resources used by the
process and can provide insight into the duration of the process. Simulation of the
process will provide you with the performance level, but also gives you the
opportunity to validate the existing process without affecting the current business
maneuvers (Fig. 1.5).

- Create a model to summarize how the collected information relates to your
  process changes.
- Once the model has been created and validated, predict the results after making
  changes in process, either positive or negative.
- Use different scenarios of resource utilization to make an improvement.

According to Naim, simulation involves a series of processes for building a
computerized model so that particular results can be achieved through the observation of the model. Simulation process includes assumption making and param-
eterization [15].

Fig. 1.5 Business process
simulation

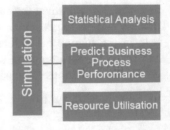

Fig. 1.6 Business process
execution

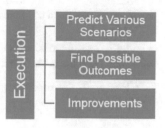

1.2.4   Execution

After the simulation process, the next task is to execute the process. In this task, the analyst can predict different scenarios as per the requirements of the business entities and cloud providers. Creating different scenarios can help to find different outcomes to choose one to make further improvements in the business process (Fig. 1.6).

- Predict different situations to ensure improvement in process changes.
- Different scenarios will help to find multiple possible outcomes.
- Experiments will allow making possible changes to improve business performance.

1.2.5   Validation and Testing

After completing all the processes, the last task is to validate and test the performance again. First thing in this task is to remove the unknown process which is making process complex and further ahead to find an unclear activity which needs to be removed (Fig. 1.7).

- Validate all the processes with results with due respect to the objective of the organization.

Fig. 1.7 Business process
validation and testing

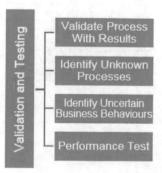

- Identify unknown processes which make process complex and need to be removed.
- Identify unclear activities which are not necessarily required.
- Finally, test the performance again to verify the process.

Clearly, all the processes will provide flow automation to a business process.

1.2.6 Tools for BPMN

There are so many tools available to design and model for the business processes such as Camunda, Bizagi Modeler, Bonita BPMN, BPMN modeler, Visual Studio, and numerous others. After consideration of few most important features, researcher has decided to use the Bizagi Modeler. All other BPMN tools have different features which are similar to Bizagi Modeler. The most important feature in Bizagi is this tool can design, execute, simulate, and publish on Web as well. Below table is the comparison of Bizagi with other simulation tools.

	Camunda	Bizagi BPMN	Bonita BPM	BPMN 2 Modeler
Features	Reporting dashboard App integration Live editing Case management Workflow management Automation multi-tenancy	Design process maps Build process applications Publish high-quality documentation Innovate drag and drop interface Based 100% on BPMN Notation Modeler collaboration services Simulate your process Publish process Multi-language	Model process Connect Customize Adapt easily Scalable	Graphical modeling Intuitive user interface Extendibility mechanism Eclipse-based
License	Open source	Proprietary	Proprietary	Proprietary
Price	Free	Free	Free	Free
Free trial	Available	Available	Available	Available
Company	Camunda	Bizagi BPMN Modeler	Bonita BPM	BPMN 2 Modeler
Platform	Windows, Linux, Mac	Windows	Windows, Linux, Mac	Windows, Linux, Mac

1.2.7 Traditional RE Versus BPMN RE

Figure 1.8 shows the long-drawn-out of traditional RE and BPMN as RE. The process starts with collecting business requirements in both types of REs, and then the next step is elicitation, where most of the subtasks are same such as documentation, analyzing current business, analyzing results, interviews, observation, and brainstorming. However, in traditional RE interviews, observations, and brainstorming are easy to complete as all employees, stakeholders, and management team in-house to contact each other. On the other hand, BPMN RE makes it difficult as this only includes cloud providers and such organization who is providing services to the third party (users), and there is no direct connection of the users. The next step in traditional RE is requirements analysis of a product or software and in BPMN its service requirements analysis. This is the main phase for BPMN as this decides what is to be used and how much it should be used to predict business performance. In other words, this phase is all about managing the resources to achieve maximum performance with due respect to cost and time.

In traditional RE, the next step is requirements change management, where they organize a meeting to make some changes to make the product better if required. And in BPMN it is called service designing and simulation. This phase designs a service framework as per the requirements of the business and cloud provider to achieve desired results with enhanced performance. Afterward, execution and implementation are the tasks for BPMN RE, which consists of creating different scenarios, making changes in business process to find out different outcomes and to decide which will be best to implement and deploy the services into the cloud.

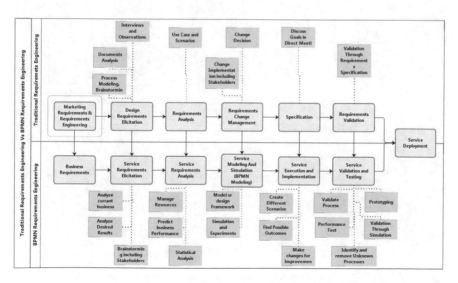

Fig. 1.8 Traditional RE versus BPMN RE

1.3   BPMN Requirements Engineering Life Cycle
for Service and Cloud Computing
(BPMN-RELC-SCC)

Traditional requirements engineering has a different process to follow to complete a product or system. This research has also defined a requirements engineering life cycle for service and cloud computing. The process starts with the service requirements elicitation, where all stakeholders and other business entities work together to decide the basic requirements of the current business to get desired results with the help of interviewing, observing, and brainstorming other employees of the organization. Next process is to specify the process requirements to achieve desired goals (Fig. 1.9).

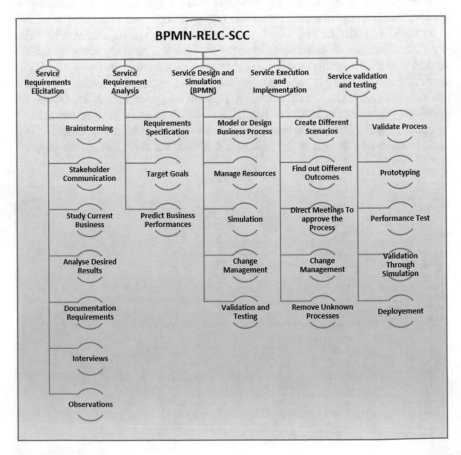

Fig. 1.9 BPMN-RELC-SCC requirements engineering life cycle processes for service and cloud computing

After elicitation and specifications of the requirements, the next task is to design the service and simulate the service to validate the process with the help of Business Process Modeling Notation (BPMN). After designing of the services, next comes the execution of the plans as per designed service. Execution of the services can help us to provide different outcomes to decide the best for further development. Finally, a process needs to be validated before deployment, which needs to be done by the BPMN simulation. All these processes of a BPMN-RELC-SCC will be explained in next subsections.

1.3.1   Service Requirements Elicitation

First and the main task of RELC-SCC is service requirements elicitation. This process consists of some sub-process such as study current business, analysis of desired results, document requirements, interviews, observations, brainstorming, and stakeholders' communications which are also depicted in the below diagram. In this task, the analyst needs to see what the current business situation is and how can we get the desired results. Brainstorming while communicating with stakeholders and interviewing all the employees will give you the depth knowledge of the current business, and keeping observation of all the processes can help to get desired results (Fig. 1.10).

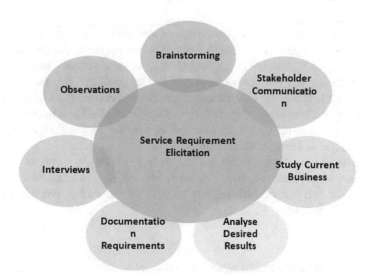

Fig. 1.10 Service requirements elicitation

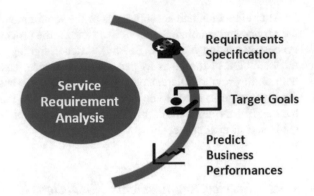

1.3.2 Service Requirements Analysis

Next process is to analyze the service requirements, which accomplish what resources required and how much required for the business. Resources managed in this task are directly connected with the goals targeted to predict business performances. To increase the business performance, analyst needs to predict different multiple business requirements (Fig. 1.11).

1.3.3 Service Design and Simulation (BPMN)

This is the main process of the RELC-SCC where BPMN RE plays an important role while designing and simulating the business process for validation and testing. An analyst can use the BPMN to design the business process and further can simulate it to analyze the performance before the deployment. First task of the BPMN service is to design the process as per the requirements of business stakeholders and cloud provider. After designing the process, business needs to manage the resources. Resources need to manage as per the business structure, cloud architecture, and business requirements.

After managing the resources, next task is to simulate the business process to analyze the performance. Multiple scenarios need to be generated to achieve the desired results and exceptional need to be selected for further enhancement and deployment (Fig. 1.12).

1.3.4 Service Execution and Implementation

Service execution and implementation is the next task after BPMN service design and simulation. After selecting a business process to deploy as a service, this needs

Fig. 1.12 Service design and simulation (BPMN)

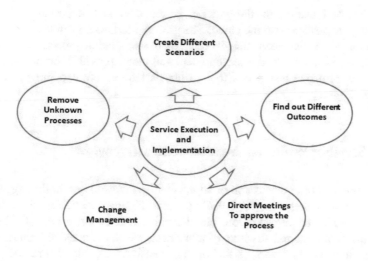

Fig. 1.13 Service execution and implementation

to be executed and implemented to decide if it needs to change or to delete any specific task which is not required. In this task, analyst creates different scenarios in selected business process, while including all stakeholders and cloud providers to check if any sub-process requires any change or is there any sub-process which is not required to be deleted (Fig. 1.13).

Fig. 1.14 Service validation
and testing

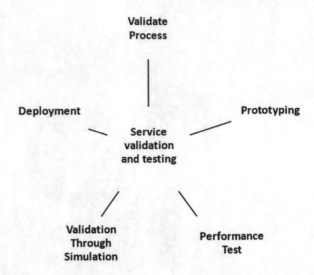

1.3.5 Service Validation and Testing

Validation and testing of the process is the final and important task before deploying the services into the cloud. The process needs to be validated through the simulation which has been already done in the simulation process. Process also needs to be validated by the stakeholders and cloud provider through the prototyping. Performance test needs to be done last time before deployment of the services (Fig. 1.14).

1.4 BPMN Combined Infrastructure Overview

This section is a brief discussion of all the entities involved within the organization and how it co-relates with the BPMN process. Figure 1.15 depicts the combined infrastructure of an organization, cloud architecture, and the role of BPMN. It shows the entire internal structure of a business, which includes all important units such as analyst, investors, stakeholders, management, cloud provider, cloud application architecture, and BPMN RE. It is clearly seen how they are internally interconnected with each other to develop the business as a cloud service application (Fig. 1.15).

Business Analyst: Business analyst plays the main role for an organization as it has direct communication between every single entity of an organization. It has direct communication between all managerial departments such as investors, stakeholders,

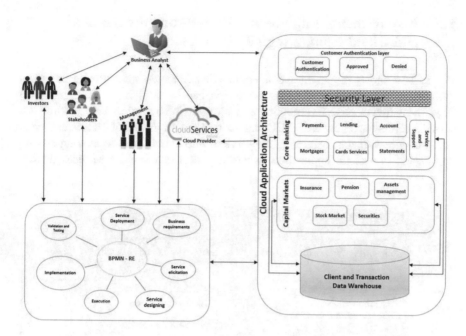

Fig. 1.15 Integrated infrastructure overview

and management. This direct interaction is very important and helpful in respect of requirements engineering of a business process. Moreover, it has a direct connection with the cloud service provider to decide the requirements of the cloud services, because cloud services need to be managed as per the business requirements. Cloud Application: Cloud application architecture displays how and what features or functions need to be on the Web services to make it easy and well structured as per the customers' perspective. It has direct communication between all the entities of the business to provide and manage the services properly, which are listed in the cloud application architecture.

BPMN: After business analyst, BPMN is the process, which decides what resources to be used in an organization and how these resources to be managed to utilize the maximum to enhance the performance. BPMN is used as the requirements engineering technique, which consists of a series of sub-process. Every sub-process has unique functionality to develop a product or service. It has direct communication between all units of the organization, with regard to delivering a process easy and comprehensible with proficiency (Fig. 1.15).

1.5 Requirements Engineering Framework for Service and Cloud Computing (REF-SCC)

Figure 1.16 shows the requirements engineering framework for service and cloud computing. REF-SCC is a framework which can be used for the designing and implementation of service computing to deploy as a Web service. This research is to provide diversity to how REF-SCC can address the migration of service and cloud computing enactments. Migrating the services into cloud refers to moving desktop application into cloud application, where any user can use the services through

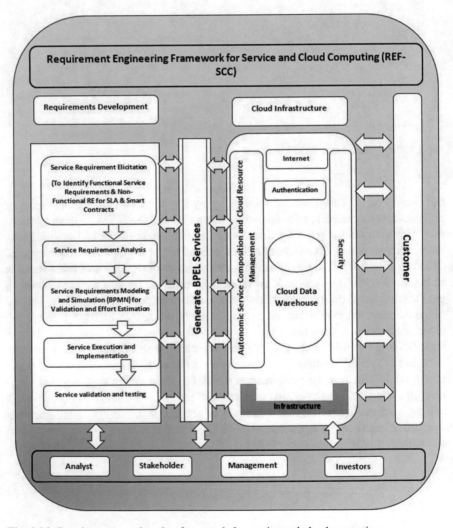

Fig. 1.16 Requirements engineering framework for service and cloud computing

Internet. Main objective of the research is to check the performance of business with respect to time, cost, and utilization, before deployment of services into cloud. The modeling and simulation through BPMN are included in the requirements engineering process of the REF-SCC as this provides validation and testing are also shown in Fig. 1.16.

In reference to Fig. 1.16, requirements development layer consists of sub-processes where service requirements need to elicit, analyze, model, simulate, execute, implement, validate, and test with the help of BPMN and with direct involvement of stakeholders, cloud providers, and other business employees. After managing the requirements and designing of the services, next task is to generate the services with the help of cloud provider.

1.6 Reference Architecture for Service and Cloud Computing

Below figure is a detailed architectural diagram for REF-FCB, which explains how every single entity is connected to each other and how process flow works. All the processes and task of the REF-FCB have been detailed below, which can be divided into three different stages to provide more clear vision, the reference architecture (Fig. 1.17) composed of the following.

Stage 1 Organizational Individuals

Below, three stages explain every task or function to perform. Stage 1 is the organizational individuals which include business analyst, investors, management, and stakeholders. Stage 2 is BPMN as a requirement engineering method, and stage 3 relates to cloud infrastructure.

- →, these arrows indicate the next process or task to be completed.
- ◄——►, these double-sided arrows define the direct communication between the processes.
- Business analyst is the main aspect of an organization to interact with all other aspects of the business.
- Investors are the financial backbone.
- Internal stakeholders such as employees, managers, and owners. External stakeholders such as suppliers, society, government, creditors, shareholders, and customers. Both stakeholders can affect or be affected by the organizational decision and policies.
- Management includes all the employees of an organization.

Stage 2 BPMN Requirements Engineering Processes

- Collecting business requirements with regard to cost, time, and efforts.

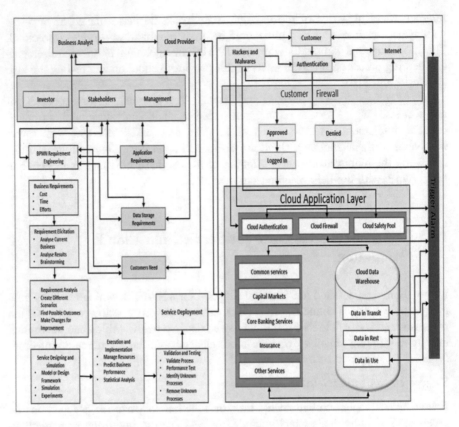

Fig. 1.17 Reference architecture for service and cloud computing

- Elicitation of business requirements with the help of analyzing the current business, analyzing the results, and brainstorming including all business individuals.
- Next process is requirements analysis, which consists of creating different scenarios to get desired results and furthermore to make changes for improvement.
- Designing, simulation, and experiments with the help of BPMN.
- Execution and implementation to predict the business performance.
- Validation and testing before deploying the services.
- Application requirements need to be managed by organizational individuals and cloud provider.
- Data storage needs to be addressed by the cloud provider as per the business requirement.
- Customers' necessities need to be identified by the organizational individuals to deploy in the cloud.

Stage 3 Cloud Application Architecture

- Internet is the connection to access the cloud Web services.
- Hackers and malware stop or abolish the services of the Internet and can also steal information.
- Customers' authentication requires id and password to access an account.
- Customer firewall can also be described as antivirus, which customers use in their respective computers or PCs.
- Cloud authentication, cloud firewall, and cloud security pool are the different security behaviors before logging into cloud to prevent the data.
- Different services can be accessed by the customer after authentication and services are totally based on the characteristics of the particular business. In this case, business relates to finance; hence, services such as capital markets, banking services, insurance, and other services can be accessed.
- Cloud data warehouse is the storage place, where all the relevant information of the customers is saved and can be used by the customer and cloud provider.
- Data in transit means data, which is traveling to data cloud storage to be saved.
- Data in rest is the data, which is saved in cloud storage and not in use.
- Data in use is the data, which is currently accessible by the customer or cloud.

Finally, there is a trigger alarm, which is directly connected to cloud provider, customer, and in between authentication processes. Trigger alarm will be activated to inform customer and cloud provider in case of any suspicious activity.

1.7 Experimental Validation

This experiment is based on the real cloud application of a credit card. There are many features available in the application such as check statement, balance, and payment. But this experiment focuses on making a payment process. Figure 1.18 shows the whole process of making payment. Process starts with login authentication process where a customer needs an id and password to log in, and wrong credentials will deny the user to log in.

After logging into account, there are all other features available. However, this research follows to make a payment. The next step is to select the payment amount such as minimum payment, payment requested, or another amount. Then, providing card details will finish the payment process. Researcher uses the aqua credit card and calculates the approximate time of completing the process to make payment. The time taken to complete a payment process is about 3 min and 30 s; however, it depends upon the consumers' experience and understanding of the application. The main aim of the researcher is to design the payment process of the credit card application and to manage the resources accordingly to maintain the performance of the credit card application, with the increment in the number of customers.

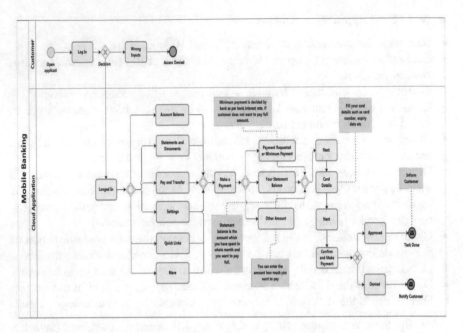

Fig. 1.18 Aqua credit card cloud application processes

1.7.1 Results

Resources used in the processes are cloud server, network bandwidth, and bank account from which payment will be made. Fifteen hypotheses have been created to achieve desired results.

- In hypotheses 1, 2, and 3, researcher has increased the number of incoming instances or customers without changing the number of resources, to find out the consequence on time, utilization, and cost.
- In hypotheses 4–7, researcher stops the increment in incoming instances and tries to manage the resources for the increased customers.
- In the hypotheses 8, 9, and 10, researcher stops increasing the resources and tries to change parameters from designed business process in the respect to manage the performance without increasing the cost.
- In hypotheses 11, 12, 13, 14, 15, after managing the resources and changing the parameters, researcher starts increasing the incoming instances or arrival counts in hypotheses 11, 12, 13, 14, and 15 without making any changes in the business process, to find out the resources we have managed in hypotheses 4–7 and the parameters we have changed in the hypotheses 8, 9, and 10 are capable enough if customers increase (Fig. 1.19).

Hypothesis	Server Utilisation %	Bandwidth Utilization %	Maximum Time	Average Time	Cloud Server Cost £	Bandwidth Cost £
H1: 1 Cloud Server, 1 Network Bandwidth, 1 Bank Account, 100 Instances	77.09	77.09	6m15s	4m3s	1584	792
H2: 1 Cloud Server, 1 Network Bandwidth, 1 Bank Account, 200 Instances	93.75	93.75	1h11m50s	25m33s	3072	1536
H3: 1 Cloud Server, 1 Network Bandwidth, 1 Bank Account, 500 Instances	99.87	99.87	6h56m15s	3h1m17s	7680	3840
H4: 2 Cloud Server, 1 Network Bandwidth, 1 Bank Account, 500 Instances	97.63	99.81	4h25m20s	1h54m51s	7680	3840
H5: 3 Cloud Server, 1 Network Bandwidth, 1 Bank Account, 500 Instances	66.48	99.77	2h45m45s	1h10m50s	7680	3840
H6: 4 Cloud Server, 1 Network Bandwidth, 1 Bank Account, 500 Instances	50.96	99.75	2h6m50s	49m14s	7680	3840
H7: 5 Cloud Server, 1 Network Bandwidth, 1 Bank Account, 500 Instances	41.22	98.62	1h48m50s	30m46s	7680	3840
H8: 5 Cloud Server, 1 Network Bandwidth, 1 Bank Account, 500 Instances (Instances Arrival time Changed to 6m)	34.62	82.83	28m	8m40s	7680	3840
H9: 5 Cloud Server, 1 Network Bandwidth, 1 Bank Account, 500 Instances (Instances Arrival time Changed to 8m)	25.99	62.19	7m5s	4m58s	7680	3840
H10: 5 Cloud Server, 1 Network Bandwidth, 1 Bank Account, 500 Instances (Instances Arrival time Changed to 10m)	20.80	49.77	7m5s	4m51s	7680	3840
H11: 5 Cloud Server, 1 Network Bandwidth, 1 Bank Account, 600 Instances	20.99	50.25	7m5s	5m1s	9296	4648
H12: 5 Cloud Server, 1 Network Bandwidth, 1 Bank Account, 700 Instances	21.30	50.99	7m5s	5m5s	10976	5488
H13: 5 Cloud Server, 1 Network Bandwidth, 1 Bank Account, 800 Instances	21.26	50.89	7m5s	5m5s	12528	6264
H14: 5 Cloud Server, 1 Network Bandwidth, 1 Bank Account, 900 Instances	21.32	51.05	7m5s	5m5s	14128	7064
H15: 5 Cloud Server, 1 Network Bandwidth, 1 Bank Account, 1000 Instances	21.61	51.75	7m5s	5m10s	15872	7936

Fig. 1.19 Result table

1.8 Conclusion

The chapter has described the series of the business process, which can be used to capture the issues or challenges in the current business operations, afterward how it can be resolved to make future improvements. This chapter also describes the importance of BPMN and its processes and how these processes help to validate the process through simulation and testing. BPMN is a standard modeling tool which can be used by different online and offline sources such as Bonitasoft, Bizagi Modeler, and Visual Paradigm.

The chapter has followed the graphical presentation to make it easily understandable. Different process stages have been described with the sub-functional task to identify the problem and to make the best possible changes to make improvement in the business process. This chapter deliberates a unique structured framework for service and cloud computing, which can be implemented and experimented by any cloud business to develop a cloud service with the help of software developers. The proposed design with BPMN can help cloud providers to manage their business processes and resources in order to give better services.

In this chapter, a financial cloud case study has been introduced to enhance the performance of the business in terms of cost, time, and utilization. The case study is connected with aqua credit card payment process, where researcher has tried to maintain the time and effort even after continuous increment in customers. The case study was explained and experimented to maintain the performance of a business process before deploying the application into cloud.

References

1. Raj P, Venkatesh V, Amirtharajan R (2013) Envisioning the cloud-induced transformations in the software engineering discipline. In: Software engineering frameworks for the cloud computing paradigm. Springer, London, pp 25–53
2. Rimal BP, Jukan A, Katsaros D, Goeleven Y (2011) Architectural requirements for cloud computing systems: an enterprise cloud approach. J Grid Comput 9(1):3–26
3. Todoran I, Seyff N, Glinz M (2013) How cloud providers elicit consumer requirements: an exploratory study of nineteen companies. In: Requirements engineering conference (RE), 2013 21st IEEE international (pp 105–114). IEEE
4. Repschlaeger J, Zarnekow R, Wind S, Turowski K (2012) Cloud requirement framework: requirements and evaluation criteria to adopt cloud solutions. In ECIS (p 42)
5. Schrödl H, Wind S (2011) Requirements engineering for cloud computing. J Commun Comput 8(9):707–715
6. Guha R (2013) Impact of semantic web and cloud computing platform on software engineering. In Software engineering frameworks for the cloud computing paradigm. Springer, London, pp 3–24
7. Harrington HJ (1991) Business process improvement: the breakthrough strategy for total quality, productivity, and competitiveness. McGraw Hill Professional, New York
8. Succi G, Predonzani P, Vernazza T (2000) Business process modeling with objects, costs and human resources. In: Bustard D, Kawalek P, Norris M (eds) Systems modeling for business process improvement. Artech House, pp 47–60
9. Lodhi A, Köppen V, Wind S, Saake G, Turowski K (2014) Business process modeling language for performance evaluation. In: 2014 47th Hawaii international conference on system sciences (HICSS). IEEE, pp 3768–3777
10. Henriksen JO (2008) Taming the complexity dragon. J Simul 2(1):3–17
11. Chwif L, Barretto M, Paul R (n.d.) On simulation model complexity. In: 2000 winter simulation conference proceedings (Cat. No.00CH37165)
12. Lodhi A, Köppen V, Saake G (2013) Business process improvement framework and representational support. In: Proceedings of the third international conference on intelligent human computer interaction (IHCI 2011), Prague, Czech Republic, August, 2011. Springer, Berlin, pp 155–167

13. Siddiqi J, Shekaran MC (1996) Requirements engineering: the emerging wisdom. IEEE Softw 13(2):15
14. Pandey D, Suman U, Ramani AK (2010) An effective requirement engineering process model for software development and requirements management. In: 2010 international conference on advances in recent technologies in communication and computing (ARTCom). IEEE, pp 287–291
15. Kheir NA (1996) System modeling and computer simulation, 2nd edn

Chapter 2
Toward an Effective Requirement Engineering Approach for Cloud Applications

Abdullah Abuhussein, Faisal Alsubaei and Sajjan Shiva

Abstract Cloud applications, also known as software as a service (SaaS), provide advantageous features that increase software adoption, accelerate upgrades, reduce the initial capital costs of software development, and provide less strenuous scalability and supportability. Developing software with these features adds new dimensions of complexity to software development that conventional software development methodologies often overlook. These complexities necessitate additional efforts in the software development process to fully utilize cloud qualities (e.g., on-demand, pay-per-use, and auto-scalability). Cloud applications can utilize one or more of these qualities based on software and business requirements. It is noteworthy that present software methodologies (traditional and agile) are deficient to a certain extent in supporting cloud application qualities. In this chapter, therefore, we propose a systematic requirement engineering approach to address this deficiency and facilitate building SaaS with cloud qualities in mind. The proposed cloud requirements engineering approach relies on identifying the number of cloud qualities to be utilized in SaaS and consequently addresses any existing shortcomings. First, we examine different deployment approaches of a cloud application. We then identify the cloud applications' qualities and highlight their importance in the cloud application development process. These qualities are used to derive additional considerations in the form of questions to guide software engineers throughout the requirement engineering process. We demonstrate how these considerations can be used in the requirement engineering process and how the proposed approach effectively produces high-quality cloud applications. This work advocates the need for a systematic approach to support cloud applications requirements engineering.

A. Abuhussein (✉)
Department of Information Systems, St. Cloud State University, St. Cloud, MN, USA
e-mail: aabuhussein@stcloudstate.edu

F. Alsubaei · S. Shiva
Computer Science Department, University of Memphis, Memphis, TN, USA
e-mail: flsubaei@memphis.edu

S. Shiva
e-mail: sshiva@memphis.edu

Keywords Software as a service · SaaS · Cloud application · Requirements engineering · Cloud computing · Software engineering · SaaS engineering · Cloud application engineering

2.1 Introduction

Software development methodologies (SDM) have evolved extensively over the past five decades. SDM applications are developed, tested, deployed, and maintained using distinct phases that structure, plan, and control the development process and the software's life cycle. These phases are referred to as the software development life cycle (SDLC), which differs based on the software development methodology (e.g., waterfall, spiral) used. There are various SDLCs, each with a series of processes to be followed that diverge from iterative to sequential to V-shaped. Later, non-traditional methodologies (i.e., agile or pattern-less), such as Scrum and Extreme Programming (XP), were introduced to promote practices such as test-driven development, refactoring, and pair programming among others. Software development methodologies (i.e., traditional and agile) are composed of five typical phases: requirements, design, implementation, testing, and maintenance. Software manufacturers generally adopt SDLCs to develop software applications depending on the size and nature of the project. Some software manufacturers employ their own software development methodology or adapt an existing one, such as the well-known Rational Unified Process (RUP), introduced by IBM in 2003 [1].

Software applications are typically delivered either as Web-based or standalone desktop applications. Since the inception of cloud computing (CC), the software-as-a-service (SaaS) paradigm has become a standard delivery model for applications in different fields. Although there has been much debate about cloud applications, scholars from research and industry commonly accept that Web-based applications from the olden days are also considered to be cloud applications [2]. However, cloud applications can be a lot more complicated than conventional Web-based applications. This is attributed to their cutting-edge and innovative concepts and technologies (e.g., multi-tenancy, load-balancing, virtualization, etc.). For example, some cloud applications share a physical machine with other tenants' applications. Also, cloud applications might request more resources and therefore scale up. These cases and many others have led to a new set of considerations that increase the complexity involved in cloud application manufacturing.

In this chapter, we explore different deployments, delivery models, architectures, and qualities of cloud application and propose a requirement engineering approach for cloud applications that considers the unique characteristics and qualities of cloud applications. The proposed approach intends to do the following:

- Complement the deficiencies of non-cloud software development methodologies.
- Alleviate the complexity arising when developing cloud-hosted applications by boosting existing software development methodologies with cloud requirements.

This chapter is organized as follows. Research-related work is discussed in the following section. In Sect. 2.3, we give a brief cloud computing and cloud application background and explore cloud application delivery and deployment models. We also demonstrate the need for a development methodology capable of successfully guiding the engineering of cloud applications. In Sect. 2.4, we shed light on some of the key reasons why organizations move to the cloud in the first place to derive a set of drivers for applications migrating to the cloud. In the same section, we discuss the cloud qualities needed for each of the identified drivers. In Sect. 2.5, we present our cloud application requirement engineering approach, and in Sect. 2.6, we describe the identified cloud application qualities and demonstrate how to use them when gathering cloud-specific requirements. In Sect. 2.7, we provide a set of enabling concepts and technologies as well as commercial and open-source solutions for each of the identified qualities. Finally, we conclude our book chapter and suggest avenues for future research on SaaS development methodologies.

2.2 Related Work

The increasing adoption and intensive use of cloud computing have induced, among other factors, a growing need to develop software methodologies that provide suitable support for the construction of cloud applications. Various scholars have made several attempts to create full and partial methodologies for the development of cloud applications [3–9]. These efforts have been geared mainly toward altering specific development methodologies such as Agile [3], whereas some others have proposed domain-specific software development approaches, such as green computing SDLC, legacy systems integration methodology [4], secure cloud application methodology [10], and methodologies to reduce development and operation costs [5]. The bulk of studies have focused on identifying some cloud application qualities, modifying existing software to include cloud application qualities, or adding a cloud application phase to an existing SDLC. However, as cloud application qualities impact the application's functionality, integrating cloud qualities as a part of the developmental life cycle promises to preserve the systematicity of the development method and guarantee software modularity and cohesiveness.

Moreover, the SaaS development industry experiences issues that cause cloud application projects to fail [11, 12]. This has led to the need for a cloud application engineering approach that supports the successful development of ubiquitously high quality, reusable, scalable, and secure cloud applications. This approach must encompass efforts from diverse areas such as cloud computing, software engineering, service-oriented architecture (SOA), human–computer interaction, project management, and security.

The software development industry has also attempted to tackle the complexity of cloud application development by creating ecosystems for SaaS development. To date, no adequate proof of success for these works exists. Also, most of these attempts represent commercial solutions and are not free [13–15].

Overall, in this chapter, we propose an approach to adapt the existing requirement engineering approach by transplanting cloud applications qualities into the process. The strength of this approach lies in the long-proven history of success that the existing software development methodologies have enjoyed and the approach conformance to standards in software engineering and the field of cloud computing.

2.3 Cloud Application Evolution

The cloud application vision is perceived primarily as a business model that emerged to provide cheap, resilient, and instantaneous software delivery. It is an innovative leap in software delivery. Despite the advancement of the development of cloud applications, the field is still considered a fertile environment for software engineering research. Cloud applications are developed, deployed, and delivered to customers using various cloud-computing-related architectures (e.g., multi-tenancy, multi-instance, etc.) and design principles (reusability, composability, scalability, etc.). These architectures and principles are relatively new to software development. Thus, adding extra considerations to existing traditional software development methodologies has become imperative. As a cloud-computing service, cloud applications can be deployed on top of public or private clouds. Adopting effective service delivery is crucial for service discovery and adoption by consumers. A fundamental reason for adopting private clouds over public clouds for application deployment is that they have an attack surface that is less exposed to threats. However, security can also be appropriately maintained in public clouds if it is planned for early and carefully. The cloud application designs and architectures, as well as the deployment and delivery models, somehow make developing cloud applications more complex than the others. This is attributed to the cloud applications' interesting and confounding features. However, a cloud application might not need to have all these new features, but perhaps just one or more. Figure 2.1 contrasts the characteristics of a traditional Web-based application and a cloud application. Although both sides of the arrows in the figure represent software that

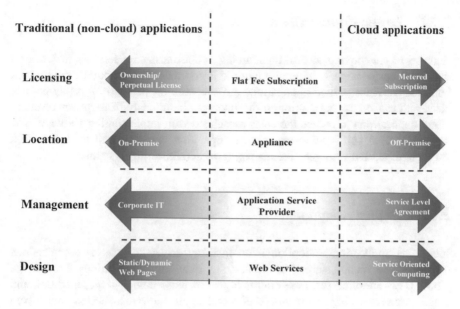

Fig. 2.1 Evolution from traditional non-cloud applications to cloud applications

is accessed through the Web, the cloud application on the right side of the figure includes more cloud qualities (e.g., scalable, automated, and pay-per-use), which in turn means more considerations in development. It is worth mentioning that not all cloud applications should comprise all cloud qualities, but if all qualities are required, cloud application development methodology must be able to accommodate them.

2.4 Key Drivers of Cloud Applications

To develop and deliver resilient, high-quality cloud applications, software engineers must understand the key motives that attract enterprises to cloud applications. This is important because these reasons are likely to impact many other related decisions when developing cloud applications. In this work, we call these reasons drivers. Drivers define the differences between cloud and non-cloud-based applications that motivate moving applications and data to the cloud. Later in this work, drivers are used to investigate and identify the desired cloud application qualities that helped attain these drivers. Considerations, concepts, and enabling technologies that made implementing these qualities possible are also researched and identified in the following sections. The key drivers of cloud applications can be described as follows:

2.4.1 Lower Capital Costs

Enterprises are moving toward outsourcing nonbusiness core services to focus on their core business goals without having to invest their own resources in non-core business activities. Cloud applications are available on demand on a pay-per-use basis and can be accessed through the Internet ubiquitously. Enterprises resort to cloud applications to reduce the costs associated with implementing software. The key point is to shift capital expenses to operating expenses and to deal with a shortage of skilled resources for managing or administering systems.

2.4.2 Higher Adoption

Increasing application's adoption is another key driver that causes organizations (i.e., software manufacturers) to think about delivering applications through the cloud. SaaS adoption rates are continuing to increase among large, medium, and small businesses alike. Consumers of cloud applications can access them from anywhere in the world and at any time. Furthermore, SaaS providers take care of backup, recovery, and technical support. However, there is no guarantee consumers will find, like, and adopt the service; therefore, service marketing and discoverability play a significant role in increasing the number of adopters.

2.4.3 Easier Upgrades/Updates

The time and effort associated with cloud application updates and upgrades are lower than in traditional software. This reduces downtime and attracts consumers to choose a cloud application. Cloud service providers carry out updates and upgrades on customers' behalf. Also, since the service is offered online, cloud application providers observe how the service is being used and collect feedback. Thus, providers can provide frequent updates with zero downtime, which means the updates are smaller, faster, and easier to test.

2.4.4 Better Scalability

For systems to be scalable, they also have to be deployable over a wide range of scales to include numbers of users and nodes, quantity of data storage, and processing rates [16]. Cloud application scalability means not just the ability to operate, but also the ability to operate efficiently over a range of configurations. Thus, it is essential to consider cloud application usage patterns and data size when planning cloud application development. Also, scalability helps enterprises to see the value of space and make decisions and plans accordingly.

2.4.5 Better Technical Support

Supportability of cloud applications relates to all the activities after service development, deployment, and delivery, including customer support, software evaluation, patching, and fixing bugs. It also includes monitoring customers and service to improve service performance and delivery. Since cloud applications are delivered to consumers online, usages can be monitored and controlled. Also, feedback and user experience can be collected and analyzed to predict anomalies and provide better support.

2.4.6 Higher Availability

Concepts such as live migration, scalability, and elasticity that emerged with the cloud made cloud-based applications more accessible to a wide range of users. Also, since the inception of the cloud, there has been a considerable increase in researching and proposing solutions to enhance the fault tolerance and self- healing properties of the cloud, which also has contributed to making cloud applications more attractive to software manufacturers and users.

Key drivers to adopting and manufacturing cloud-based applications and the cloud qualities that enable them are illustrated in Fig. 2.2. When engineering cloud applications, every cloud application quality from the 18 qualities needs to be carefully planned during the requirement engineering process. The reason for this is that consumers usually do not know how to technically explain what they need, and it is overwhelming for software engineers to remember all the cloud qualities and details when gathering the requirements. As mentioned earlier, a cloud application can be as simple as a traditional Web-based application or as complicated as a pure cloud application with more cloud application qualities.

2.5 Cloud Applications Requirements Engineering

When developing applications, both practitioners and experts regard requirements engineering as the most important phase. One key reason is that it might encounter the most common and time-consuming errors as well as the most expensive ones to repair [17]. The requirement phase in any software development project is carried

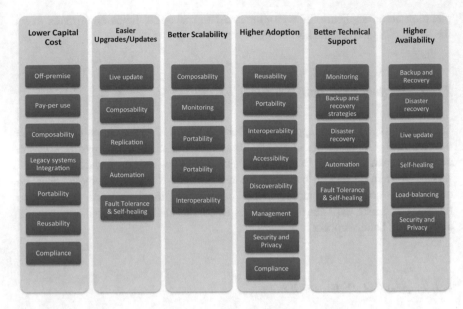

Fig. 2.2 Key drivers of cloud applications mapped to qualities

out as three processes, viz eliciting the requirements of a feasibility study; specifying the requirements; and validating the requirements. These processes and their activities are illustrated in the top portion of Fig. 2.3. Because clouds are used as deployment and delivery mediums, there will inevitably be additional tasks to be planned for as part of the requirement elicitation process. These are the tasks associated with cloud-related qualities. As a result, the requirements will be divided into three groups instead of two as follows:

- Application's Functional Requirements: These describe what the applications should do.
- Non-Functional Requirements: These specify criteria that can be used to define the operations of a system, rather than specific functions.
- Cloud Application Requirements: These specify additional functional requirements to be added due to the cloud's use as a medium to deploy and deliver the software service.

In this work, we present a four-step process to derive these requirements as follows: (1) Identify SaaS drivers, (2) extract SaaS qualities that correspond to the drivers, (3) derive SaaS considerations from the qualities, and (4) use these considerations to specify the cloud-specific requirements. These are shown in the lower portion of Fig. 2.3.

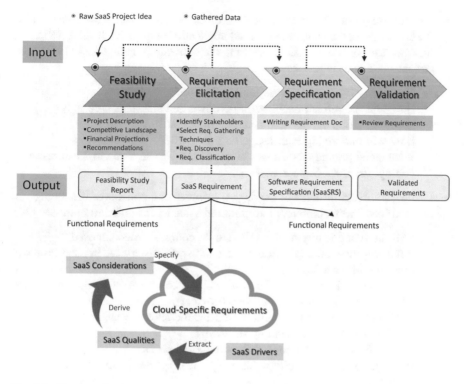

Fig. 2.3 Cloud application requirement engineering process

2.6   Cloud Application Qualities and Requirements

In this section, we present questions that software engineers should consider when eliciting additional cloud-related requirements using the cloud qualities illustrated in Fig. 2.2. First, we define each of the desired cloud application qualities/attributes and follow the definitions with a set of various considerations in the form of questions, as follows:

- Off-premise: A measure of the degree to which application location can affect control over the application and its data. This is important because it specifies the hardware and platform on which the software resides, and which is not usually owned by users [18].

  - What will the IaaS/PaaS configuration and features be?
  - Where will the IaaS be located?
  - What will the response time (latency) be?
  - Who will own, manage, monitor, and control the cloud application's underlying infrastructure and platforms?
  - What is the application deployment model (e.g., private, public hybrid)?

- Pay-per-use (or on-demand): A measure of the degree to which users can control the cost of using the application. Cloud applications cost a small subscription fee on a per user or usage basis every time the software is being used [19]. It also defines on-demand requests and usage.

  - How many users will access the application?
  - How will the infrastructure scale up/down and charge accordingly?
  - How will cloud application usage be measured?
  - How will underlying IaaS usage be measured?
  - What cloud pricing models will be used (e.g., Pay as You Go, Feature-Based Pricing, Free, Ad-Supported)?

- Composability: A measure of the degree to which components of the service can be combined and recombined into different systems for different purposes [20].

  - Will the functionality of the software be composed/decomposed?
  - What will the communication models between subservices be, and how will encryption be carried out?
  - Who owns the service? Can the subservice move and run on another infrastructure owned by a different provider?
  - How will usage patterns be collected for the subservices, and what are the thresholds for each to scale up or down?
  - Who will monitor and control the composed services?
  - How to verify reliability, security, and privacy in the composing components?
  - Will the underlying infrastructure support replication of services and/or their components?
  - What will the message exchange pattern be among cloud application components?

- Legacy systems integration: A measure of the degree to which legacy applications can be integrated and provided to users as a service [21]. Software encapsulation or containerization enable this quality.

  - Will it be possible to transform current systems into cloud applications rather than building from scratch?
  - Will it be possible to maintain the same legacy system functionality?
  - Will it be possible to migrate old (existing) data and their formats?
  - Will the legacy application and its APIs be able to handle users workload?
  - Will the legacy system be able to address security privacy issues emerging from using the cloud?
  - How will the legacy system be connected to GUI and other components?
  - Will the legacy system be able to run the code concurrently?

- Portability: A measure of the degree to which application transports and adapts to a new environment to save the cost of redevelopment [22]. The purpose is to provide a service with the ability to run across multiple clouds' infrastructures from different cloud environments.

- Will the portability of a cloud application and its components be considered during design and development?
- Are there any portability barriers?
- Will a portability test plan be considered?
- Will the cloud application and its components be moved from one cloud provider to another?
- How will compliance issues in the ported application and data be handled?

- Monitoring: A measure of the degree to which a cloud application and its subservices behavior are self-monitored in terms of the quality of the service (QoS), performance, security, etc. [23].

  - Will the SaaS environment have monitors to monitor performance, scaling user experience and health?
  - Will the cloud application provide a dashboard to monitor status, report (analytics), and visualize service health?
  - Who will monitor, control, and manage cloud application monitors (e.g., provider, end user, both)?
  - Will the monitoring of the cloud application be automated?
  - Who will be responsible for monitoring service health when service is composed?
  - Will the billing of the cloud application infrastructure also be monitored, (e.g., overcharges, outstanding balance, etc.)?
  - Will the resources (e.g., storage and CPU) of the cloud application be monitored?
  - Will the cloud application monitor and report on bug fixes and upgrades?

- Backup and recovery strategies: A measure of the degree to which the application and/or its components can be efficiently backed up and recovered [24].

  - Will the cloud application include a backup plan and facility?
  - Will backup/restoration be automated? How often?
  - Will the cloud application allow restoring the data of one user without affecting another's data (granular backup and restore)?
  - Will redundant copies of backups be stored in different off-site locations?

- Disaster recovery (DR): A measure of the degree to which the application and its components recover to the latest working state after an unexpected shutdown or hardware failure [25].

  - Will the cloud application include a disaster recovery plan and facility?
  - Will the disaster recovery process be automated?
  - How quickly does the cloud application recover from failure (time to recover)?
  - Will the cloud application be hosted on multiple, secure, disaster-tolerant data centers?
  - What procedures will be followed when one or multiple infrastructures are down (part of the service but not the whole)?

- Automation: A measure of the degree to which the application and its components are automated. This includes application provisioning, deployment, and management [26].

  - Which parts of the cloud application will be automated (e.g., discovery, provision, backup, billing, self-healing, rollback, compensation)?
  - Will the service requests service be automated?
  - Will the service upgrade be automated?
  - Will the service charge be automated?
  - Will the monitoring of service be automated?
  - Will the backup process be automated?

- Fault Tolerance and Self-healing: A measure of the degree to which a cloud application detects the improper operation of software applications and transactions, and then initiates corrective action without disrupting users [27].

  - Will the cloud application self-heal?
  - Will service be capable of exception handling?
  - Will the service support transactions and rollback?
  - Who is responsible for service failure caused by a subservice of the application maintained by multiple cloud infrastructures? (i.e., governance)
  - Will the cloud application include multiple instances of the same services/ components to improve performance?
  - What measures will be implemented to ensure high availability?
  - What measures will be implemented to ensure fault tolerance?
  - Will the cloud application include load-balancing techniques?

- Live Update: A measure of the degree to which a cloud application can effectively be updated without requiring the service to go down [28].

  - Will the cloud application or its components require a live update?
  - Will the cloud application support live updates?
  - How will cloud application updates be scheduled?

- Reusability: A measure of the degree to which a cloud application can effectively be used in constructing new systems [29].

  - Will the software or its components be reusable?
  - Who will own the subservice? Can the subservice be reused and run on another infrastructure owned and managed by a different provider?
  - Will the service be reused in another region? How many different compliances should the service conform to?
  - What software licensing will there be (e.g., open-source, GNU, MIT, etc.)?
  - Will the application be integrated with third-party components?

- Interoperability: A measure of the degree to which the cloud application and/or its components can effectively be integrated, despite differences in language, interface, protocol, and execution platform [30].

- How will the subservices conform to multiple compliances, and how general should they be (will they cover multiple countries)?
- Will all services be accessible through APIs?

- Accessibility: A measure of the degree to which the cloud application can be available globally to serve customers spanning large enterprises and small business alike.

  - Will the cloud application require a content delivery network service for distant geographical areas (e.g., commercial, self-configuring, or private)?
  - Can the cloud application be configured to allow for choosing server(s) located near clients?

- Discoverability: A measure of the degree to which interested users can discover a cloud application with minimal or without human intervention on the part of the service provider during the entire service adoption lifecycle [7].

  - Will the cloud application be self-discovered, configured?
  - Will the cloud application have metadata about its capabilities and behavior in a service registry?
  - Will service description in the registry be abstract and loosely coupled with the underlying infrastructure, service logic, and technology?
  - Will service metadata be updated automatically when service is updated?
  - How will discovery authentication and access control be managed?

- Service Management: A measure of the degree to which the cloud application can be managed, monitored, controlled, and customized [31].

  - Will the cloud application be customizable (e.g., user interfaces, theme, user experiences, etc.)?
  - Will the cloud application allow for role changes (profiling)?
  - Will the cloud application include a feature to provide technical support?
  - Will the cloud application include measures for cloud economics (e.g., cost-effectiveness, revenue, and cost of operating the cloud application)?
  - Will the cloud application allow for the authorization of tasks?

- Security and Privacy: A measure of the degree to which security and privacy of the cloud application and its components and subservience can be addressed [32–34].

  - Will the cloud application include device and user authentication and access control?
  - Will the cloud application require malware detection?
  - Do cloud application users require access to application logs?
  - Will data in the cloud application require encryption at the client, or while in transit, at rest, and in process?
  - Will the cloud application be delivered through a secure channel (e.g., https, VPN, etc.)?

- Compliance: A measure of the degree to which conformance to standards, laws, and regulations is effectively addressed in the cloud application [35].

- How will the different compliances affect moving the application and data from one region to another?
- Can the application conform to multiple compliances?
- Does the cloud application need to conform to domain-specific compliance (e.g., HIPAA, FERPA, etc.)?
- Does the cloud application require conformance to global compliance (e.g., ISO, AICPA SOC)?
- Does the cloud application require conformance to local compliance (e.g., GPDR in Europe, HIPPA in the USA)?
- Will the cloud application use standard technologies/protocols?
- Will the cloud application be audited to verify its conformance to SLA?

2.7 Enabling Technologies for SaaS Qualities

In this section, we investigate and classify the concepts, enabling technologies, and open-source and commercial tools (refer to Table 2.1) that software engineers can use when implementing cloud applications. Column one lists all the qualities identified in this chapter. Columns two and three represent the concepts, enabled concepts, and technologies, respectively.

Table 2.1 Cloud application qualities, concepts, and enabling technologies

SaaS quality	Concepts	Enabling technologies and examples
Off premises	SaaS management	• SaaS management platforms (SMP) • SaaS asset management (SAM) • Cloud data management interface (CDMI) [36] • Cloud orchestration platforms • Cloud monitors
	Load-balancing	• Application load balancer • Network load balancer
Pay-per-use	Multi-tenancy	• Virtualization
	Cloud pricing models [37]	• Pay as you go, feature-based pricing, per storage pricing, tiered user pricing, per user pricing, flat rate pricing/subscription, "roll your own," Freemium, Ad-supported
Composability	Serverless computing	• Stateless computing: Fn project [38], Serverless Framework [39]
	Containerization	• Building containers: dockers • Managing containers: Kubernetes
	Load-balancing	• Application load-balancing

(continued)

Table 2.1 (continued)

SaaS quality	Concepts	Enabling technologies and examples
		• Network load-balancing
	Application programming interface (API)	• API management tools: Azure API management [40] • Simple object access protocol (SOAP) • Representational state transfer (REST)
Integrate legacy systems	Virtualization	• IaaS (infrastructure as a service): EC2 [41], Azure compute [42] • PaaS (platform as a service)
	Containerization	• Building containers: dockers • Managing containers: Kubernetes
	API	• API management tools: Azure API management • SOAP • REST
Portability	Standardization	• Service catalogs
	SoA	• Statelessness architecture • Service discovery: Web services description language (WSDL), SOAP, XML: working with WSDLs [43] • Service encapsulation: OO programming
	Live migration	• Migration as a service: RiverMeadow [44], AWS Migration Hub [45], Azure cloud migration [46]
	Containerization	• OS containers • PaaS framework: OpenShift [47]
	Multicloud toolkits	• Avoid lock-in: Apache Libcloud [48], Apache jclouds [49], Fog [50]
Monitoring	Infrastructure monitoring	• vRealize Hyperic [51], AWS cloud watch [52], Rackspace monitoring [53]
	Dashboard and reporting	
	Middleware monitoring	
	Application monitoring	
	Logging and reporting	
Backup and recovery strategies	Cloud-to-cloud backup	• AWS backup and restore [54] • Azure backup [55]
	Live migration	• Migration as a service: RiverMeadow [27], AWS migration Hub [28], Azure cloud migration [29]
	Snapshots and cloning	• Snapshots and cloning admin [56]
	Granular backup and restore	• AWS backup and restore [54] • Azure backup [55]
Disaster recovery (DR) and Business continuity (BC)	End-to-end DR	• AWS disaster recovery [57] • Azure disaster recovery [58]
	DR patterns	• Disaster recovery scenarios for applications [59]

(continued)

Table 2.1 (continued)

SaaS quality	Concepts	Enabling technologies and examples
Automation	Discoverability Automation	• Snow for SaaS [60]
	Provisioning automation	• SaaS self-provisioning solutions [61]
	Analytics Automation	• SaasNow [62]
	Monitoring automation	• SaaS performance monitoring [63], vRealize Hyperic [34], AWS cloud watch [35], Rackspace monitoring [36]
	Security automation	• Sqreen [64]
Fault tolerance and self-healing	Resiliency	• Redundancy and replication
	Error handling	• Try catch method • Error logging
	Fail safe	• AWS disaster recovery [57] • Azure disaster recovery [58]
Live update	Automated update	• RiverMeadow [27]]
	Manual update	• In-place updates [65] • Blue-green deployments [65]
Reusability	Attractiveness	• Analytics: Google analytics [66] • CRM: Salesforce [67] • Video marketing: Wistia [68]
	SoA [69, 70]	• Statelessness architecture • Service discovery • Service autonomy • Loose coupling • Granularity • Composability
Interoperability	API	• API management tools: Azure API management [40] • SOAP • REST
	Loose coupling	• RunMyProcess [71]
	Pipeline builder	• Harness [72]
	SoA	• Statelessness architecture • Service discovery: WSDL, SOAP, XML: working with WSDLs [43] • Service encapsulation: OO programming
	Strategies to avoid lock-in	• Apache Libcloud [31], Apache jclouds [32], Fog [33]

(continued)

Table 2.1 (continued)

SaaS quality	Concepts	Enabling technologies and examples
Accessibility	User experience (UX) [73]	• Frictionless signups • Laser focus on your target audience • Simple onboarding • Very easy-to-use UI • Personable • Beautiful design • Support is readily available
	UX design	• InVisionApp [74]
	Unified access management [75]	• SSO • Multifactor authentication • Unified directory • Monitoring and reporting
	Content delivery networks and edge computing	• Section [76]
Discoverability	Visibility techniques [77]	• Outsourcing, paid advertising, portfolios, free advertising, live networking, recommendations, referrals, shock and awe, social networking, cold contact, cross-promotion, curation
	API directories	• APIs.guru [78]
	SaaS directories	• FinancesOnline [79]
Management	Subscription management	• Servicebot [80]
	UX management	• Screenshot feedback: feedback.js [81]
	Subscription and billing management	• Chargify [82]
	Performance management	• Ideagen [83]
Security and privacy	Security as a service	• Sqreen [64]
	Real-time monitors	• CloudPassage Halo [84] • CipherCloud [85]
	Multifactor authentication and single sign-on	• Okta SaaS Authentication [86]
	Penetration testing	• 4ARMED penetration testing [87]
	Auditing and assurance	• Deloitte [88]
Compliance	Monitor real-time compliance	• Intello [89]
	Compliance audit reports	• CyberGuard compliance [90] • Reciprocity [91]

2.8 Conclusion

In this chapter, we have presented an approach to augment cloud qualities in existing requirement engineering processes. The suggested augmentation enables software engineers to gather and specify requirements with cloud qualities in mind, which makes identifying and remembering cloud-related features while gathering and articulating cloud application requirements less cumbersome. Although the cloud application qualities have been described, it is understood that various qualities of a cloud-based application may be chosen based on customer requirements without departing from the spirit and scope of the requirement engineering process. Accordingly, other cloud qualities that may emerge in the future are within the scope of the proposed work. This work has also presented sets of questions that software engineers can use when eliciting cloud-specific requirements. This will hopefully remove the burden of ignorance and forgetfulness from software engineers' shoulders and guide them successfully throughout the requirement engineering process. This process outlined is for the use of software engineers, business, and clients for integrating cloud-specific requirements. This area of software engineering is worth pursuing. In the future, we plan to build an ecosystem to walk cloud application stakeholders through the requirement engineering process based on the approach presented in this work.

References

1. TP026B R (2017) Rational unified process. www.ibm.com/developerworks/rational/library/content/03July/10 00/1251/1251_bestpractices_TP026B Pdf
2. Hudli AV, Shivaradhya B, Hudli RV (2009) Level-4 SaaS applications for Healthcare Industry. In: Proceedings of the 2nd Bangalore annual compute conference. ACM, New York, NY, USA, pp 19:1–19:4
3. Mohagheghi P, Saether T (2011) Software engineering challenges for migration to the service cloud paradigm: ongoing work in the REMICS project. In: Proceedings of the 2011 IEEE world congress on services. IEEE Computer Society, Washington, DC, USA, pp 507–514
4. Chauhan NS, Saxena A (2013) A green software development life cycle for cloud computing. IT Prof 15:28–34. https://doi.org/10.1109/MITP.2013.6
5. Mahmood Z, Saeed S (2013) Software engineering frameworks for the cloud computing paradigm. Springer, London
6. Kao T, Mao C, Chang C, Chang K (2012) Cloud SSDLC: cloud security governance deployment framework in secure system development life cycle. In: 2012 IEEE 11th international conference on trust, security and privacy in computing and communications, pp 1143–1148
7. Zack WH, Kommalapati H (2019) The SaaS development lifecycle. In: InfoQ. https://www.infoq.com/articles/SaaS-Lifecycle. Accessed 20 May 2019
8. La HJ, Kim SD (2009) A systematic process for developing high quality SaaS cloud services. In: Jaatun MG, Zhao G, Rong C (eds) Cloud computing. Springer, Berlin, pp 278–289
9. Aldhahari E, Abuhussein A, Shiva S (2015) Leveraging crowdsourcing in cloud application development. ACTA Press, Calgary

10. Casola V, De Benedictis A, Rak M, Rios E (2016) Security-by-design in clouds: a security-SLA driven methodology to build secure cloud applications. Proc Comput Sci 97:53–62. https://doi.org/10.1016/j.procs.2016.08.280
11. Chhabra B, Verma D, Taneja B (2019) Software engineering issues from the cloud application perspective, vol 5
12. Aleem S, Ahmed F, Batool R, Khattak A (2019) Empirical investigation of key factors for SaaS architecture dimension. IEEE Trans Cloud Comput 1–1. https://doi.org/10.1109/tcc.2019.2906299
13. Rishabsoft (2019) SaaS application development services. https://www.rishabhsoft.com/cloud/saas-app-development. Accessed 20 May 2019
14. Apprenda (2019) SaaS (software-as-a-service) development platform. In: Apprenda. https://apprenda.com/library/software-on-demand/saas-softwareasaservice-development-platform/. Accessed 29 Apr 2019
15. Suffescom (2019) SAAS Application Development Services | SAAS Application Development Company. https://www.suffescom.com/saas-application-development-services. Accessed 29 Apr 2019
16. Jogalekar P, Woodside M (2000) Evaluating the scalability of distributed systems. IEEE Trans Parallel Distrib Syst 11:589–603. https://doi.org/10.1109/71.862209
17. Wieringa R (2001) Software requirements engineering: the need for systems engineering and literacy. Requir Eng 6:132–134. https://doi.org/10.1007/s007660170010
18. Natis YV, Gall N, Cearley DW, Leong L, Desisto RP, Lheureux BJ, Smith DM, Plummer DC (2008) Cloud, SaaS, hosting and other off-premises computing models, vol 6
19. Mell P, Grance T (2011) The NIST definition of cloud computing. National Institute of Standards and Technology, Gaithersburg
20. Weisel EW (2004) Models, composability, and validity. PhD Thesis, Old Dominion University, USA
21. Parnami P, Jain A, Sharma N (2019) Toward adapting metamodeling approach for legacy to cloud migration. In: Hu Y-C, Tiwari S, Mishra KK, Trivedi MC (eds) Ambient communications and computer systems. Springer, Singapore, pp 275–284
22. Mooney JD (1990) Strategies for supporting application portability. Computer 23:59–70. https://doi.org/10.1109/2.60881
23. Benedetti F, Cocco AD, Marinelli C, Pichetti L (2017) Monitoring resources in a cloud-computing environment
24. Sato T, He F, Oki E, Kurimoto T, Urushidani S (2018) Implementation and testing of failure recovery based on backup resource sharing model for distributed cloud computing system. In: 2018 IEEE 7th international conference on cloud networking (CloudNet), pp 1–3
25. Alhazmi OH (2016) A cloud-based adaptive disaster recovery optimization model. Comput Inf Sci 9:58. https://doi.org/10.5539/cis.v9n2p58
26. Wettinger J, Binz T, Breitenbücher U, Kopp O, Leymann F (2015) Streamlining cloud management automation by unifying the invocation of scripts and services based on TOSCA. Cloud Technol Concepts Methodol Tools Appl 2240–2261. https://doi.org/10.4018/978-1-4666-6539-2.ch106
27. Park J, Yoo G, Lee E (2008) A reconfiguration framework for self-healing software. In: 2008 international conference on convergence and hybrid information technology, pp 83–91
28. Zhang X, Zheng X, Wang Z, Li Q, Fu J, Zhang Y, Shen Y (2019) Fast and scalable VMM live upgrade in large cloud infrastructure. In: Proceedings of the twenty-fourth international conference on architectural support for programming languages and operating systems. ACM, New York, NY, USA, pp 93–105
29. Prieto-Diaz R (1993) Status report: software reusability. IEEE Softw 10:61–66. https://doi.org/10.1109/52.210605
30. Wegner P (1996) Interoperability. ACM Comput Surv 28:285–287. https://doi.org/10.1145/234313.234424

31. Brogi A, Canciani A, Soldani J (2015) Modelling and analysing cloud application management. In: Dustdar S, Leymann F, Villari M (eds) service oriented and cloud computing. Springer International Publishing, Berlin, pp 19–33

32. Abuhussein A, Bedi H, Shiva S (2012) Evaluating security and privacy in cloud computing services: a Stakeholder's perspective. In: 2012 international conference for internet technology and secured transactions, pp 388–395

33. Abuhussein A, Alsubaei F, Shiva S, Sheldon FT (2016) Evaluating security and privacy in cloud services. In: 2016 IEEE 40th annual computer software and applications conference (COMPSAC), pp 683–686

34. Abuhussein A, Shiva S, Sheldon FT (2016) CSSR: cloud services security recommender. In: 2016 IEEE world congress on services (SERVICES), pp 48–55

35. Singh S, Sidhu J (2017) Compliance-based multi-dimensional trust evaluation system for determining trustworthiness of cloud service providers. Future Gener Comput Syst 67:109–132. https://doi.org/10.1016/j.future.2016.07.013

36. SNIA (2019) Cloud data management interface (CDMI). https://www.snia.org/cdmi. Accessed 27 Apr 2019

37. Incredo LLC (2016) What are the most successful SAAS pricing models and what you may be doing wrong. https://www.incredo.co/blog/successful-saas-pricing-models. Accessed 27 Apr 2019

38. FnProject (2019) The container native serverless framework. https://fnproject.io/. Accessed 27 Apr 2019

39. Serverless (2019) The serverless application framework powered by AWS Lambda, API Gateway, and more. In: Serverless. https://serverless.com/. Accessed 27 Apr 2019

40. Microsoft (2019) Microsoft Azure, API Management: Establish API Gateways | Microsoft Azure. https://azure.microsoft.com/en-us/services/api-management/. Accessed 27 Apr 2019

41. Microsoft (2019) Amazon EC2. https://aws.amazon.com/ec2/?nc2=h_m1. Accessed 27 Apr 2019

42. Microsoft (2019) Virtual machines: Linux and Azure virtual machines | https://azure.microsoft.com/en-us/services/virtual-machines/. Accessed 27 Apr 2019

43. Soapui (2019) Working with WSDL files | Documentation | SoapUI. https://www.soapui.org/soap-and-wsdl/working-with-wsdls.html. Accessed 27 Apr 2019

44. RiverMeadow (2019) Inc RS cloud migration software—cloud migration tools | RiverMeadow. https://www.rivermeadow.com/rivermeadow-saas. Accessed 27 Apr 2019

45. Microsoft (2019) AWS Migration Hub—Amazon Web Services. In: Amazon Web Services Inc. https://aws.amazon.com/migration-hub/. Accessed 27 Apr 2019

46. Microsoft (2019) Cloud Migration Services—Azure Migration Center | Microsoft Azure. https://azure.microsoft.com/en-us/migration/. Accessed 27 Apr 2019

47. OpenShift (2019) Container application platform by Red Hat, Built on Docker and Kubernetes. https://www.openshift.com. Accessed 27 Apr 2019

48. Apache (2019) Apache Libcloud standard Python library that abstracts away differences among multiple cloud provider APIs | Apache Libcloud. https://libcloud.apache.org/. Accessed 27 Apr 2019

49. Apache (2019) Apache jclouds: home. http://jclouds.apache.org/. Accessed 27 Apr 2019

50. FogIo (2019) The Ruby Cloud Services Library. http://fog.io/. Accessed 27 Apr 2019

51. VMWare (2019) Application monitoring | vRealize Hyperic. In: VMWare. https://www.vmware.com/products/vrealize-hyperic.html. Accessed 27 Apr 2019

52. Microsoft (2019) Amazon CloudWatch—application and infrastructure monitoring. In: Amazon Web Services Inc. https://aws.amazon.com/cloudwatch/. Accessed 27 Apr 2019

53. Rackspace (2019) Custom infrastructure monitoring | Rackspace. In: Rackspace hosting. https://www.rackspace.com/en-us/cloud/monitoring. Accessed 27 Apr 2019

54. Microsoft (2019) Backup and data protection solutions | Amazon Web Services. In: Amazon Web Services Inc. https://aws.amazon.com/backup-restore/. Accessed 28 Apr 2019

55. Microsoft (2019) Cloud backup—online backup software | Microsoft Azure. https://azure.microsoft.com/en-us/services/backup/. Accessed 28 Apr 2019

56. Oracle (2019) Administering PaaS Services. In: Oracle Help Center. https://docs.oracle.com/en/cloud/paas/psmon/snapshots-and-clones.html#GUID-98A389EF-C271-411E-9C8E-6F3E068C7D2A. Accessed 28 Apr 2019

57. Microsoft (2019) Disaster recovery cloud computing service—Amazon Web Services (AWS). In: Amazon Web Services Inc. https://aws.amazon.com/disaster-recovery/. Accessed 28 Apr 2019

58. Microsoft (2019) Azure disaster recovery service—Azure site recovery | Microsoft Azure. https://azure.microsoft.com/en-us/services/site-recovery/. Accessed 28 Apr 2019

59. Google (2019) Disaster recovery scenarios for applications | architectures. In: Google cloud. https://cloud.google.com/solutions/dr-scenarios-for-applications. Accessed 28 Apr 2019

60. Snow Software (2018) Snow for SaaS. In: Snow Software. https://www.snowsoftware.com/int/snow-saas. Accessed 28 Apr 2019

61. Apprenda (2019) SaaS self-provisioning, https://apprenda.com/library/software-on-demand/saas-selfprovisioning/. Accessed 28 Apr 2019

62. Saas Now (2019) SaaS Now self-service BI and analytics from the cloud. https://www.saasnow.com/. Accessed 28 Apr 2019

63. Catchpoint (2019) Digital experience monitoring. https://www.catchpoint.com/. Accessed 28 Apr 2019

64. Sqreen (2019) Application security management platform. In: Sqreen. https://www.sqreen.com/. Accessed 28 Apr 2019

65. OpenShift (2019) Upgrade methods and strategies | Upgrading clusters | OpenShift container platform 3.9. https://docs.openshift.com/container-platform/3.9/upgrading/index.html. Accessed 28 Apr 2019

66. Google (2019) Analytics tools and solutions for your business—Google analytics. https://marketingplatform.google.com/about/analytics/. Accessed 28 Apr 2019

67. Salesforce.com (2019) The customer success platform to grow your business. https://www.salesforce.com/. Accessed 28 Apr 2019

68. Wistia (2019) Video hosting for business. In: Wistia. https://wistia.com. Accessed 28 Apr 2019

69. Service-Architecture.com (2019) Service-Oriented Architecture (SOA) Definition. www.service-architecture.com/articles/web-services/service-oriented_architecture_soa_definition.html. Accessed 28 Apr 2019

70. Abuhussein A, Bedi H, Shiva S (2014) Exploring security and privacy risks of SoA solutions deployed on the cloud. In: Proceedings of the international conference on grid computing and applications (GCA). The steering committee of the world congress in computer science, Computer…, p 1

71. RunMyProcess (2019) Workflow automation. www.runmyprocess.com/workflow-automation/. Accessed 28 Apr 2019

72. Harness (2019) Build pipelines in minutes. In: Harness. https://harness.io/build-pipelines-in-minutes/. Accessed 28 Apr 2019

73. Chargify (2019) SaaS UX Bible: 7 must have user experience principles from the experts at InVision. https://www.chargify.com/blog/saas-ux-bible-must-have-ux-principles-from-invision/. Accessed 28 Apr 2019

74. InVision (2019) Digital product design, workflow and collaboration. In: InVision. https://www.invisionapp.com/. Accessed 28 Apr 2019

75. OneLogin (2019) Unified access management solution. In: OneLogin. https://www.onelogin.com/lp/product-demo?v=bf&utm_term=watch&headline=OneLoginTMUnifiedAccessManagementSolution&_bt=325941279821&_bk=+unified+access+management&_bm=b&_bn=g&utm_source=GOOGLE&utm_medium=cpc&gclid=Cj0KCQjwnpXmBRDUARIsAEo71tTJj8C50b43L_JwA4rN7elLgNmj5fyI0fpfpU7iiNMMYilO8i_gycQaAkFFEALw_wcB. Accessed 28 Apr 2019

76. Edge Compute (2019) Edge compute platform—global edge delivery | section. https://www.section.io/. Accessed 28 Apr 2019

77. Taprun (2019) Discoverability—how to find clients and customers | software, SAAS pricing strategy. https://taprun.com/discoverability/#cold-contacts. Accessed 28 Apr 2019
78. APIs.guru (2019) API tooling development: GraphQL, OpenAPI | APIs.guru. In: APIsguru—Wikipedia WEB APIs. https://APIs.guru/. Accessed 28 Apr 2019
79. FinanceOnline (2019) B2B directory—Trusted SaaS software reviews. In: Financesonline.com. https://financesonline.com/. Accessed 28 Apr 2019
80. Servicebot (2019) Subscription billing made for SaaS companies. https://www.servicebot.io/. Accessed 28 Apr 2019
81. Hertzen.com (2019) Hertzen N von feedback.js. In: Niklas Von Hertzen. http://hertzen.com/experiments/jsfeedback/. Accessed 28 Apr 2019
82. Chargify (2019) Recurring billing | subscription billing software—Chargify. https://www.chargify.com/. Accessed 28 Apr 2019
83. Ideagen.com (2019) Performance management systems | Ideagen Plc. https://www.ideagen.com/solutions/performance-management?utm_source=Google&utm_medium=cpc&utm_campaign=Performance+Management. Accessed 28 Apr 2019
84. CloudPassage (2019) Halo—cloud security tool | CloudPassage. https://www.cloudpassage.com/product/. Accessed 28 Apr 2019
85. CipherCloud (2019) CASB + Plaform. In: CipherCloud. https://www.ciphercloud.com/casb. Accessed 28 Apr 2019
86. OKTA (2019) SaaS authentication. In: Okta. https://www.okta.com/blog/tag/saas-authentication/. Accessed 28 Apr 2019
87. Armed (2019) Limited CI for 4ARMED L. In: 4ARMED Cloud Secur. Prof. Serv. https://www.4armed.com/assess/saas-penetration-testing/. Accessed 28 Apr 2019
88. Deloitte US (2019) Deloitte audit and assurance services | Deloitte US. In: Deloitte U.S. www2.deloitte.com/us/en/pages/audit/solutions/deloitte-audit.html. Accessed 28 Apr 2019
89. Intello (2019) Intelligent SaaS operations. https://www.intello.io/?gclid=Cj0KCQjwnpXmBRDUARIsAEo71tTvA5rWWloz5AkZfGvf2rubynETvu2Jj-Fd_T3DfCjFIQ7MrFVgESsaAi2VEALw_wcB. Accessed 28 Apr 2019
90. CGcompliance.com (2019) LLP CC Software as a Service (SaaS) | CyberGuard compliance. https://www.cgcompliance.com/industries/software-as-a-service-saas. Accessed 28 Apr 2019
91. GRC Software (2019) GRC tools and solutions. In: Reciprocity. https://reciprocitylabs.com/. Accessed 28 Apr 2019

Chapter 3
Requirements Engineering for Large-Scale Big Data Applications

Thalita Vergilio, Muthu Ramachandran and Duncan Mullier

Abstract As the use of smartphone proliferates, and human interaction through social media is intensified around the globe, the amount of data available to process is greater than ever before. As consequence, the design and implementation of systems capable of handling such vast amounts of data in acceptable timescales has moved to the forefront of academic and industry-based research. This research represents a unique contribution to the field of software engineering for Big Data in the form of an investigation of the big data architectures of three well-known real-world companies: Facebook, Twitter and Netflix. The purpose of this investigation is to gather significant non-functional requirements for real-world big data systems, with an aim to addressing these requirements in the design of our own unique reference architecture for big data processing in the cloud: MC-BDP (Multi-Cloud Big Data Processing). MC-BDP represents an evolution of the PaaS-BDP (Platform as a Service for Big Data Processing) architectural pattern, previously developed by the authors. However, its presentation is not within the scope of this study. The scope of this comparative study is limited to the examination of academic papers, technical blogs, presentations, source code and documentation officially published by the companies under investigation. Ten non-functional requirements are identified and discussed in the context of these companies' architectures: batch data, stream data, late and out-of-order data, processing guarantees, integration and extensibility, distribution and scalability, cloud support and elasticity, fault tolerance, flow control and flexibility and technology agnosticism. They are followed by the conclusion and considerations for future work.

T. Vergilio (✉) · M. Ramachandran · D. Mullier
School of Built Environment, Engineering, and Computing,
Leeds Beckett University, Leeds, UK
e-mail: T.Vergilio@leedsbeckett.ac.uk

M. Ramachandran
e-mail: m.ramachandran@leedsbeckett.ac.uk

D. Mullier
e-mail: d.mullier@leedsbeckett.ac.uk

© Springer Nature Switzerland AG 2020
M. Ramachandran and Z. Mahmood (eds.), *Software Engineering in the Era of Cloud Computing*, Computer Communications and Networks,
https://doi.org/10.1007/978-3-030-33624-0_3

Keywords Big data · Requirements engineering · Batch · Stream · Scalability · Fault tolerance · Flow control · Technology agnosticism · Processing guarantees

3.1   Introduction

Big data is defined as data that challenges existing technology for being too large in volume, too fast or too varied in structure. Big data is also characterised by its complexity, with associated issues and problems that challenge current data science processes and methods [1]. Large internet-based companies have the biggest and most complex data, which explains their leading role in the development of state-of-the-art big data technology. It has been reported that, in one minute of internet usage, 1 million people log into Facebook, 87,500 new tweets are posted and 694,444 h of videos are watched on Netflix [2]. Processing these vast amounts of data presents challenges not only in terms of developing the most appropriate algorithms to best inform these companies and their future decisions, but also in terms of assembling the technology needed to perform these calculations in a timely manner.

This paper contributes to the existing knowledge in the area of software engineering for Big Data by performing a search of the existing literature published by three major companies, and an outline of the strategies devised by them to cope with the technological challenges posed by big data in their production systems. Non-functional requirements are important quality attributes which influence the architectural design of a system [3]. Ten non-functional requirements for big data systems are identified and discussed in the context of these real-world implementations. These requirements are used to guide the design and development of new reference architecture for big data processing in the cloud: MC-BDP. The presentation, evaluation and discussion of MC-BDP are addressed in a different publication.

The companies targeted for this study are Facebook, Twitter and Netflix. The methodology used in this comparative study is explained in Sect. 3.2, which covers the scope of this research, the criteria used for selecting the target companies, as well as definitions for the non-functional requirements under examination. Section 3.3 discusses related work, and Sect. 3.4 addresses each non-functional requirement and discusses how they are implemented by the three companies in their production systems. Finally, Sect. 3.5 presents the conclusion and considerations for future work.

3.2 Research Methodology Using Systematic Literature Review

This section describes this research's methodology using a systematic literature review, as illustrated in Fig. 3.1. The first step is the definition of the scope of this research. Since this research is literature-based, its scope is defined in terms of which literature sources to use. The next step is an explanation of the criteria used to select the companies under examination from a pool of potential matches. The third step consists of explaining the comparison criteria used to evaluate the architectures of the selected companies. It describes the ten non-functional requirements for large-scale big data applications which form the focus of this study. Finally, the last step is an evaluation of the different approaches taken by these companies to implement the aforementioned requirements.

3.2.1 Scope

The scope of this paper is limited to academic papers, technical documentation and presentations or blog posts officially published by the companies under evaluation. Although the authors recognise that direct observation of the systems under evaluation by way of a set of case studies would have yielded more reliable, and perhaps more valuable results, time and resource constraints limited the scope of the present study. Aware of this limitation, this research endeavoured to use only sources officially published or endorsed by the companies under evaluation, in an effort to be true to the systems under examination. Where no information could be found with regard to specific criteria or particular aspects of the systems under examination, this is clearly stated by the authors.

Fig. 3.1 Summary of research methodology

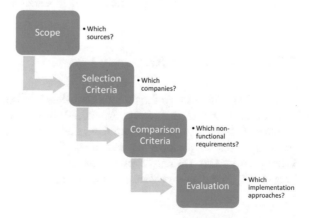

Table 3.1 Classification and summary of source materials

Company	Academic paper	Technical blog	Presentation	Code/documentation
Facebook	2	2	1	1
Twitter	3	2	1	3
Netflix	1	8	2	0

Table 3.1 shows a summary of the materials used as source for this research, classified by type. Six academic papers were found describing the systems used by the three companies under study. Additionally, twelve technical blog articles, four presentations and the source code or documentation of four systems were used as source for this study. Twitter had the strongest academic presence at the time of writing, with two papers available through the ACM Digital Library [4, 5], and one paper available through the IEEE Xplore Digital Library [6]. Two relevant academic papers were found for Facebook, one available through the ACM Digital Library [7], and one available through the IEEE Xplore Digital Library [8]. Finally, only one academic paper was found at the time of writing describing the big data systems at Netflix: [9], available through the ACM Digital Library.

Although Netflix was underrepresented in terms of academic sources when compared to the other two companies evaluated, it had the highest number of relevant non-academic sources: eight technical blog articles and two presentations. Both Facebook and Twitter, in comparison, had two technical blog articles and one presentation relevant to this research. Finally, in terms of source code and documentation available for public peruse, Twitter had three entries, corresponding to the source code for Scalding [10], Heron [11] and Storm [12]. The code for Facebook's Scribe [13] was available as open source through the company's archive repository. At the time of writing, none of the Netflix systems assessed by this study were open source.

This section presented the scope of the evaluation conducted in this research, which was limited to academic papers, technical blog articles, presentations and source code/documentation published by Facebook, Netflix and Twitter. The next section explains the selection criteria used to select the three target companies.

3.2.2 Selection Criteria

This section explains how the three companies: Facebook, Twitter and Netflix were selected as target of this study. An initial survey of big data architectures was conducted, limited to peer-reviewed academic papers. Three search engines were primarily used to perform the searches: Google Scholar, IEEE Xplore Digital Library and ACM Digital Library. The initial survey searched for terms such as "big data", "big data processing", "big data software" and "big data architecture".

In the interest of thoroughness, synonyms were used to replace key terms where appropriate, e.g. "system" for "software".

The first classification which became apparent was in terms of who developed the solutions presented. The results found comprised technologies developed

(1) by academia,
(2) by real-world big data companies,
(3) by industry experts as open-source projects, or
(4) by a combination of the above.

This research focuses on category number 2.

A further classification can be drawn from the academic papers reviewed, this time in terms of how the contributions presented were evaluated. Three cases were encountered:

(A) cases where there is no empirical evaluation of the proposed solution,
(B) cases where the empirical evaluation of the proposed solution is purely experimental and
(C) cases where peer-reviewed published material was found describing the results of implementing the proposed solution in large-scale commercial big data settings.

In order to select suitable companies to include in this study, the focus of this research was limited to category C.

Three companies were selected within the criteria characterised above: Facebook, Twitter and Netflix. These were selected from a wider pool of qualifying companies which included Microsoft [14, 15], Google [16, 17] and Santander [18]. The rationale for choosing the three aforementioned companies is based on the quantity, quality and clarity of the information encountered, as well as availability of technical material online such as project documentation and architectural diagrams.

3.2.3 Definitions

This section explains how the ten non-functional requirements: batch data, stream data, late and out-of-order data, processing guarantees, integration and extensibility, distribution and scalability, cloud support and elasticity, fault tolerance, flow control and flexibility and technology agnosticism were identified as non-functional requirements for this study. It then provides definitions for each requirement.

The ten non-functional requirements selected for this study were based on the initial literature survey of official academic publications explained in Sect. 3.2.2. As with the previous selection, focus was given to solutions developed by real-world big data companies. However, it is worth noting that these requirements are widely addressed in open source, as well as purely academic solutions [19–22]. Likewise, they are highlighted in non-academic sources such as commercial solutions and cloud-based services [23–25].

3.2.3.1 Batch Data

This requirement refers to the processing of data which is finite and usually large in volume, e.g. data archived in distributed file systems or databases. An important requirement for real-world big data systems is that they must be capable of processing large amounts of finite, usually historical data. Figure 3.2 illustrates a typical case for batch data processing.

As we can see in Fig. 3.2, large amounts of data are first collected in static storage spaces such as, for example, distributed databases, file systems, logs or data warehouses. It is then processed in finite batches by powerful, usually distributed technology. The name batch processing comes from this approach to data processing whereby the data is collected into finite batches before it is processed.

This section described the non-functional requirement for a large-scale big data system to be capable of processing batch data, defined as data which is finite, usually historical and large in volume. The next section describes the non-functional requirement for a large-scale big data system to be capable of processing stream data.

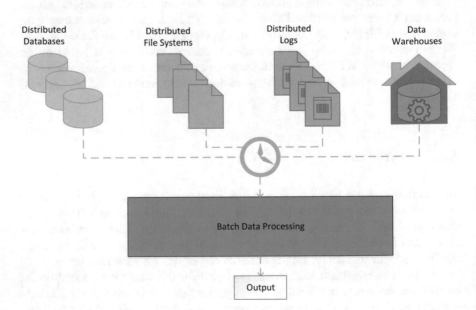

Fig. 3.2 Batch data processing

3.2.3.2 Stream Data

This requirement refers to the processing of data which is potentially infinite and usually flowing at high velocity. For example, real-world big data systems are generally required to capture and process user activity or monitoring data in real time, or close to real time. Figure 3.3 illustrates a typical case for stream data processing.

In Fig. 3.3, we can see that data is collected from a variety of sources such as, for example, smart homes, application logs or the Internet of things (IoT). It is then processed in real time, or as close to real time as the technology allows. This approach is called stream processing because the incoming data is very large, potentially infinite, so processing cannot wait until all the data is available before it starts. Instead, processing is ongoing. It takes place at defined intervals and emits results at defined intervals. Differently from batch processing, completion is not a concept that is used in the context of stream processing, since the data source is potentially infinite. This, however, does not mean that accuracy is compromised, as it remains not only possible, but indeed a desirable quality of mature streaming systems, as demonstrated by Akidau et al. [17].

The capacity to process stream data was identified as a non-functional requirement not only within the architectures of the three companies evaluated, as discussed in detail in Sect. 3.4.2, but also of other large-scale big data companies such as Microsoft's library for large-scale stream analytics, Trill, used in Azure Stream Analytics and ads reporting for the Bing search engine [26], Google's Dataflow,

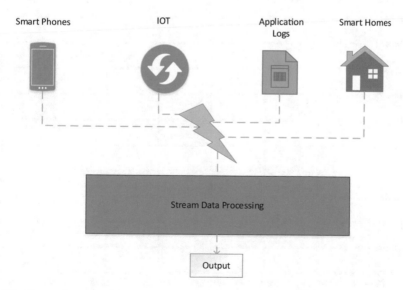

Fig. 3.3 Stream data processing

used for statistics calculations for abuse detection, billing, anomaly detection and others [17] and LinkedIn's Samza, currently used in production and deployed to more than 10,000 containers for anomaly detection, performance monitoring, notifications, real-time analysis and others [27]. The literature review conducted as part of this research therefore concluded that the capacity to process stream data is a fundamental requirement of large-scale big data architectures, and, as stream technology develops, it becomes capable of catering for a larger number of use-cases previously consigned to batch processing, as discussed comprehensively in [28].

This section described the non-functional requirement for large-scale big data systems to be capable of processing stream data, defined as data which is potentially infinite in size and usually arriving at high velocity. The next section describes the non-functional requirement for a large-scale stream big data system to be capable of processing late and out-of-order data.

3.2.3.3   Late and Out-of-Order Data

This requirement relates to stream processing and refers to the processing of data which arrives late or in a different order from that in which it was emitted. Streaming data from mobile users, for example, could be delayed if the user loses reception for a moment. In order to handle late and out-of-order data, a system must have been designed with this requirement in mind. Figure 3.4 illustrates late and

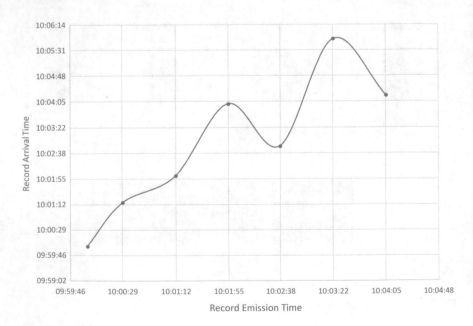

Fig. 3.4 Late and out-of-order data

out-of-order data. The records emitted at 10:01:55 and 10:03:22 are significantly delayed. The record emitted at 10:02:38 actually arrives before the one emitted at 10:01:55. Similarly, the record emitted at 10:04:05 arrives before the record emitted at 10:03:22. Late and out-of-order records such as the ones depicted are common with real-time user data which is subject to network delays.

Figure 3.5 shows a summary of strategies for dealing with late and out-of-order data. The windowing strategy defines how the data is grouped into windows of time to enable the processing of otherwise infinite data. At a minimum, the period (how frequently each window starts) and duration (how long each window lasts for) must be defined. Thus, a scenario where the period is longer than the duration characterises sampling, whereas one where the duration is longer than the period characterises tumbling or sliding windows. Where the period and duration are the same length, the window is considered a fixed window. The triggering strategy defines how often results are emitted, e.g. at the end of every window, at the end of every windows plus a defined tolerance, at fixed intervals, etc. The state strategy defines what data is persisted during stream data processing and how long for. It is useful for computing aggregates and can be defined at key, window or application level. Finally, the watermark strategy defines how late the data is expected to be and usually signals the application to start processing at a time all data is believed to have been received. These four strategies combined define how late and out-of-order data is handled by a stream application.

As an example, a system may be configured to process a simple count of distinct words entered into a search engine. The windowing strategy is defined to use sliding windows of 10 s, starting every 5 s. The triggering strategy is configured to emit (partial) results every 5 s and to accumulate more data as it arrives to emit in the next 5 s. The state strategy is configured to use per-window state. Finally, the watermark strategy is configured to expect all data to have arrived within 5 min of emission. Figure 3.6 summarises this sample configuration.

Any data that is less than 5 min late is incorporated into the calculations. Partial results for a window of 10 s are emitted after 5 s, with subsequent emissions

Fig. 3.5 Strategies for dealing with late and out-of-order data

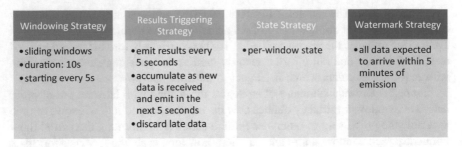

Windowing Strategy	Results Triggering Strategy	State Strategy	Watermark Strategy
• sliding windows • duration: 10s • starting every 5s	• emit results every 5 seconds • accumulate as new data is received and emit in the next 5 seconds • discard late data	• per-window state	• all data expected to arrive within 5 minutes of emission

Fig. 3.6 Sample configuration for dealing with late and out-of-order data

(adjustments) occurring every 10 s, until the watermark is achieved. Any data arriving later than the watermark is discarded.

This section described the non-functional requirement for streaming systems being capable of processing data which arrives late and out of order. The next section describes the non-functional requirement of ensuring that a distributed system honours one of three processing guarantees.

3.2.3.4 Processing Guarantees

This requirement refers to the processing guarantees that a distributed stream system offers, i.e. exactly once, at least once and at most once. It determines whether processing tasks assigned to workers are replayed in case of system failure [29]. While exactly once processing is ideal, it comes at a cost which could translate into increased latency. This requirement is used to evaluate how different systems and different use-cases warranted different compromises in terms of latency and processing guarantees. Figure 3.7 illustrates the three types of processing guarantees.

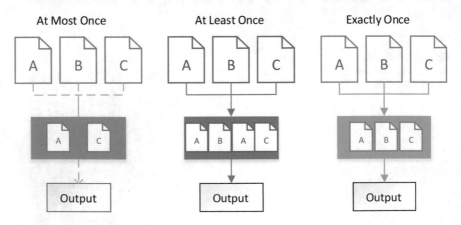

Fig. 3.7 Processing guarantees

The first type of processing guarantee illustrated in Fig. 3.7, at most once, focuses on avoiding reprocessing of data, to the detriment of duplication. In the event of worker failure, the data processing task assigned to that worker will not be restarted, resulting in data loss, as illustrated in Fig. 3.7. The second type of processing guarantee, at least once, focuses on avoiding data loss, even if it means that processing task (and results) is duplicated. Finally, the third type of processing guarantee, exactly once, is a combination of the former two: it ensures that there is no data loss, and it also ensures that there is no duplication. Although the exactly once processing guarantee is the most accurate, it is not always the most desirable, as it is more costly resource-wise to achieve when compared to the other two. In order to ensure that data is processed exactly once, an external checkpointing system is usually employed to ensure that each task can be replayed from where it is left off (or as close as possible to that) in the event of node failure. These checkpoints can be expensive and involve additional disc and networking resources which may not be desirable in every particular use-case. The less processing duplication desired, i.e. the stricter the exactly once guarantee required, the higher the checkpointing frequency needed, since each worker must output the state of the processing task several times throughout its execution.

This section described the non-functional requirement for processing guarantees, defined as assurances provided by a distributed parallel system with regards to how many times incoming data will be processed regardless of possible node or transmission failures. The next section describes the non-functional requirement for integration and extensibility.

3.2.3.5 Integration and Extensibility

This requirement refers to how well the systems presented integrate with existing services and components. It also refers to provisions made to facilitate the extension of the existing architecture to incorporate different components in the future. For illustration, Fig. 3.8 is a simplified diagram representing Heron's architecture. Heron was designed by Twitter to be fully compatible with Storm, their previous big data framework for stream processing. As Fig. 3.8 shows, the Heron API

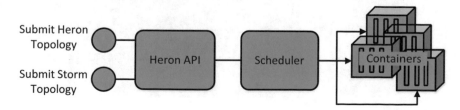

Fig. 3.8 Twitter Heron simplified architecture

accepts both Heron and Storm topologies, thus facilitating the integration of the
new system with legacy processing code defined as topologies.

This section described the non-functional requirement for integration and
extensibility, defined as the capacity of a system to integrate with existing services
and components, as well as the provisions made to facilitate the extension of the
existing architecture to incorporate different components in future. The next section
defines the non-functional requirement of distribution and scalability.

3.2.3.6 Distribution and Scalability

This requirement refers to how easily the data processing can be distributed
amongst different machines, located in different data centres, in a multi-clustered
architecture. Dynamic scaling, which addresses the possibility of adding or
removing nodes to a running system without downtime, is also part of this
requirement. Figure 3.9 illustrates the processing of stream big data by a
container-based architecture using a pipe analogy: the length of the pipe represents
the time each container takes to process data, and the diameter of the pipe represents
the number of containers processing the data.

The wider the diameter of the pipe, the more containers there are processing the
data, so the pipe is shorter and the queue is reduced, since data is processed faster.
An example of horizontal scaling would be to launch more data processing containers running on the same physical infrastructure (same number of nodes, of same
capacity). This is illustrated in Fig. 3.10.

At some point, however, horizontal scaling fails to translate into faster processing, and it is necessary to commission more nodes to provide more processing
capacity at infrastructure level. This is known as vertical scaling, and, using the
previous analogy, it is the equivalent of adding more pipes. As consequence, the
data flows faster through the pipes and the queue is reduced, as illustrated in
Fig. 3.11.

This section described the non-functional requirement of distribution and scalability, defined as how easily the data processing can be distributed amongst

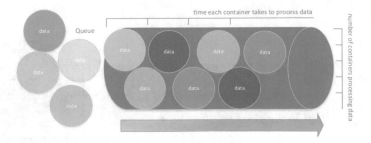

Fig. 3.9 Pipe analogy for container-based stream big data processing pipeline

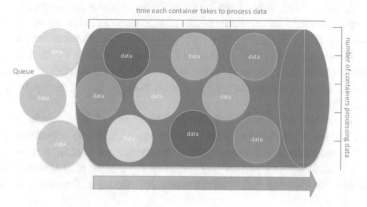

Fig. 3.10 Horizontal scaling

Fig. 3.11 Vertical scaling

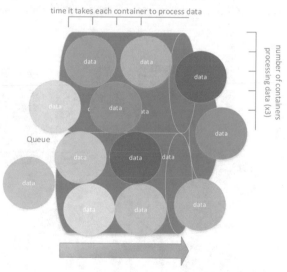

different machines, located in different data centres, in a multi-clustered architecture. The next section describes the non-functional requirement of cloud support and elasticity.

3.2.3.7   Cloud Support and Elasticity

This requirement refers to the ease with which the architecture (or part of it) can be moved into the cloud to take advantage of the many benefits associated with its economies of scale. Elasticity in particular is a cloud property which allows a system to scale up and down according to demand. Since the user only pays for resources actually used, there is less wastage and it is theoretically cheaper than running the entire infrastructure locally with enough idle capacity to cover for eventual spikes.

Since being unable to easily switch between cloud providers represents a risk to cloud consumers [30], support for a multi-cloud architecture which mitigates the risk of vendor lock-in and allows cloud consumers to transfer resources across providers is addressed as part of this requirement. Figure 3.12 shows MC-BDP's multi-tenant multi-cloud infrastructure enabled through the use of container technology.

This section described the non-functional requirement of cloud support and elasticity, defined as the ease with which the architecture (or part of it) can be moved into the cloud to take advantage of the many benefits associated with its economies of scale, in particular with respect to elasticity, defined as the capacity of a cloud system to scale up and down according to demand. The next section describes the non-functional requirement of fault tolerance.

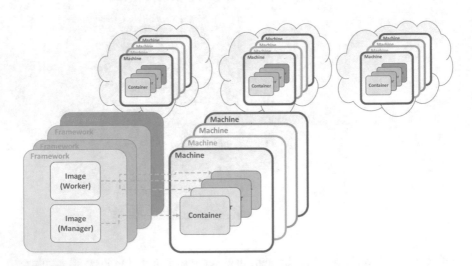

Fig. 3.12 Multi-tenant multi-cloud infrastructure enabled by container technology

3.2.3.8   Fault Tolerance

This requirement refers to provisions made at design time, so the system can continue to operate should one or more nodes fail. Ideally, production systems should recover gracefully, with minimal effect (if at all) on the user experience. Fault tolerance can be observed at different levels, as illustrated in Fig. 3.13. Using MC-BDP's container-based architecture as an example, a container could become unresponsive, as shown in A, and the container orchestrator would be expected to provide fault tolerance (i.e. relaunch the task in another container). Similarly, a node where several containers are running could become unresponsive, as in B, requiring fault tolerance at both container and node levels (i.e. launch an equivalent node and relaunch the containers that were running on the lost node). A more serious scenario is depicted in C, where a cloud provider's entire region fails. This requires a number of nodes and other resources to be recreated, and all the containers running in those nodes to be relaunched. Finally, D illustrates a multi-region failure for a single provider where all resources previously running on that cloud need to be relaunched. Although uncommon, multiple availability zone failures do sometimes occur [31], as do multiple region outages, as exemplified by a DNS disruption that affected Azure customers in all regions in 2016 [32].

3.2.3.9   Flow Control

This requirement refers to scenarios where the data source emits records faster than the system can consume. Strategies for dealing with backpressure, e.g. dropping

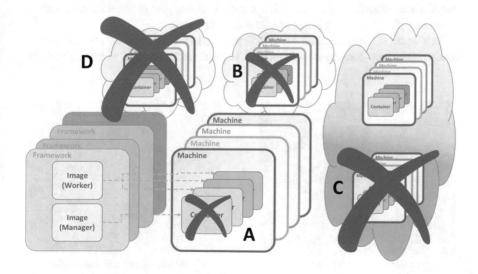

Fig. 3.13 Fault tolerance at different levels

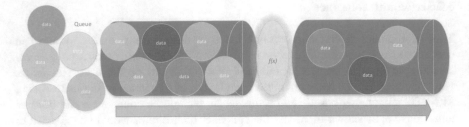

Fig. 3.14 Flow control

records, sampling, combining, applying source backpressure, etc., are generally required from real-world big data systems. Figure 3.14 illustrates flow control defined by an abstract function $f(x)$, which results in fewer records flowing down the stream.

3.2.3.10 Flexibility and Technology Agnosticism

This requirement refers to the extent to which the architecture provides the implementer with the option to use different technology in place of existing components. A modular architecture, for example, allows separate components to be replaced or upgraded with no detrimental effect to the functioning of the system as a whole. Figure 3.15 shows the MC-BDP reference architecture developed by the authors, together with a sample prototype implementation. MC-BDP is an example of a highly flexible reference architecture designed with technology agnosticism in mind. Each module depicted in the concrete prototype implementation can be replaced with a technologically equivalent component.

This section presented the methodology used in this research and included a subsection where the ten non-functional requirements referred to throughout this

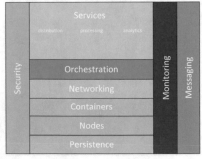
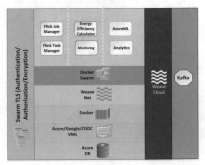

MC-BDP Reference Architecture   MC-BDP Prototype Implementation

Fig. 3.15 MC-BDP reference architecture and prototype implementation

chapter were defined. The next section discusses related work and is followed by Sect. 3.4, which looks at how the three target companies, Facebook, Twitter and Netflix, addressed the aforementioned requirements.

3.3 Related Work

This section is divided in three parts: Sect. 3.3.1 discusses academic developments in the area of requirements engineering for large-scale big data systems, Sect. 3.3.2 examines the literature around big data architectures and finally Sect. 3.3.3 reviews the gap in the literature which the current research endeavours to bridge.

3.3.1 Requirements Engineering for Big Data

Although requirements engineering is an established area of academic research, requirements engineering for big data is an incipient field and has tended to concentrate on infrastructure requirements, to the detriment of other aspects [33]. This section discusses work which attempted to bring such aspects to the forefront of requirements engineering research. Section 3.3.1.1 addresses work which attempted to incorporate the 3 Vs of big data (volume, velocity and variety) into traditional requirements engineering. Section 3.3.1.2 looks at proposals to devise a requirements engineering context model for the domain of big data.

3.3.1.1 Incorporating the 3 Vs of Big Data into Requirements Engineering

This section addresses related work which integrates key aspects of big data into the field of requirements engineering. Big data is traditionally defined as displaying the three characteristics of volume, velocity and variety. Being able to integrate these characteristics into a systematic process for the specification of requirements has been highlighted as one of the research challenges in the field [34]. Moreover, these characteristics must be represented in requirements notation in order to ensure they are adequately captured [33]. Noorwali et al. [35] proposed an approach to integrate big data characteristics into quality requirements. However, at the time of writing, their approach has not yet been evaluated empirically. The current research uses known large-scale industry implementations to identify ten non-functional requirements for the specific domain of big data processing.

3.3.1.2 Devising a Context Model for Big Data Requirements Engineering

This section discusses related work which proposes a new context model for big data in the field of requirements engineering. Eridaputra et al. used the goal-oriented requirements engineering (GORE) method to model requirements and propose a new requirement model based on the characteristics of big data and its challenges. Their research was empirically evaluated through a case study set at a government agency where 26 functional and 10 non-functional requirements were obtained from the model, and further validated by stakeholders as accurate [36]. Al-Najran and Dahanayake developed a new requirements specification framework for the domain of data collection which incorporates requirements engineering for big data [37]. This framework was empirically evaluated through quantitative experiments to measure the relevance of Twitter feeds [38].

Madhavji et al. introduced a new context model of big data software engineering where not only computer science and software engineering research were taken into account, but also big data software engineering practice, corporate decision-making and business and client scenarios [33]. Arruda and Madhavji subsequently identified a lack of known artefact models to support requirements engineering process design and project understanding and proposed the creation of a requirements engineering artefact model for big data end-user applications (BD-REAM) [39]. Based on this initial study, Arruda later developed a context model for big data software engineering and introduced a requirements engineering artefact model containing artefacts such as development practice, corporate decision-making and research, as well as the relationships and cardinalities between them. At the time of writing, this research was still in early stages of development and had not been empirically evaluated [34].

This research is similar to Madhavji et al.'s, in that it looks beyond theoretical contributions in the computer science and software engineering fields in an effort to incorporate recent developments from the industry into its review of non-functional requirements for the domain of large-scale big data applications. Since big data is a very active area of development not only in academia, but also (and perhaps even more) commercially, a thorough specification of requirements for big data ought to include both spheres. Furthermore, the current study looks at real-world implementations of non-functional requirements by some of the largest big data corporations and discusses the particular use-cases that led to some of their key design decisions.

This section discussed related work in the area of requirements engineering for big data and briefly presented a number of functional services proposed by this research for big data requirements engineering in a multi-cloud environment. The next section addresses the literature related to big data architectures.

3.3.2 Big Data Architectures

The literature addressing the challenges presented by big data is vast, and the majority of new solutions, architectures and frameworks proposed acknowledge and aim to fulfil one or more of the non-functional requirements identified in Sect. 3.2. This section therefore discusses related work by introducing a classification based on the type of contribution proposed:

3.3.2.1 Evaluation or Unique Application of Widely Adopted Existing Technologies for Big Data Processing

This class consists of research which leveraged existing, widely adopted technologies for big data processing and either applied it to a original case or evaluated it in a unique way.

Examples of papers which present an evaluation of existing technology are Spark's evaluation by Shoro and Soomro using a Twitter-based case study [40] and Kiran et al.'s implementation of the Lambda architecture using Amazon cloud resources [41].

Examples of unique applications of widely adopted existing technology for big data processing are Sun et al.'s use of Hadoop, Spark and MySQL to process big data related to spacecraft testing [42] and Naik's use of Docker Swarm to create a distributed containerised architecture for data processing using Hadoop and Pachyderm [41].

3.3.2.2 New Technologies for the Processing of Big Data

This class consists of research which proposed entirely new technologies for the processing of big data.

Examples of contributions within this category are Borealis, a distributed stream processing engine developed by a consortium involving Brandeis University, Brown University and the MIT [19], Millwheel, Google's distributed stream processing system based on the concept of low watermarks which was later open-sourced as Apache Beam [16] and Storm, one of the most popular open-source stream processing frameworks, originally developed by Twitter [4].

3.3.2.3 Original Architectural Proposals Where Existing Technologies Are Used or Recommended

This class consists of research which proposed an entirely new architecture based on existing technologies.

AllJoyn Lambda is an example of an architecture which makes use of MongoDB and Storm as part of architecture for smart environments in IoT [43]. Basanta-Val et al. propose a new time-critical architecture which utilises Spark and Storm for data processing [44]. An ETL-based approach to big data processing is proposed by Guerreiro et al. This proposal utilises Spark, SparkSQL and MongoDB technologies [45]. Finally, an example of a proposed architecture where the choice of big data technology is left open to the implementer is [46].

This section discussed related work by introducing a classification for research on big data architectures. The next section identifies the gap in the literature addressed by this research.

3.3.3 Gap in the Literature

Significant advances have been made in the fields of big data recently, and it continues to develop further as devices and applications produce more and more data. However, although new frameworks and technologies are developed and launched into the market at a very fast pace, matching them with system requirements continues to present a challenge [34]. Arruda highlights the need for addressing big data-specific characteristics in the definition, analysis and specification of both functional and non-functional requirements [34]. This research aims to bridge this gap by abstracting from functional and concentrating on non-functional requirements for the domain of big data which are common to well-known large-scale big data implementations.

Eridaputra et al. call for new methods to model requirements for big data applications [36], and Madhavji et al. identify the need to develop new techniques to assess the impact of architectural design decisions on functional and non-functional requirements [33]. This research is a step in the recommended direction, as it looks at implementations of non-functional requirements for big data by real-world companies and discusses the design decisions that resulted in specific implementations.

Madhavji et al. also drew attention to the lack of academic work on reference architectures and patterns for big data applications, and how these reference architectures can be translated into existing technologies, frameworks and tools to yield concrete deployments [33]. The current research bridges this gap by proposing new reference architecture for the domain of stream data processing and by providing a prototype implementation based on open-source technology.

This section discussed the gap in the literature which motivated the authors to undertake the current research. The next section examines the literature published by the three companies under study to understand how the non-functional requirements defined in Sect. 3.2 are addressed in their real-world implementations.

3.4 Requirements Engineering for Big Data

Systematic approaches to identifying functional services and non-functional requirements are necessary to design and build big data systems. The traditional requirements engineering approaches are unsatisfactory when it comes to identifying requirements for service-oriented systems [34, 35, 47, 48]. This research bridges this gap by identifying five functional services for big data requirements engineering in Sect. 3.4.1 and by providing a comparison of three large companies' approaches to implementing non-functional requirements for big data in Sect. 3.4.2.

3.4.1 Identification of Functional Services for Big Data

The non-functional requirements identified in this study, together with a gap analysis exercise based on the literature survey, were used to identify a number of functional services for big data requirements engineering in a multi-cloud environment. These functional services are illustrated in Fig. 3.16.

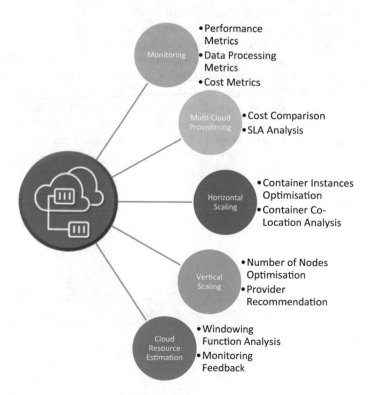

Fig. 3.16 Identified functional services for big data requirements engineering in a multi-cloud environment

A monitoring service provides performance metrics (CPU, memory and network usage), as well as data processing metrics (number of records ingested, number of records processed, percentage of data loss). It also provides cost expended in terms of resources utilised from each provider and cost estimations based on past usage. A multi-cloud provisioning service compares offerings from different providers in terms of cost and other preconfigured SLAs. A horizontal scaling service provides optimisations for the number of container instances running in the cluster, as well as container co-location analysis and recommendations. A vertical scaling service provides optimisations for the number of nodes in the cluster and offers provider comparison and recommendations based on weighted desired qualities. Finally, the cloud resource estimation service adjusts the initial estimations entered by designers of stream processing systems before a given configuration is run, based on the windowing function selected. It also communicates with the monitoring service to adjust estimations for running systems. Figure 3.17 summarises the resource estimation process provided by MC-Compose, a cloud resource estimation service developed as part of this research.

The resource estimation process summarised in Fig. 3.17 starts with a resource estimate for CPU, memory and network consumption submitted by a user. This is based on the assumption that the system is unknown or has not yet been deployed to production. The windowing function, used to render the potentially infinite stream of data finite for processing, is then selected. It consists of a period, which represents how frequently a processing window starts, and a duration, which represents the duration of each processing window. Strategies such as sampling can be implemented by selecting a period higher than the duration, whereas sliding windows can be implemented by selecting a higher duration than period. MC-Compose takes into account the windowing function selected and adjusts the resource estimation entered by the user, who is prompted to accept the adjusted requirements, or manually override them. Once the system is deployed, the monitoring service depicted in Fig. 3.16 sends metrics to MC-Compose, which are used to further

Fig. 3.17 MC-compose resource estimation process summary

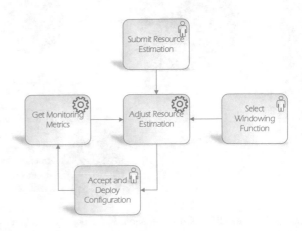

refine the resource estimation calculations and thus improve the recommendations made to the user.

This section summarised five functional services for big data in a multi-cloud environment, identified as part of this research. The next section discusses the approaches of the three companies selected by this study to implementing the ten non-functional requirements identified in Sect. 3.2.3.

3.4.2 Non-functional Requirements

This section presents ten non-functional requirements discussed in the literature published by the three companies selected in Sect. 3.2.2. It then examines how they implemented these requirements and compares the different solutions.

Figure 3.18 shows a summarised view of the ten non-functional requirements discussed in this section, organised hierarchically. At the top level, there are three main requirements: the capacity to process batch data, the capacity to process stream data, the capacity to distribute processing tasks across several machines and to scale up that number and the capacity to seamlessly integrate with existing and future technology. Three other requirements are related to stream processing: the capacity to handle late and out-of-order data, the capacity to offer one of three processing guarantees and the capacity to offer strategies for flow control. Two requirements are related to distribution and scalability: the capacity to offer cloud support and elasticity, so distribution can expand to encompass the ream of the cloud and the capacity to offer fault tolerance, usually by taking snapshots of the state of data processing and relaunching the failed task on a healthy node. Finally, the capacity of a system to have its components substituted for equivalent technology is related to the requirement for integration and extensibility.

The remainder of this section explores how the three companies selected for this study implemented these ten requirements and compares their different solutions.

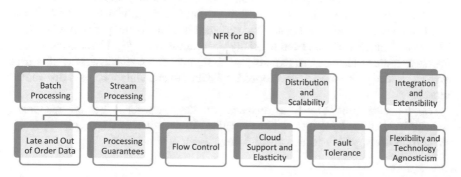

Fig. 3.18 Non-functional requirements for large-scale big data systems

3.4.2.1 Batch Data

This requirement refers to the capability to process data which is finite and usually large in volume.

In terms of size, both Facebook and Twitter estimate that the finite data they hold on disc reaches hundreds of petabytes, with a daily processing volume of tens of petabytes [49]. Netflix's big data is one order of magnitude smaller, with tens of petabytes in store and daily reads of approximately 3 petabytes [50].

With regard to how this requirement is addressed, Facebook uses a combination of three independent, but communicating systems to manage its stored data: an Operational Data Store (ODS), Scuba, Hive and Laser [7].

Twitter's batch data is stored in Hadoop clusters and traditional databases and is processed using Scalding and Presto [49]. Scalding is a Scala library developed in-house to facilitate the specification of map-reduce jobs [10]. Presto, on the other hand, was originally developed by Facebook. It was open-sourced in 2013 [51] and has since been adopted not only by Twitter, but also by Netflix [52].

Differently from the previous two companies, Netflix's Hadoop installation is cloud-based, and it uses an in-house developed system called Genie to manage query jobs submitted via Hadoop, Hive or Pig. Data is also persisted in Amazon S3 databases [53].

3.4.2.2 Stream Data

This requirement refers to the capability to process data which is potentially infinite and usually flowing at high velocity.

Stream processing at Facebook is done by a suite of in-house developed applications: Puma, Swift and Stylus. Puma is a stream processing application with a SQL-like query language optimised for compiled queries. Swift is a much simpler application, used for checkpointing. Finally, Stylus is a stream processing framework which combines stateful or stateless units of processing into more complex DAGs [7].

Storm, one of the most popular stream processing frameworks in use today, was developed by Twitter [4]. Less than five years after the initial release of Storm, however, Twitter announced that it had replaced it with a better performing system, Heron, and that Storm had been officially decommissioned [6]. Heron uses Mesos, an open-source cluster management tool designed for large clusters. It also uses Aurora, a Mesos framework developed by Twitter to schedule jobs on a distributed cluster.

Netflix also uses Mesos to manage its large cluster of cloud resources. Scheduling is done by a custom library called Fenzo, whereas stream processing is done by Mantis, which is also custom-developed.

3.4.2.3 Late and Out-of-Order Data

This requirement relates to stream processing and refers to the capability to process data which arrives late or in a different order from that in which it was emitted. All three streaming architectures utilise the concept of windows of data to transform infinite streaming data into finite windows that can be processed individually [5, 7, 54].

For handling late and out-of-order data, Facebook's Stylus utilises low watermarks. No mention was found in Twitter Heron's academic paper of whether it provides a mechanism for dealing with late or out-of-order data. However, looking at the source code for the Heron API, the BaseWindowedBolt class, merged into the master project in 2017, has a method called withLag(), which allows the developer to specify the maximum amount of time by which a record can be out of order [55].

No mention was found in documentation published by Netflix of Mantis's strategy for dealing with late and out-of-order data. Because the source code for Mantis is proprietary, further investigation was limited.

3.4.2.4 Processing Guarantees

This requirement refers to a stream system's capability to offer processing guarantees, i.e. exactly once, at least once and at most once. Exactly once semantics involves some level of checkpointing to persist state. There is therefore an inherent latency cost associated with it, which is why not all use-cases are implemented this way.

An example of a use-case where exactly once semantics is not a requirement is Facebook's Scuba system. Since the data is intended to be sampled, completeness of the data is not a requirement. Duplication, however, is not acceptable. In this case, at most once is a more fitting processing guarantee than exactly once [7], since it is in line with sampling and does not allow duplicate records to occur. Facebook also has use cases where exactly once processing guarantees are required. These are catered for by Stylus, a real-time system designed with optimisations to provide at least once processing semantics through the use of checkpointing [7].

At Twitter, both Storm and its successor, Heron, offered at least once and at most once guarantees. Identified as a shortcoming by Kulkarni et al. [5], the lack of exactly once semantics in Heron was subsequently addressed and implemented as "effectively once semantics". Effectively once semantics means that data may be processed more than once (the processing would undergo a rewind in case of failure), but it is only delivered once [56].

Netflix uses Kafka as its stream platform and messaging system [57], which means it provides inherent support for exactly once processing through idempotency and atomic transactions [58]. Additionally, at least once and at most once processing guarantees are also supported by Kafka [59].

3.4.2.5 Integration and Extensibility

This requirement refers to the capability to integrate with existing services and components. It also refers to provisions made to facilitate the extension of the existing architecture to incorporate different components in future.

Although Facebook's real-time architecture is composed of many systems, they are integrated thanks to Scribe. Scribe works as a messaging system: all of Facebook's streaming systems write to Scribe, and they also read from Scribe. This allows for the creation of complex pipelines to cater for a multitude of use-cases [7]. In terms of extensibility, any service developed to use Scribe as data source and data output could integrate seamlessly with Facebook's architecture.

As part of a process to make Heron open source, Twitter introduced a number of improvements to make it more flexible and adaptable to different infrastructures and use-cases. By adopting a general-purpose modular architecture, Heron achieved significant decoupling between its internal components and increased its potential for adoption and extension by other companies [6].

Netflix's high-level architecture is somewhat rigid in that there is no alternative to using Mesos as an orchestration and cluster management tool or AWS as a cloud provider [60]. Additionally, Titus must run as a single framework on top of Mesos. This limitation, however, was introduced by design. With Titus running as a single framework on Mesos, it can allocate tasks more efficiently and has full visibility of resources across the entire cluster [9]. Because Titus is a proprietary system designed by Netflix and optimised to fulfil its own use cases, it was initially tightly coupled to Netflix's infrastructure. It has, however, evolved into a more generic product since being open-sourced in April 2018 [61].

3.4.2.6 Distribution and Scalability

This requirement refers to the capability to distribute data processing amongst different machines, located in different data centres, in a multi-clustered architecture. Dynamic scaling, which addresses the possibility of adding or removing nodes to a running system without any downtime, is also addressed as part of this requirement.

Scalability was one of the driving factors behind the development of Scribe as a messaging system at Facebook. Similarly to Kafka, Scribe can be scaled up by increasing the number of buckets (brokers) running, thus increasing the level of parallelism [7]. There is no mechanism in place for dynamic scaling of Puma and Stylus systems [7].

At Twitter, Heron was developed as a more efficient and scalable alternative to Storm. Heron uses an in-house developed proprietary framework called Dhalion to help determine whether the cluster needs to be scaled up or down [62].

As Netflix's architecture is cloud-based, it is inherently elastic and scalable. Fenzo is responsible for dynamically scaling resources by adding or removing EC2 nodes to the Mesos infrastructure as needed [54].

3.4.2.7   Cloud Support and Elasticity

This requirement refers to the capability to move the architecture (or part of it) into the cloud to take advantage of the many benefits associated with its economies of scale.

Based on the material examined, Neflix's architecture is the only which is predominantly cloud-based. Having started with services running on AWS virtual machines, they are now undergoing a shift towards a container-based approach, with a few services now running in containers on AWS infrastructure [9]. Twitter has also undergone a shift towards a containerised architecture, albeit not cloud-based, with the development and implementation of Heron. As containers become more widespread, the risk of vendor lock-in is lowered, since containers enable the decoupling of the processing framework from the infrastructure they run in. Future migration to a safer multi-cloud set-up is not only possible, but desirable [63].

3.4.2.8   Fault Tolerance

This requirement refers to the capability of a system to continue to operate should one or more nodes fail. Ideally, the system should recover gracefully, with minimal repercussions for the user experience.

Fault tolerance is a requirement of Facebook's real-time systems, currently implemented through node independence and by using a persistent messaging system for all inter-system communication. Scribe, Facebook's messaging system, persists data to disc and is backed by Swift, a stream platform designed to provide checkpointing [7].

At Twitter, fault tolerance is addressed at different levels. At architectural level, a modular distributed architecture provides better fault tolerance than a monolithic design. At container level, resource provisioning and job scheduling are decoupled, with the scheduler being responsible for monitoring the status of running containers and for trying to restart any failed ones, along with the processes they were running. At JVM level, Heron limits task processing to one per JVM. This way, should failure occur, it is much easier to isolate the failed task and the JVM where it was running [6]. At topology level, the management of running topologies is decentralised, with one Topology Master per topology, which means failure of one topology does not affect others [5].

As Netflix's production systems are cloud-based, fault tolerance is addressed from the perspective of a cloud consumer. The Active-Active project was launched by Netflix with the aim of achieving fault tolerance through isolation and redundancy by deploying services to the USA across two AWS regions: US-East-1 and US-West-2 [64]. This project was later expanded to incorporate the EU-West-1 region, as European locations were still subjected to single points of failure [65]. With this latest development, traffic could be routed between any of the three regions across the globe, increasing the resilience of Netflix's architecture.

3.4.2.9   Flow Control

This requirement refers to the capability to handle scenarios where the data source emits records faster than the system can consume.

All real-time systems at Facebook read and write to Scribe. As described by Chen et al., this central use of a persistent messaging system makes Facebook's real-time architecture resilient to backpressure. Since nodes are independent, if one node slows down, the job is simply allocated to a different node, instead of the slowing down the whole pipeline [7]. The exact strategy used by Scribe to implement flow control is not made explicit in the paper.

Heron was designed with a flow control mechanism as an improvement over Storm, where producers dropped data if consumers were too busy to receive it. When Heron is in backpressure mode, the Stream Manager limits incoming data through the furthest upstream component (the spout) in order to slow down the flow of data throughout the topology. The data processing speed is thus reduced to the speed of the slowest component. Once backpressure is relieved and Heron exits backpressure mode, the spout is set back to emit records at its normal rate [5].

At Netflix, Mantis jobs are written using ReactiveX, a collection of powerful open-source reactive libraries for the JVM [66]. RxJava, one of the libraries in ReactiveX originally developed by Netflix, offers a variety of strategies for dealing with backpressure such as, for example, the concept of a cold observable which only starts emitting data if it is being observed, at a rate controlled by the observer. For hot observables which emit data regardless of whether or not they are being observed, RxJava provides the options to buffer, sample, debounce or window the incoming data [67].

3.4.2.10   Flexibility and Technology Agnosticism

This criterion refers to the capability of architecture to use interchangeable technology in place of existing components.

Out of the three architectures investigated, Facebook's set-up is the least flexible and the least technologically agnostic. With the exception of Hive and its ODS, built on HBase [68], Facebook's data systems were developed in-house to cater for very specific use-cases. This is perhaps the reason why, at the time of writing, only Scribe has been made open source [69]. It is worth noting, however, that the Scribe project was not developed further, and the source code has been archived [13].

Heron's modular architecture is flexible by design, and the technologies chosen for Twitter's particular implementation, Aurora and Mesos, are not compulsory for other implementations. Heron's flexibility is evidenced by its adoption by large-scale companies such as Microsoft [70], and its technology agnosticism is evidenced by its successful implementation on a Kubernetes (instead of Mesos) cluster [71].

At programming level, Netflix is an active participant of the Reactive Streams initiative, which aims to standardise reactive libraries with an aim to rendering them interoperable. Considering that JDK 9, released in September 2017, is also compatible with Reactive Streams, there is potential for Mantis's jobs to be defined in standard Java in the future.

At cloud infrastructure level, the use of containers as a deployment abstraction reduces the tight coupling between Netflix's artefacts and specific virtual machine offerings provided by AWS. This is defined by Leung et al. [9] as a shift to a more application-centric deployment. It is worth noting, however, that, at the time of writing, Netflix officially relies on a single cloud provider: AWS, despite there being indication that they would have started to evaluate Google Cloud in an effort towards achieving a multi-cloud strategy [72].

At architecture level, because Titus was only recently open-sourced, this study did not evaluate whether essential parts of its architecture such as the Mantis, Fenzo or the Mesos cluster could be replaced with an equivalent. It is expected, however, that its transition to open source could attract important contributions from the community and enhance its flexibility and technology agnosticism.

3.4.2.11 Summary and Applications

This section provides a summary of the implementation approaches of the ten non-functional requirements by the three companies selected. Additionally, it introduces direct applications of the current study: the design and development of MC-BDP, a new reference architecture for large-scale stream big data processing.

As continuation of this research, the non-functional requirements discussed in this study were used to guide the design and implementation of new reference architecture for big data processing in the cloud: MC-BDP. MC-BDP is an evolution of the PaaS-BDP architectural pattern originally proposed by the authors. While PaaS-BDP introduced a framework-agnostic programming model and enabled different frameworks to share a pool of location and provider-independent resources [63], MC-BDP expands this model by explicitly prescribing a multi-tenant environment where nodes are deployed to multiple clouds. Figure 3.19 shows a summary of how Facebook, Twitter and Netflix implemented the ten non-functional requirements discussed in this research. The last column shows MC-BDP, the proposed reference architecture.

MC-BDP was subsequently evaluated via a simulated energy efficiency case study where a prototype was developed using open-source technology to calculate the Power Usage Effectiveness (PUE) of a data centre at Leeds Beckett University. The components of this prototype implementation were deployed to the OSDC, Azure and Google clouds. Based on the non-functional requirements discussed in the current study, three hypotheses were formulated and verified empirically:

H1. MC-BDP is scalable across clouds.
H2. MC-BDP is fault-tolerant across clouds.
H3. MC-BDP's provision for technology agnosticism does not incur a significant increase in processing overhead.

Criteria / Architecture	Facebook	Twitter	Netflix	MC-BDP
Batch Data	ODS/Scuba/Hive/Laser	Scalding/Presto	Genie/Hadoop/Hive/Pig	flexible (integration framework)
Stream Data	Puma/Swift/Stylus	Heron	Mantis/Fenzo	flexible (integration framework)
Late and Out-of-Order Data	yes (low watermarks)	yes (record lag)	no information found	yes (low watermarks)
Processing Guarantees	exactly once*	effectively once	exactly once	exactly once
Integration and Extensibility	with Facebook technology	with a range of technologies (modular architecture)	with Netflix technology	with a range of technologies (modular architecture)
Distribution and Scalability	multiple data centres lack of dynamic scaling	multiple data centres Dhalion provides dyn. scaling	multiple AWS regions Fenzo provides dyn. scaling	multiple clouds framework provides dyn. scaling
Cloud Support and Elasticity	no	no	yes	yes
Multi-Cloud Support	no	no	no	yes
Fault Tolerance	architectural and node levels	architectural, container, JVM and topology levels	architectural level (cloud consumer perspective)	architectural, container and framework levels
Flow Control	via persistent messaging system (Scribe)	via Heron's backpressure mode	programmatically via ReactiveX	Kafka + big data framework
Flexibility and Technology Agnosticism	low (proprietary non-generic components)	high (modular architecture)	high at programming level (Reactive Streams)	high (modular architecture)

* not by all systems

Fig. 3.19 Summary of non-functional requirements for big data and implementations

This section examined how the three companies selected for this study: Facebook, Twitter and Netflix implemented the ten non-functional requirements defined in Sect. 3.2.2. Additionally, it introduced two instances where the present study was applied to inform the design and development of further contributions: MC-BDP and MC-Compose. A full presentation and discussion of MC-BDP and MC-Compose, however, lie outside the scope of this chapter and will be the subject of a future publication. The next section presents the conclusion to this work and suggestions for future work.

3.5 Conclusion and Future Work

This study presented the results of a literature search for non-functional requirements relevant to real-world big data implementations. Three companies were selected for this comparative study: Facebook, Twitter and Netflix. Their specific implementations of the non-functional requirements selected were compared and discussed in detail and are summarised in this section.

Facebook and Twitter process the largest volume of data, with Twitter having the lowest requirement for latency. Differently from Facebook, these two architectures were also explicitly designed to handle late and out-of-order data. In terms of processing guarantees, all three architectures support exactly once semantics.

Although the existing systems at Facebook and Netflix are integrated, they were not designed as a unified modular framework. Heron, on the other hand, was developed by Twitter as an improvement over Storm, which suffered from bottlenecks and single points of failure. Heron's modular architecture makes it more flexible and technologically agnostic, as well as a stronger candidate for adoption by other companies when compared to systems developed by the other two companies.

Differently from Facebook and Twitter, which provide mechanisms for scalability and fault tolerance in their infrastructures, Netflix approaches this concept from a cloud consumer's perspective, since its architecture is cloud-based. Netflix's deployments are distributed over multiple regions, although support for multi-cloud is still lacking.

All three architectures provide mechanisms for flow control. Facebook and Twitter control backpressure from an infrastructure level, whereas Netflix provides methods and constructs to achieve this programmatically.

The authors recognise that more thorough results could have been obtained if our approach had included direct observation of the systems under evaluation by way of a set of case studies. However, due to time and resource constraints, the scope of the present study was limited to published sources.

Future work shall involve a prototype implementation of the MC-BDP reference architecture and its subsequent evaluation in terms of a minimum of three of the non-functional requirements for large-scale big data applications identified in this study. Additionally, this research aims to develop one or more of the functional services for big data requirements engineering in a multi-cloud environment described in the previous section. An example of such service is MC-Compose, a cloud resource estimation service for stream big data systems which adjusts user-entered estimations based on the windowing function selected and on monitoring feedback.

Acknowledgements This work made use of the **Open Science Data Cloud (OSDC)** which is an Open Commons Consortium (OCC)-sponsored project. Cloud computing resources were provided by **Google Cloud** and **Microsoft Azure** for Research awards. Container and cloud native technologies were provided by **Weaveworks**.

References

1. Cao L (2017) Data science: challenges and directions. Commun ACM 60(8):59–68
2. Desjardins J (2019) What Happens in an internet minute in 2019?. Visual capitalist, 13 Mar 2019. Available: https://www.visualcapitalist.com/what-happens-in-an-internet-minute-in-2019/. Accessed 22 Mar 2019
3. Chung L, Prado Leite JC (2009) Conceptual modeling: foundations and applications. In: Borgida AT, Chaudhri VK, Giorgini P, Yu ES (eds). Springer, Berlin, pp 363–379
4. Toshniwal A et al (2014) Storm@Twitter. In: Proceedings of the 2014 ACM SIGMOD international conference on management of data, New York, NY, USA, pp 147–156
5. Kulkarni S, et al (2015) Twitter heron: stream processing at scale. In: Proceedings of the 2015 ACM SIGMOD international conference on management of data, New York, NY, USA, pp 239–250
6. Fu M et al (2017) Twitter Heron: towards extensible streaming engines. In: 2017 IEEE 33rd international conference on data engineering (ICDE), 2017, pp 1165–1172
7. Chen et al GJ (2016) Realtime data processing at Facebook. In: Proceedings of the 2016 international conference on management of data, New York, NY, USA, pp 1087–1098
8. Bronson N, Lento T, Wiener JL (2015) Open data challenges at Facebook. In: 2015 IEEE 31st international conference on data engineering, 2015, pp 1516–1519

9. Leung A, Spyker A, Bozarth T (2017) Titus: introducing containers to the Netflix cloud. Queue 15(5):30:53–77
10. Twitter, Inc. (2018) Scalding: a scala API for cascading
11. Heron Documentation (2019) Heron documentation—Heron's architecture. Available https://apache.github.io/incubator-heron/docs/concepts/architecture/. Accessed 02 Jun 2019
12. Goetz PT, Lim J, Patil K, Brahmbhatt P (2019) Apache storm. The Apache Software Foundation
13. Scribe (2014) Facebook archive
14. Eliot S (2010) Microsoft cosmos: petabytes perfectly processed perfunctorily, 11 May 2010. Available https://blogs.msdn.microsoft.com/seliot/2010/11/05/microsoft-cosmos-petabytes-perfectly-processed-perfunctorily/. Accessed 24 Jan 2018
15. Bernstein P, Bykov S, Geller A, Kliot G, Thelin J (2014) Orleans: distributed virtual actors for programmability and scalability
16. Akidau T et al (2013) MillWheel: fault-tolerant stream processing at internet scale. Proc VLDB Endow 6(11):1033–1044
17. Akidau T et al (2015) The dataflow model: a practical approach to balancing correctness, latency, and cost in massive-scale, unbounded, out-of-order data processing. Proc VLDB Endow 8:1792–1803
18. Cheng B, Longo S, Cirillo F, Bauer M, Kovacs E (2015) Building a big data platform for smart cities: experience and lessons from Santander. In: 2015 IEEE International Congress on Big Data, pp 592–599
19. Abadi DJ et al (2005) The design of the borealis stream processing engine. CIDR 5:277–289
20. Loesing S, Hentschel M, Kraska T, Kossmann D (2012) Stormy: an elastic and highly available streaming service in the cloud. p 55
21. Alexandrov A et al (2014) The stratosphere platform for big data analytics. VLDB J 23(6), pp 939–964
22. Zhu JY, Xu J, Li VOK (2016) A four-layer architecture for online and historical big data analytics. pp 634–639
23. Amazon EMR—Amazon Web Services (2019) Amazon EMR. Available: https://aws.amazon.com/emr/. Accessed 15 Mar 2019
24. Azure HDInsight—Hadoop, Spark, & Kafka Service | Microsoft Azure (2019) HDInsight. Available: https://azure.microsoft.com/en-gb/services/hdinsight/. Accessed 15 Mar 2019
25. Big Data Analytics Infrastructure Solutions | IBM (2019) IBM big data analytics solutions. Available: https://www.ibm.com/it-infrastructure/solutions/big-data. Accessed 15 Mar 2019
26. Chandramouli B, Goldstein J, Barnett M, Terwilliger JF (2015) Trill: engineering a library for diverse analytics. IEEE Data Eng Bull 38:51–60
27. Noghabi SA et al (2017) Samza: stateful scalable stream processing at LinkedIn. Proc VLDB Endow 10(12):1634–1645
28. Akidau T, Chernyak S, Lax R (2018) Streaming systems: the what, where, when, and how of large-scale data processing, 1st edn. O'Reilly Media, Beijing
29. Akber SMA, Lin C, Chen H, Zhang F, Jin H (2017) Exploring the impact of processing guarantees on performance of stream data processing. In: 2017 IEEE 17th international conference on communication technology (ICCT), pp 1286–1290
30. Satzger B, Hummer W, Inzinger C, Leitner P, Dustdar S (2013) Winds of change: from vendor lock-into the meta cloud. IEEE Internet Comput 17(1):69–73
31. Brodkin J (2011) Amazon EC2 outage calls "availability zones" into question. Network World, 21 Apr 2011. Available: https://www.networkworld.com/article/2202805/cloud-computing/amazon-ec2-outage-calls--availability-zones--into-question.html. Accessed 22 Feb 2019
32. Dayaratna A (2016) Microsoft azure recovers from multi-region azure DNS service disruption. Cloud Computing Today, 15 Sep 2016. Available: https://cloud-computing-today.com/2016/09/15/microsoft-azure-recovers-from-multi-region-azure-dns-service-disruption/. Accessed 22-Feb-2019

33. Madhavji NH, Miranskyy A, Kontogiannis K (2015) Big picture of big data software engineering: with example research challenges. In: 2015 IEEE/ACM 1st international workshop on big data software engineering, 2015, pp 11–14
34. Arruda D (2018) Requirements engineering in the context of big data applications. ACM SIGSOFT Softw Eng Notes 43(1):1–6
35. Noorwali I, Arruda D, Madhavji NH (2016) Understanding quality requirements in the context of big data systems. In: 2016 IEEE/ACM 2nd international workshop on big data software engineering (BIGDSE), pp 76–79
36. Eridaputra H, Hendradjaya B, Sunindyo WD (2014) Modeling the requirements for big data application using goal oriented approach. In: 2014 international conference on data and software engineering (ICODSE), pp 1–6
37. Al-Najran N, Dahanayake A (2015) A requirements specification framework for big data collection and capture. In: New trends in databases and information systems, pp 12–19
38. Al-Najran N (2015) A requirements specification framework for big data collection and capture. In: Masters of Science in Software Engineering, Prince Sultan University, Riyadh
39. Arruda D, Madhavji NH (2017) Towards a requirements engineering artefact model in the context of big data software development projects: research in progress. In: 2017 IEEE international conference on big data (big data), pp 2314–2319
40. Shoro AG, Soomro TR (2015) Big data analysis: apache spark perspective. Glob J Comput Sci Technol 15(1)
41. Kiran M, Murphy P, Monga I, Dugan J, Baveja SS (2015) Lambda architecture for cost effective batch and speed big data processing. pp 2785–2792
42. Sun B, Zhang L, Chen Y (2017) Design of big data processing system for spacecraft testing experiment. In: 2017 7th IEEE international symposium on microwave, antenna, propagation, and EMC technologies (MAPE), pp 164–167
43. Villari M, Celesti A, Fazio M, Puliafito A (2014) AllJoyn Lambda: an architecture for the management of smart environments in IoT. pp 9–14
44. Basanta-Val P, Audsley NC, Wellings A, Gray I, Fernandez-Garcia N (2016) Architecting time-critical big-data systems. IEEE Trans Big Data 99:1–1
45. Guerreiro G, Figueiras P, Silva R, Costa R, Jardim-Goncalves R (2016) An architecture for big data processing on intelligent transportation systems. An application scenario on highway traffic flows. pp 65–72
46. Costa C, Santos MY (2016) BASIS: a big data architecture for smart cities. pp 1247–1256
47. Ramachandran M (2013) Business requirements engineering for developing cloud computing services. In: Mahmood Z, Saeed S (eds) Software engineering frameworks for the cloud computing paradigm. Springer, London, pp 123–143
48. Ramachandran M, Mahmood Z (eds) (2017) Requirements engineering for service and cloud computing. Springer International Publishing, Berlin
49. Krishnan S (2016) Discovery and consumption of analytics data at Twitter. 29 Jun 2016
50. Gianos T, Weeks D (2016) Petabytes scale analytics infrastructure @Netflix. Presented at the QCon, San Francisco, 11 Aug 2016
51. Pearce J (2013) 2013: a year of open source at Facebook. Facebook Code, 20 Dec 2013. Available: https://code.facebook.com/posts/604847252884576/2013-a-year-of-open-source-at-facebook/. Accessed 12 Feb 2018
52. Tse E, Luo Z, Yigitbasi N (2014) Using presto in our big data platform on AWS. The Netflix Tech Blog, 10 Jul 2014
53. Krishnan S, Tse E (2013) Hadoop platform as a service in the cloud. The Netflix Tech Blog, 10 Jan 2013
54. Schmaus B, Carey C, Joshi N, Mahilani N, Podila S (2016) Stream-processing with Mantis. Netflix TechBlog, 14 Mar 2016
55. Peng B (2017) [ISSUE-1124]—windows bolt support #2241. Twitter Inc.
56. Heron Documentation (2019) Heron delivery semantics

57. Wu S et al (2016) The Netflix Tech Blog: evolution of the Netflix data pipeline. 15 Feb 2016. Available: http://techblog.netflix.com/2016/02/evolution-of-netflix-data-pipeline.html. Accessed 30 Oct 2016
58. Woodie A (2017) A peek inside Kafka's new "exactly once" feature. Datanami, 07 Mar 2017
59. Dobbelaere P, Esmaili KS (2017) Kafka versus RabbitMQ: a comparative study of two industry reference publish/subscribe implementations: industry paper. In: Proceedings of the 11th ACM international conference on distributed and event-based systems, New York, NY, USA, pp 227–238
60. Titus (2018) Titus documentation. Available: https://netflix.github.io/titus/. Accessed 18 Mar 2019
61. Joshi A et al (2018) Titus, the Netflix container management platform, is now open source. Medium, 18 Apr 2018
62. Graham B (2017) From rivulets to rivers: elastic stream processing in Heron. 16 Mar 2017
63. Vergilio T, Ramachandran M (2018) PaaS-BDP—a multi-cloud architectural pattern for big data processing on a platform-as-a-service model. In: Proceedings of the 3rd international conference on complexity, future information systems and risk, Madeira
64. Meshenberg R, Gopalani N, Kosewski L (2013) Active-active for multi-regional resiliency. Netflix TechBlog, 02 Dec 2013
65. Stout P (2016) Global cloud—active-active and beyond. Netflix TechBlog, 30 Mar 2016
66. Christiansen B, Husain J (2013) Reactive programming in the Netflix API with RxJava. Netflix TechBlog, 04 Dec 2013
67. Gross D, Karnok D (2016) Backpressure. ReactiveX/RxJava Wiki, 27 Jun 2016. Available: https://github.com/ReactiveX/RxJava/wiki/Backpressure. Accessed 15 Feb 2018
68. Tang L (2012) Facebook's large scale monitoring system built on HBase. In: Presented at the strata conference + Hadoop world, New York, NY, USA, 24 Oct 2012
69. Johnson R (2018) Facebook's scribe technology now open source. Facebook Code, 24 Oct 2008
70. Ramasamy K (2016) Open sourcing Twitter Heron. Twitter Engineering Blog, 25 May 2016
71. Kellogg C (2017) The Heron stream processing engine on Google Kubernetes Engine. Streamlio, 28 Nov 2017
72. McLaughlin K (2018) Netflix, long an AWS customer, tests waters on Google cloud. The Information, 17 Apr 2018. Available: https://www.theinformation.com/articles/netflix-long-an-aws-customer-tests-waters-on-google-cloud. Accessed 18 Mar 2019

Chapter 4
Migrating from Monoliths to Cloud-Based Microservices: A Banking Industry Example

Alan Megargel, Venky Shankararaman and David K. Walker

Abstract As organizations are beginning to place cloud computing at the heart of their digital transformation strategy, it is important that they adopt appropriate architectures and development methodologies to leverage the full benefits of the cloud paradigm. A mere "lift and move" approach, where traditional monolith applications are moved to the cloud will not support the demands of digital services. While monolithic applications may be easier to develop and control, they are inflexible to change to become more suitable for cloud environments. Microservices architecture, which adopts some of the concepts and principles of service-oriented architecture, provides a number of benefits, when developing an enterprise application, over a monolithic architecture. Microservices architecture offers agility, faster development and deployment cycles, scalability of selected functionality and the ability to develop solutions using a mixture of technologies. Microservices architecture aims to decompose a monolithic application into a set of independent services which communicate with each other through open APIs or highly scalable messaging. In short, microservices architecture is more suited for building agile and scalable cloud-based solutions. This chapter provides a practice-based view and comparison between the monolithic and microservices styles of application architecture in the context of cloud computing vision and proposes a methodology for transitioning from monoliths to cloud-based microservices.

Keywords Microservices architecture · Monolithic architecture · Cloud-based · Microservice identification · Migration · Microservices

A. Megargel (✉) · V. Shankararaman
School of Information Systems, Singapore Management University, Singapore, Singapore
e-mail: alanmegargel@smu.edu.sg

V. Shankararaman
e-mail: venks@smu.edu.sg

D. K. Walker
Independent Consultant, Singapore, Singapore
e-mail: davidkwalker@me.com

© Springer Nature Switzerland AG 2020
M. Ramachandran and Z. Mahmood (eds.), *Software Engineering in the Era of Cloud Computing*, Computer Communications and Networks,
https://doi.org/10.1007/978-3-030-33624-0_4

85

4.1   Introduction

Digital transformation requires organizations to be nimble and adopt accelerated innovation methods which enable the delivery of new digital services to customers, partners and employees. To achieve this, organizations are looking towards building flexible cloud-based applications, whereby it is easier to add and update digital services as requirements and technologies change. Legacy monolithic applications might be operationally acceptable on a day-to-day basis but these applications are not well suited for building digital services. Traditional monolithic architecture and software development methods remain a stumbling block for driving digital transformation. In order to efficiently drive digital transformation, organizations are exploring a new software development methodology and architecture, "cloud-based microservices architecture", whereby IT solutions can be organized around granular business capabilities which can be rapidly assembled to create new cloud-based digital experience applications.

The "microservices" architecture is a style and method of developing software applications more quickly by building them as collections of independent, small, modular services. Organizations are currently faced with two challenges, namely how to build new applications using a microservices architecture and how to migrate from a monolith to a cloud-based microservices architecture. This chapter provides practical guidance and a methodical approach to address these two challenges.

A single monolith is typically composed of tens or hundreds of business functions, which are deployed together in one software release. Microservices, on the other hand, typically encapsulate a single business function which can be scaled separately and deployed separately. It is possible to develop a large enterprise application for cloud deployment by assembling and orchestrating a set of microservices, as an alternative to developing a monolith.

In this chapter, we first discuss the challenges of monolithic applications in terms of technology stack, scalability, change management and deployment. We then propose a microservices architecture, as an alternative, and provide a comparison with the monolith. In the next section, we discuss a methodical approach to transitioning from a monolith application to a cloud-based microservices application, both from the perspective of building new solutions from scratch and migrating existing solutions built as monoliths. Finally, we conclude with a summary and ideas for future work.

4.2   Monolithic Applications: Background and Challenges

A monolithic application, or "Monolith", describes a legacy style of application architecture which does not consider modularity as a design principle. Originally, the term "monolithic" was used to describe large mainframe applications [1], which are self-contained and become increasingly complex to maintain as the number of functions they support increases over many years of version updates. Following the mainframe era, incarnations of the monolithic architecture style emerged, namely client–server architecture and three-tier web application architecture [2]. A common characteristic of all forms of monolithic application is the presence of three distinct architecture layers: the user interface layer, the business logic layer and the database layer. A simplified illustration of these three layers, or tiers, is provided in Fig. 4.1.

The defining characteristic of a monolith is that all of the business logic is developed and deployed together onto the middle tier, typically hosted on an application server. More broadly, beyond mainframes, a monolith can be described as any block of code that includes multiple functions. Business logic is coded into functions, each fulfilling a specific business capability, e.g. order management or account maintenance. The first release of an application might include several tens of functions, and with subsequent releases, the application might grow to include several hundreds of functions [3]. Before discussing their issues it is only fair to state that monoliths, especially mainframe systems, are highly performing in terms of response time and throughput and are highly resilient and reliable [4]. Many established banks are still relying on 1970s mainframe technology for their core banking systems [4]. However, monoliths are not suitable for cloud deployment due to several reasons which are explained below.

Technology Stack

In a monolith, the functions which implement business logic are all typically written using the same programming language as was popular and relevant at the time the original application was developed. Mainframe applications, especially legacy systems for example, are written using the COBOL language, and any extensions or subsequent version releases of the application must also be written in COBOL. Developers are locked into the original technology stack, and as such are not free to develop new functions using modern application frameworks or languages suitable for cloud deployment [2]. As the number of functions increases, a monolith becomes more complex and requires a larger team of developers to support the application [3].

Fig. 4.1 Monolithic application (or "monolith")

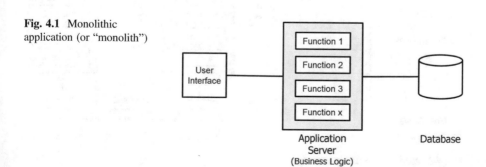

The functions implemented in a monolith, all developed using the same programming language, must interact with each other using native method calls and are therefore tightly coupled [3]. For cloud deployment, loosely coupled functions are more suitable [5]. Loose coupling of functions within a monolith is not possible; for example, it is not possible for function-A to interact with function-B using native method calls if the two functions are deployed onto different servers.

Scalability

Functions within a monolith collectively share the resources (CPU and memory) available on the host system. The amount of system recourses consumed by each function varies depending on demand. A high-demand function, for example one that has a high number of requests via the user interface, or one that is computationally intensive, may at one point consume all of the available resources on the host machine. Therefore, scalability within a single monolith is limited [3].

Vertically scaling the monolith by increasing the system memory would be an option, but a high-demand function would eventually consume the additional memory as well. Since functions within a monolith are tightly coupled and cannot be individually deployed in separate systems, as mentioned in the previous section, the best and most widely used option would be to scale the monolith horizontally.

Horizontal scaling of a monolith, as illustrated in Fig. 4.2, involves adding whole new redundant servers [6], as many as necessary, in order to handle any number of incoming requests through the user interface. A load balancer is needed in order to split the load of incoming requests evenly between the servers. Session replication between servers is needed so that a single user session can span across

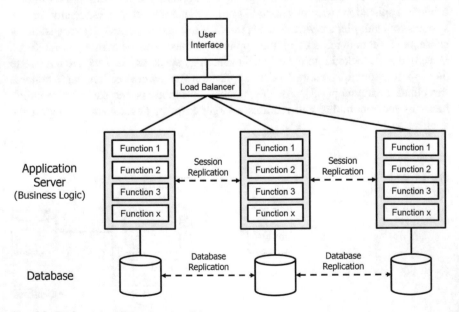

Fig. 4.2 Horizontal scaling of a monolith

servers, or alternatively "sticky" session can be configured to ensure that all requests from the same user are routed consistently to the same server. Either way, database replication is needed in order to ensure that all redundant instances of the database are kept current. Horizontal scaling adds cost and complexity to a monolith, making it impractical for cloud deployment.

Change Management

As mentioned above, a monolith becomes more complex as the number of functions increases over time, requiring a larger support team. Functions which interact using native method calls are tightly coupled and interdependent and therefore are susceptible to change. A change to one function might impact any other function which interacts with that function. Due to these interdependencies, testing only the function which has changed would be insufficient; rather, the entire application should be retested to ensure there is no impact due to the change. Retesting an entire application implies that all test cases need to be regression tested, ensuring that tests which are expected to pass still pass and tests which are expected to fail still fail.

Because a change to any function requires the entire application to be retested, change management processes for monoliths are complex. Test cases need to be maintained. Regression tests need to be planned and scheduled. Test results need to be reviewed. Any test failures cause the entire application to revert back to the development team for bug fixing. Monoliths are typically managed using a waterfall software development lifecycle (SDLC) methodology [7], which requires the entire application to be promoted through a sequence of states, namely development, system integration testing (SIT), user acceptance testing (UAT) and production. Typically large enterprises such as banks have specialized change management teams who plan, schedule and execute changes. Incident management teams report that 80–90 per cent of production problems occur due to improperly tested changes as the root cause [4], even with rigorous testing practices in place. Due to the risk of production problems, banks may schedule the redeployment of monoliths to occur only once per month, even for routine enhancements. The careful and rigorous testing practices implemented for monoliths can inhibit the time-to-market of new customer experience-driven innovations.

Deployment

Individual functions within a monolith cannot be individually deployed; rather, the entire application must be deployed. The deployment package for a monolith is typically one large file. For example, the deployment package for a java web application is a single web application resource (WAR) file. Other types of single file deployment archives include java archive (JAR), enterprise java bean (EJB), tape archive (TAR) for Linux/Unix and dynamic link library (DLL) for Windows. The deployment archive for a monolith increases in size as the number of functions within the monolith increases.

Individual functions within a monolith cannot be individually restarted after deployment; rather, the entire application must be restarted. The implication of this is that the entire application would be unavailable to users while it is being restarted, unless the application is deployed in a high availability (HA) configuration of servers, in which case the application could be deployed and restarted on one HA server at a time. Large monoliths, with hundreds of functions, can take 20–40 min to restart [3].

The size of the deployment archive, together with long restart times, plus the fact that the entire application must be redeployed each time there is a change, makes the use of modern DevOps methods and tools challenging if not impractical for large monoliths. This would be the case for cloud deployments as well as for on-premises deployments of monoliths.

4.3   Microservices: A Cloud-Based Alternative

Microservices are "a variant of the service-oriented architecture (SOA) architectural style that structures an application as a collection of loosely coupled services" [8]. A microservice encapsulates a function, or a business capability [9], which owns its own data [10], and can be independently deployed and independently scaled [3]. Microservices can encapsulate business entities (e.g. Product, Customer, Account) or can encapsulate business activities which orchestrate multiple business entities (e.g. Credit Evaluation, Trade Settlement) [5, 11].

An atomic microservice [5] is a fine-grained service which encapsulates the functionality and data of a single business entity such as Product. In this example, the Product service owns product data; that is, if any other service requires product data, it must access it via the Product service interface. The Product service exposes functions or operations via its interface such as GET product data, POST (create) new product data, PUT (update) existing product data and DELETE product data. Atomic microservices represent the smallest reusable software modules which cannot usefully be further subdivided or decomposed [5].

A composite microservice [5] is a course-grained service which encapsulates the functionality of a single business activity, such as fund transfer. In this example, the fund transfer service orchestrates an end-to-end process by invoking the operations of several atomic microservices in a sequence which fulfils the business activity, as illustrated in Fig. 4.3. Composite microservices can also perform transaction management (e.g. commit or rollback) across the orchestration.

In a microservices layered architecture, the service integration and orchestration responsibilities previously handled by an enterprise service bus (ESB) are now transferred to and dispersed among composite microservices [5]. Alternatively, the communication between microservices can be event-based [12], in which case the end-to-end process logic is distributed among the microservices. With this latter event-based approach, the end-to-end process logic can be reconstructed using service discovery and architecture recovery tools [13].

Microservices architecture (MSA) principles are similar to those established for SOA, with some additions. With regard to legacy issues associated with monoliths, the objectives of SOA and MSA are similar [12]. Both SOA and MSA aim to

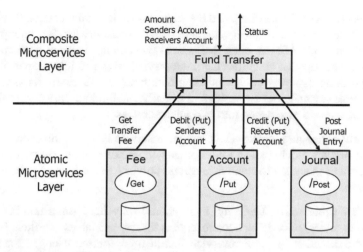

Fig. 4.3 Microservice layers (with fund transfer example)

transform inflexible legacy architectures into services-based architectures which are more flexible and agile for developing new innovative digital solutions [5]. In complex organizations like traditional banks, SOA maturity is key to overcoming legacy systems as an inhibiter for digital banking transformation [4]. While SOA is a key enabler for the agility of on-premises solutions in the presence of monolithic legacy systems, this architecture does not translate well onto the cloud where monolith implementations are impractical. As such, MSA is now a key enabler for the agility of cloud-based solutions, provided that microservices are designed at the right level of encapsulation or boundary context [9]. A set of MSA boundary-setting design principles are provided as follows:

MSA Boundary-Setting Design Principles

P1. Do one thing well Microservices should be highly cohesive [9, 14] in that they encapsulate elements (methods and data) that belong together. A microservice has a specific responsibility enforced by explicit boundaries. It is the only source of a function or truth; that is, the microservice is designed to be the single place to get, add or change a "thing". A microservice should "do one thing well".

P2. No bigger than a squad Each microservice is small enough that it can be built and maintained by a squad (small team) working independently [3]. A single quad team should comfortably own a microservice, whereby the full context of the microservice is able to be understood by a single person. The microservice should ideally be less than a few hundred lines of code or zero code using a GUI-driven designer studio. Smaller microservices are optimised to be rewritten/refactored.

P3. Grouping like data Data and its operations set boundaries. The functional boundary of a microservice is based on the data that it owns, the operations it performs (e.g. REST resources), and the views it provides on that data [9, 14]. Data that is closely related belongs under the same microservice; for example, data needed for a single API call often (but not always) belongs to a single microservice. If putting data together simplifies the microservice APIs and interactions, then that is a good thing. Conversely, if separating data does not adversely impact APIs or code complexity and does not result in a trivially small microservice, then that data might make sense to separate into two microservices.

P4. Do not share data stores Only one microservice is to own its underlying data [9]. This implies moving away from normalized and centralised shared data stores. Microservices that need to share data can do so via API interaction or event-based interaction.

P5. A few tables only Typically, there should only be a small number of data stores (e.g. tables) underlying a microservice; that is, 1–3 tables are often the range. Data store selection for a microservice should be optimised using fit for purpose styles, e.g. in-memory data grid, relational database (SQL) or key-value pair (No-SQL).

P6. Independent technology selection Unlike monoliths, the small size of services allows for flexibility in technology selection. Often a business requirement or constraint may dictate a specific technology choice. In other cases, technology choice may be driven by engineering skills, preference and familiarity.

P7. Independent release cadence Microservices should be loosely coupled [14] and therefore should have their own release cadence and evolve independently. It should always be possible to deploy a microservice without redeploying any other microservices. Microservices that must always be released together could be redesigned and merged into one microservice.

P8. Limit chatty microservices Any interdependence between atomic microservices should be removed. If two or more microservices are constantly chatty (interacting), then that is a strong indication of tight coupling [9, 14], and these microservices should be merged into one. Note: If principle P1 is followed ("do one thing well"), then there should be no chatty interdependent microservices.

Cloud Deployment of Microservices

Following the above MSA boundary-setting design principles, highly cohesive and loosely coupled microservices are more practical for cloud deployment as compared to monoliths. Microservices can be deployed independently and can be scaled independently and are small enough in size that automated build, test and deploy scripts can be implemented using agile DevOps methods and tools [7].

The small size of the deployment objects also makes containerization practical, using Docker or similar technology [15], whereby each microservice is deployed inside a separate virtual machine image which then can be run (instantiated) any number of times on any number of different host systems as self-contained light-weight containers which can be scaled out elastically. For example, a high-demand Product microservice can be instantiated into another active-active load-balanced container during the peak load period, and then, the redundant container can be later removed as the load subsides.

Cloud-based microservices are exposed to internal user interfaces and external third-party applications via an API Gateway [3]. An API Gateway provides a single point of entry into the microservices, as well as a single point of control. Features of an API Gateway include (a) user authentication, (b) user authorization to access specific microservices, (c) transformation between various data formats (e.g. JSON, XML), translation between various transport protocols (e.g. HTTP, AMQP) and (d) scripting for aggregating or orchestrating multiple microservices in order to reduce network traffic. Figure 4.4 illustrates a microservices-based architecture.

Challenges of Microservices Deployment

The complexity of a microservices-based architecture increases over time as the number of deployed microservices increases [3, 15]. Monitoring and management tools are needed in order to: (a) monitor the run-time status of microservices and restart any which have stopped, (b) monitor the loading on microservices and manage the elastic scaling of active-active load-balanced containers (instances) accordingly and (c) provide a framework for microservice discovery so that, for example, a composite microservice can locate a newly redeployed atomic microservice which gets assigned a new IP address.

Another complexity arises when interdependent microservices are located on different host systems across a wide area network (WAN); for example, microservice 'A' requests data from microservice 'B'. In such cases, synchronous request/reply interactions would cause high network traffic across the WAN. A better approach would be to use an asynchronous event-based interaction [12] across the WAN, whereby microservice 'B' publishes data, whenever it becomes available (i.e. the event) to all microservices which have subscribed to that data.

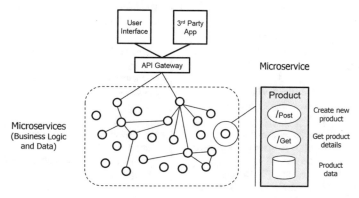

Fig. 4.4 Microservices-based architecture

Similarly, if the same service 'A' was instantiated on multiple host systems across the WAN, there arises complexity and design challenges around how to ensure availability and/or eventual consistency [16] of data across the WAN.

Monolith Versus Microservices Feature Comparison

Based on what has been discussed so far in this chapter, a feature comparison between monoliths and microservices is summarized in Table 4.1.

4.4 Building Cloud-Based Applications

Many established enterprises which are encumbered with inflexible monolithic systems are beginning to transition to a microservices architecture. Newly created enterprises have an option to build a cloud-based microservices architecture from day one, rather than to buy or build monolithic systems. In such greenfield scenarios, one of the main challenges faced by architects is the identification of candidate microservices which are highly cohesive and loosely coupled [9, 14].

Without reference to any existing monolith which can be used as a starting point for microservices decomposition, architects can take a top-down approach starting

Table 4.1 Monolith versus microservices feature comparison

Feature	Monolith	Microservices
Technology stack	• Locked into original technology stack and framework • All functions developed with one programming language	• Each microservice can be developed using a different technology, fit for purpose or based on developer preference
Scalability	• Functions within a monolith cannot be scaled independently • Horizontal scaling of the entire monolith is necessary	• Each microservice can be scaled independently via containers • Tools are needed for monitoring and managing containers
Change management	• For any small change, the entire monolith needs to be retested • Change/testing processes are complex and time consuming	• Microservices are small and can be tested quickly • Microservices have independent release cadences
Deployment	• Deployment file is large, slow to startup, may incur downtime • Use of agile DevOps methods and tools is not practical	• Microservices can be deployed independently • Use of agile DevOps methods and tools is appropriate

with a set of business requirements, then deriving a set of business process models and/or business capability models [9] and finally decomposing those models into a set of microservice candidates. Capability-based services can be distinguished in layers as shown in Table 4.2 [9].

Even without an existing monolith as a reference, a bottom-up approach for microservices identification can be used, provided there exists a data model of the target application in the form of a Unified Modeling Language (UML) compliant Entity Relationship Diagram (ERD) and use cases. Service Cutter [14] is a tool which can assist architects in identifying microservice candidates which are highly cohesive and loosely coupled. Using the ERD and use cases as inputs, the Service Cutter tool extracts the "building blocks" of an application, referred to as "nanoentities", which are to be encapsulated and owned by microservices. These nanoentities are:

- Data: which is exclusively owned and maintained/manipulated by a microservice,
- Operations: which are the business rules/logic exclusively provided by a microservice,
- Artefacts: which are a "collection of data and operations results transformed into a specific format", e.g. a business report which is exclusively provided by a service [14].

Using a predefined "coupling criteria", the relationship between each pair of nanoentities in the model is scored, and finally, a clustering algorithm is used to identify the candidate microservices [14].

Another source of information which can support bottom-up microservices identification, in the absence of a monolith as a reference, is industry specific models. The banking industry, for example, has produced a number of widely used information models. The Banking Industry Architecture Network (BIAN) is a consortium of over 30 banks, technology vendors and universities, which have collaborated on a service decomposition framework for banks [17, 18]. The BIAN

Table 4.2 Capability-based service types

Service type	Description
Business process service	Stateful services which orchestrate automated composite business and data services, including human interaction
Composite business service	Automated services which provide business logic, by orchestrating atomic business and data services
Composite data service	Automated services which provide data, by orchestrating atomic business and data services
Atomic business service	Automated services which provide atomic business logic functionality, e.g. a pricing calculator
Atomic data service	Automated services with provide atomic data manipulation functionality, e.g. CRUD (create, read, update, delete) product information

Service Landscape, as it is called, is a decomposition of a generic universal bank (retail, corporate and investment banking) into a finite set of service domains which cannot be useably further decomposed. As shown in Fig. 4.5, the framework is organized into three levels: (a) business area, (b) business domain and (c) service domain. The "Loan" service domain, for example, can encapsulate all of the business logic and data for loans. Even in the absence of a data model, this framework can be a good starting point for architects to identify candidate microservices.

Technology vendor supplied data warehouse models are another source of information which can support bottom-up microservices identification. The two most widely used data warehouse models in the banking industry are: 1) Teradata Financial Services Logical Data Model (FSLDM) and 2) IBM Information Framework (IFW) Banking Data Model. Each of these vendors supplied information models comes out-of-the-box with a set of core banking entities, also referred to as "subject areas" as illustrated in Fig. 4.6. Data warehouses are organized into subject areas in order to support "subject area experts" i.e. data scientists/analysts whom are tasked to help bank management make decisions around; product, party (customer), channel, campaign and others. The FSLDM and IFW banking industry

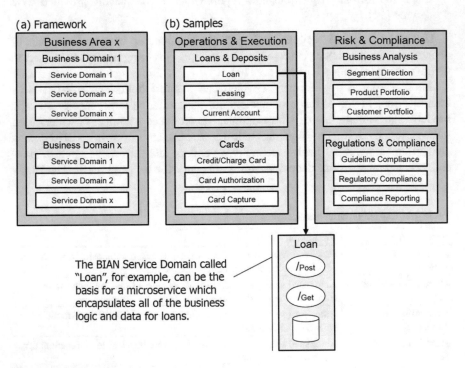

Fig. 4.5 BIAN service landscape (sample)

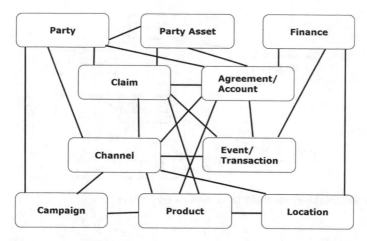

Fig. 4.6 Teradata FSLDM subject areas

models are improved overtime as requirements from many banks are incorporated and therefore have become industry standards [18]. While these standard subject areas are course-grained at a high level of abstraction, they suggest a good baseline for further decomposition into more fine-grained microservices.

Each subject area has a default set of attributes which are extensible. Figure 4.7 shows a worked example for the Account/Agreement subject area.

Each subject area has an extensible set of relationships with other subject areas. Figure 4.8 illustrates a worked example for the campaign subject area. These

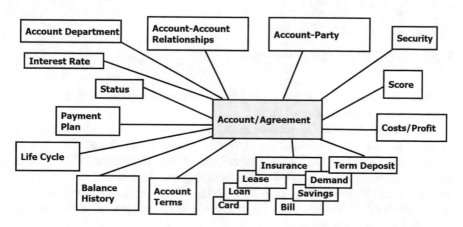

Fig. 4.7 Attributes of FSLDM account/agreement subject area

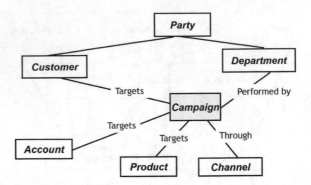

Fig. 4.8 Relationships around FSLDM campaign subject area

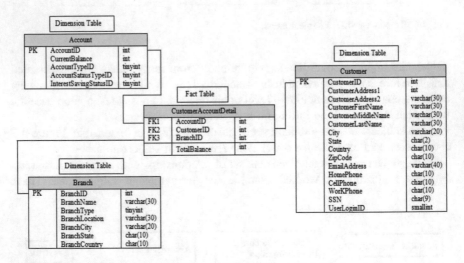

Fig. 4.9 IBM IFW banking data model (sample)

relationship maps are useful for identifying composite services [9], as well as inter-process communications [3].

Each vendor supplied information model comes with a baseline ERD as shown in Fig. 4.9. Data warehouse ERDs are typically implemented using a "star schema", whereby a normalized "fact" table is related to multiple de-normalized "dimension" tables. Dimension tables are useful for identifying atomic data services [9] or data nanoentities [14]. Fact tables are useful for identifying artefact nanoentities [14]. The "Customer" dimension table, for example, can encapsulate all of the business logic and data for customers.

Data warehouse models exist for other industries as well, for example healthcare—Health Catalyst Enterprise Data Model and Oracle Healthcare Data Model. In the absence of an existing monolith to reference, it is a challenge for

architects to decompose the functional boundaries of an enterprise into microservices at an optimum level of cohesiveness and loose coupling [12].
Industry-specific models offer a good starting point.

The development cycle for a cloud-based application starts with a microservice
identification phase. At the end of this phase, the architect will have produced a
library of microservice interface definitions, typically in the form of a Swagger
Definition File for REST-based microservices or a Web Service Description
Language (WSDL) for SOAP-based microservices [19]. Interface definitions serve
as a software specification, enabling concurrent development by the back-end
microservice development team and the front-end user interface development team.

In a microservices-based application, each individual microservice can be
developed using a different programming language, can be developed and maintained by a small team and can be deployed and scaled separately. A GUI-driven
development tool such as TIBCO BusinessWorks Container Edition (BWCE)
enables rapid development of container-ready microservices, involving very little or
zero coding. REST-based microservices can be tested using Swagger or Postman.
SOAP-based microservices can be tested using SOAPUI. Container-ready
microservices can be built into a Docker image [20], together with the required
lightweight operating system, run-time libraries and database drivers. Docker
images are deployed and run as lightweight virtual machine (VM) containers within
the target cloud environment [20]. Microservices are small enough that DevOps
tools such as Jenkins can be used to automate the build, test and deploy steps [7].
Kubernetes is a popular Docker cluster management suite which covers service
discovery, monitoring, orchestration, load balancing and cluster scheduling [15].
The microservice development lifecycle, annotated with some popular tools, is
shown in Fig. 4.10.

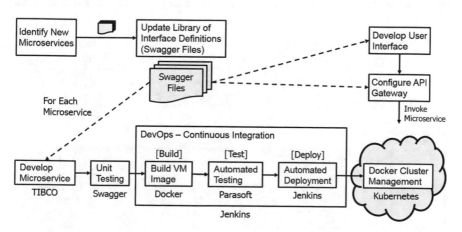

Fig. 4.10 Microservice development lifecycle

4.5 Transitioning from Monoliths to Cloud-Based Microservices

Migrating from legacy monoliths to a services-oriented architecture has been a long-standing challenge [21], up to and including the recent microservices era [13]. In the pre-microservices era, the best outcome of an SOA migration was to provide a layer of abstraction (i.e. a services layer) in front of the legacy monolith, in order to provide a more flexible architecture while extending the lifespan of the legacy system [22]. There are several case studies of SOA migrations in banking [21, 23].

In the microservices era, the ultimate goal for established enterprises is to replace their on-premises legacy monoliths with a functionally equivalent collection of cloud-based microservices which can be independently developed, deployed and scaled. Only one case study could be found of monolith to microservices migration in banking [6], and in this case, the bank did not decommission its legacy monolith.

One of the main migration challenges involves reverse engineering of the legacy monolith in order to identify service candidates [24]. If the source code and/or database schema are not available for analysis, capturing and analysing the run-time interaction at the monolith interface (API) can help to identify service candidates, as illustrated in Fig. 4.11. Service identification can also be aided by referring to industry models as discussed in the previous section. Various phases or steps and post-migration benefits are now discussed below.

4.5.1 Migration Phases

In this section, we offer a phased approach for migrating from a monolith to a cloud-based microservices architecture, as shown in Fig. 4.12 and detailed in the

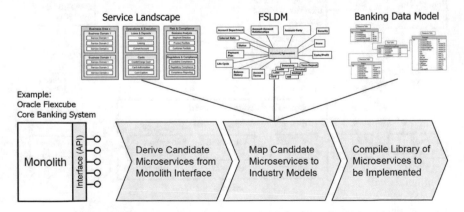

Fig. 4.11 Microservice identification

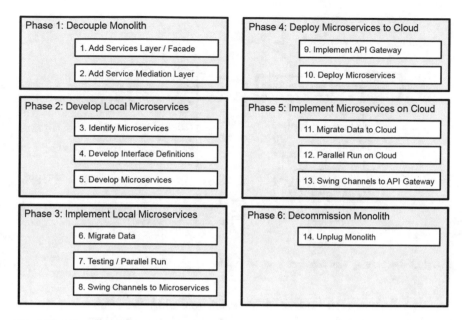

Fig. 4.12 Migration phases

section which follows. The migration phases presented here are based on an actual core banking system migration conducted in an academic setting under a project referred to as SMU tBank [25], whereby an Oracle Flexcube retail banking system was directly replaced by over 200 microservices.

Phase 1: Decouple Monolith

A common approach for decoupling the front-end presentation layer from the bank-end business logic layer is to introduce a facade layer [3, 26] between the user interface and the monolith, in order to prepare for the eventual transition away from the monolith. Initially, each facade implements "pass-through" logic (i.e. no data transformation) which reflects the underlying monolith interface, such that any existing user interfaces do not require code changes, and are then physically decoupled from the monolith. To cater for any new user interfaces (e.g. banking channels), each facade may then be refactored into a service which implements the target microservice interface definition, if already identified, such that the service "adapter" performs a data transformation back to the underlying monolith interface. The facade/services layer is illustrated in Fig. 4.13.

A service mediation layer [24] is then introduced above the facade/services layer, to provide run-time control over the channel-to-service mapping. For example, if Service X invokes the monolith interface for "getAccountBalance", and Service Y invokes the equivalent microservice for "getAccountBalance", and both services use the same request/reply fields as specified in the service interface definition, then through run-time control, channel Z can be reassigned to consumer

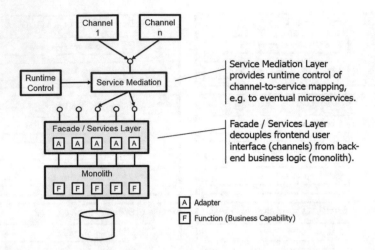

Fig. 4.13 Decoupling user interface from monolith

Service Y (microservice) instead of Service X (monolith). With this capability, it is possible to reassign, i.e. "swing" the entire set of channels to consume microservices, in one shot, without having to change a single line of code in any of the channels. The service mediation layer is illustrated in Fig. 4.13. The service mediation layer also provides monitoring, logging and security features.

Phase 2: Develop Local Microservices

Decompose the monolith into separate microservices. This may involve reverse engineering the monolith in order to identify candidate microservices [8], as illustrated in Fig. 4.12. Service identification is both the most tedious step and the most critical step in the entire migration process. It is important to realize the optimum level of cohesion and loose coupling for each microservice.

Develop a library of microservice interface definitions in the form of Swagger files (or WSDL files), which can then be imported into any number of standards compliant microservices development and testing tools. Employ a design time governance tool to manage the microservices design lifecycle and to make the interface definitions available to developers.

Develop and unit test the microservices, as illustrated in Fig. 4.12. The microservice should implement the equivalent business logic and the equivalent data schema as the original function within the monolith. While each microservice can be developed by a small team, a complex monolith such as a core banking system may be decomposed into several hundred microservices. Therefore, this is the most resource intensive step in the entire migration process. GUI-driven development, standards-based testing tools and DevOps continuous integration (build, test and deploy) tools enable rapid development of microservices as illustrated in Fig. 4.10.

Phase 3: Implement Local Microservices

Once the microservices are developed, unit tested and deployed locally, i.e. on-premises, then the channel-to-service mapping can be changed independently or in batches. For each microservice, the following steps are repeated: (1) migrate the data from the monolith to the microservice, (2) conduct a parallel run, such that the channel invokes both the monolith and microservice, and the resulting data is reconciled between the two and (3) change the channel-to-service mapping to reassign, i.e. "swing" the channel to invoke the microservice instead of the monolith. This process can be repeated systematically, until all of the channels are invoking only microservices. At any point in time, any channel-to-service mapping can be temporarily reassigned back to the monolith, in case of a bug. Service mediation capability enables channels to swing back and forth between the monolith and the microservice without changing any code or configurations on the channel. This capability is illustrated in Fig. 4.14 (annotations 1 and 2).

Phase 4: Deploy Microservices to Cloud

Implement an API Gateway in the target cloud environment to provide a single point of entry and a simple point of control for microservices invocation. Deploy the microservices from the on-premises environment to the cloud environment. Deploy any necessary microservices management and monitoring tools onto the cloud. Conduct end-to-end testing to ensure each microservice can be invoked externally via the API Gateway.

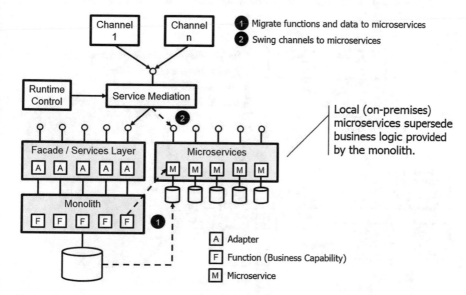

Fig. 4.14 Migrating to local (on-premises) microservices

Phase 5: Implement Microservices on Cloud

Once the microservices have been implemented locally, i.e. on-premises, then the channel-to-service mapping can be changed independently or in batches. For each microservice, the following steps are repeated: (1) migrate the data from the on-premises microservice to the cloud-based microservice, (2) conduct a parallel run, such that the channel invokes both the on-premises microservice and cloud-based microservice, and the resulting data is reconciled between the two and (3) change the channel-to-service mapping to reassign, i.e. "swing" the channel to invoke the cloud-based microservice instead of the on-premises microservice. This process can be repeated systematically, until all of the channels are invoking only cloud-based microservices. At any point in time, any channel-to-service mapping can be temporarily reassigned back to the on-premises microservice, in case of a bug. Service mediation capability enables channels to swing back and forth between the on-premises microservice and the cloud-based microservice without changing any code or configurations on the channel. This capability is illustrated in Fig. 4.15 (annotations 3 and 4).

Phase 6: Decommission Monolith

At this point, or even after Phase 3, the monolith is no longer used and can be decommissioned, i.e. taken off line. The on-premises environment then becomes a staging area for microservices development and testing. Channel applications in a UAT environment can be mapped to invoke the on-premises microservices. Existing channels can be systematically refactored to invoke the API Gateway directly, instead of via the service mediation layer. New channels and third-party

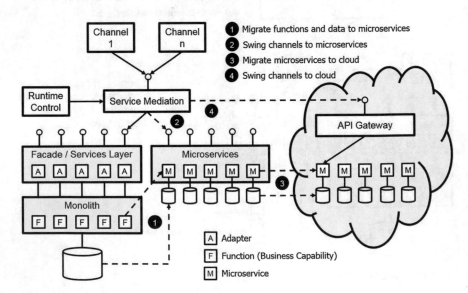

Fig. 4.15 End-to-end migration from monolith to cloud-based microservices

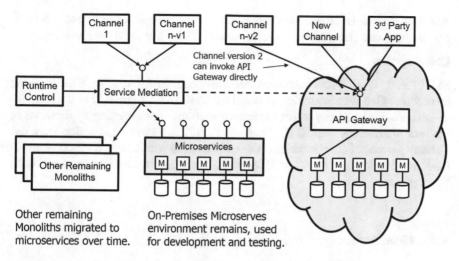

Fig. 4.16 Final configuration of on-premises and cloud-based environments

apps can invoke the API Gateway directly. The service mediation layer would remain until all of the other remaining monoliths, and any future acquired monoliths, are eventually migrated to cloud-based microservices. Figure 4.16 shows the final configuration.

4.5.2 Post-migration Benefits

For the case of SMU tBank [24] which the above migration phases are based upon, a number of benefits have been realized as follows:

Performance

Average response time as measured at the service mediation logging point improved from 200 ms (monolith) to 40 ms (microservice), the difference being the database technology used. With Oracle Flexcube core banking system, we were locked into using the heavy footprint Oracle database. And, for our database intensive microservices, we selected MySQL ndbcluster replication engine which operates in-memory efficiently.

Reuse/Agility

During one stage of the SMU tBank development, three student teams developed four banking channels (Teller, Internet Banking, Mobile Banking and ATM-simulation) concurrently during one school semester, without creating any new business logic or database tables. This was possible due to their reuse of existing microservices, which were developed during the previous semester.

Collaboration

SMU tBank cloud-based microservices are available for use by other learning institutions. One such institution has used the SMU tBank Open API as the basis for student projects, whereby student teams develop their own banking applications or FinTech alternatives. The SMU tBank Open API has attracted attention from our industry partners. Future work includes collaborating with a large Swiss investment bank to develop a library of BIAN/IFX compliant microservices for wealth management.

4.6   Conclusion

When organizations continue their digital transformation efforts, they should consider an agile style of application architecture which enables the rapid delivery of new cloud-based digital services. Microservices architecture is seen as a key enabler towards this effort. The main tenet of this architecture is to develop software applications more quickly by building them as collections of independent, small, modular services. A primary benefit of this architecture is to empower decentralized governance that allows small, independent teams to innovate faster, thus improving time-to-market of new digital services.

This chapter contributes to the software engineering community by filling a gap in the literature around best practices and methodologies for decomposing monoliths and transitioning to cloud-based microservices. This chapter presented two approaches: (a) blank slate approach, whereby applications are developed completely from cloud-based microservices from day one and (b) migration approach, whereby existing monoliths are decomposed into cloud-based microservices, and transitioned function by function onto the cloud, until the original monolith can be literally unplugged. Though the context and examples presented in this chapter relate to the banking domain, the method is generic enough to be applied to other domains such as e-commerce, supply chain and logistics, health care. The blank slate approach is best suited for building new applications, and the migration approach is best suited for transitioning from existing monoliths to a microservices architecture. The migration methodology presented in this chapter is more detailed compared to the blank state approach. Our future work will focus on further identifying and refining the steps for developing microservices-based enterprise solutions from a blank slate.

References

1. Khadka R, Saeidi A, Jansen S, Hage J, Haas GP (2013) Migrating a large scale legacy application to SOA: challenges and lessons learned. In: 2013 20th working conference on reverse engineering (WCRE). IEEE, pp 425–432
2. Shankararaman V, Megargel A (2013) Enterprise integration: architectural approaches. In: Service-driven approaches to architecture and enterprise integration, vol 67
3. Lloyd W, Ramesh S, Chinthalapati S, Ly L, Pallickara S (2018) Serverless computing: an investigation of factors influencing microservice performance. In: 2018 IEEE international conference on cloud engineering (IC2E). IEEE, pp 159–169
4. Peinl R, Holzschuher F, Pfitzer F (2016) Docker cluster management for the cloud-survey results and own solution. J Grid Comput 14(2):265–282
5. Wikipedia (2019) Microservices. Available: https://en.wikipedia.org/wiki/Microservices
6. Megargel A, Shankararaman V, Fan TP (2018) SOA maturity influence on digital banking transformation. IDRBT J Bank Technol 2(2):1
7. Kohlmann F, Alt, R (2009) Aligning service maps-a methodological approach from the financial industry. In: 2009 42 Hawaii international conference on system sciences. IEEE, pp 1–10
8. Winter A, Ziemann J (2007) Model-based migration to service-oriented architectures. In: International workshop on SOA maintenance and evolution. CSMR, pp 107–110
9. Indrasiri K, Siriwardena P (2018) The case for microservices. In: Microservices for the enterprise: Springer, Berlin pp 1–18
10. Pardon G, Pautasso C (2017) Consistent disaster recovery for microservices: the CAB theorem. In: IEEE cloud computing
11. Wikipedia (2019) Monolithic application. Available: https://en.wikipedia.org/wiki/Monolithic_application
12. Sun Y, Nanda S, Jaeger T (2015) Security-as-a-service for microservices-based cloud applications. In: 2015 IEEE 7th international conference on cloud computing technology and science (CloudCom). IEEE, pp 50–57
13. Richardson C, Smith F (2016) Microservices: from design to deployment. Nginx Inc., pp 24–31
14. Palihawadana S, Wijeweera C, Sanjitha M, Liyanage V, Perera I, Meedeniya D (2017) Tool support for traceability management of software artefacts with DevOps practices. In: 2017 Moratuwa engineering research conference (MERCon). IEEE, pp 129–134
15. Malavalli D, Sathappan S (2015) Scalable microservice based architecture for enabling DMTF profiles. In: 2015 11th international conference on network and service management (CNSM). IEEE, pp 428–432
16. Ząbkowski T, Karwowski W, Karpio K, Orłowski A (2012) Trends in modern banking systems development. Inf Syst Manag XVI:82
17. Caetano A, Silva AR, Tribolet J (2010) Business process decomposition-an approach based on the principle of separation of concerns. Enterp Model Inf Syst Archit (EMISAJ) 5(1):44–57
18. Knoche H, Hasselbring W (2018) Using microservices for legacy software modernization. IEEE Softw 35(3):44–49
19. Di Francesco P, Lago P, Malavolta I (2018) Migrating towards microservice architectures: an industrial survey. In: 2018 IEEE international conference on software architecture (ICSA). IEEE, pp 29–2909
20. Cerny T, Donahoo MJ, Trnka M (2018) Contextual understanding of microservice architecture: current and future directions. ACM SIGAPP Appl Comput Rev 17(4):29–45
21. Dragoni N, Dustdar S, Larsen ST, Mazzara M (2017) Microservices: migration of a mission critical system. arXiv preprint arXiv:1704.04173

22. Galinium M, Shahbaz N (2009) Factors affecting success in migration of legacy systems to service-oriented architecture (SOA). School of Economic and Management, Lund University, Lund
23. Frey FJ, Hentrich C, Zdun U (2015) Capability-based service identification in service-oriented legacy modernization. In: Proceedings of the 18th European conference on pattern languages of program. ACM, p 10
24. Megargel A (2018) Digital banking: overcoming barriers to entry (Doctoral dissertation). Retrieved from Singapore Management University. https://ink.library.smu.edu.sg
25. Gysel M, Kölbener L, Giersche W, Zimmermann O (2016) Service cutter: a systematic approach to service decomposition. In European conference on service-oriented and cloud computing, Springer, pp 185–200
26. Natis YV (2017) Core architecture principles for digital business and the IoT—part 1: Modernize. Gartner Publication G00324415

Chapter 5
Cloud-Enabled Domain-Based Software Development

**Selma Suloglu, M. Cagri Kaya, Anil Cetinkaya, Alper Karamanlioglu
and Ali H. Dogru**

Abstract A cloud-based software development framework is presented that does
not require programmer capabilities. The development starts with a graphical
modelling of the process model, defining the top-level flow for the application.
Such a flow coordinates the functional units that are components or services linked
to the process again through graphical means such as drag and drop. Variability
affects all processes and functional constituents, being the principal specification
requires for the application under development. The idea has been partially
implemented in a commercial setting and is in its assessment phase. This framework needs to be domain-specific for successful deployment of user ideas without
programming-level input. As a platform, the suggested environment allows the
setting up of different development environments for different domains. A user
community can construct new domains by defining reference architectures, process
models and other assets for the application developers. Consequently, there is a
possibility of a market place shaping up where such assets can be offered and
consumed, subject to an administration for security and optionally commercial
purposes. Open, free or paid marketplaces can be created based on administrative
policies.

Keywords Cloud computing · End-User Development · IaaS · PaaS · SaaS ·
Software ecosystem

S. Suloglu
Department of Software Engineering, Rochester Institute of Technology,
Rochester, NY, USA

M. C. Kaya (✉) · A. Cetinkaya · A. Karamanlioglu · A. H. Dogru
Department of Computer Engineering, Middle East Technical University, Ankara, Turkey
e-mail: mckaya@ceng.metu.edu.tr

© Springer Nature Switzerland AG 2020
M. Ramachandran and Z. Mahmood (eds.), *Software Engineering in the Era
of Cloud Computing*, Computer Communications and Networks,
https://doi.org/10.1007/978-3-030-33624-0_5

109

5.1 Introduction

Software development approaches, with their ever-continuing improvements, experience big changes as an outcome of existing motivations. Sometimes our pioneers investigate paradigms and their ramifications ripple down to technologies or more frequently, practitioners invent technologies that evolve into systematic conceptions. In either case, a new trend in development or architectural deployment gets established for a considerable length of time and interest. Innovations seem to develop mostly in a bottom-up manner, favouring the second of those motivations. Recently, some new relevant technologies have been emerging with their snowball effects; such as cloud computing, Internet of Things (IoT), cyber-physical systems (CPS) and machine learning. These have appeared with promising capabilities; however, they also influence each resulting with interesting synergies.

Software Ecosystems is one such new phenomenon that probably owes its existence to cloud technologies. A similar concept is software stores for mobile applications. There may also be a catalogue of products offered to the buyer, similar to e-commerce stores. However, there is also the support for the offer side where software developers utilize. There are developers/sellers and users/buyers involved. Also, the whole site can be managed by an authority which regulates this market and optionally makes profits through a percentage out of related transactions. The authority may also be involved in the testing, security, quality assessment and enforcement of the products.

A similar environment is considered in this research. However, easier and faster development of software is targeted for an accelerated market. There may also be a huge demand for applications that are envisioned by users who are not programmers. The targeted development environment should make it possible for a non-programmer to deliver software products through a "no-code" software development method. Hence, the envision–produce–use cycles will be accelerated to gain momentum to such a market, hopefully.

Developers will also be able to market their sub-products besides complete applications. These need to be developed any way to be able to deliver complete products. The development paradigm is a compositional one—in that a software system will be composed of components that are built with the purpose of being composed into different products. Actually, the proposed tool will allow the creation of different domains that is a set of components availing development in a specific application field. Once a domain has matured [1], new applications will be created through the selection, modification and integration of existing components. These two main activities to define a domain and to develop new applications correspond so far to the domain engineering and product engineering modes of the software product line architecture. Figure 5.1 depicts the interaction of different users with the development platform.

Different users are interacting with the platform in Fig. 5.1 create domains, create applications in a specific domain and use the applications. The users depicted

Fig. 5.1 User interactions
with the development
platform

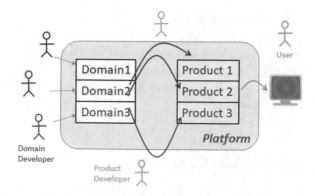

on the left are the domain developers. They can create or modify software assets.
The users with the product developer role will utilize these domain assets, logging
into one existing domain, to construct applications. Finally, those applications will
be used in the common understanding of Software as a Service.

A key capability of this new environment, however, adds to the value of the
known architectures. Variability is taken as the primary means for information input
to the environment, especially in the product engineering phase. The other constituents of the architecture, hence the parts of the solution, will be configured as
automated as possible through the ripple-down path of the variability propagation.

Variability should be handled with special care, as the single most important
specification element for the system under development. The other structures to be
configured as a result of variability resolution include the top-level coordination in
the form of a process model, a set of components and a set of connectors. The
offered mechanisms come with configurable process models, components and
connectors. Variability is hierarchically managed within its interaction with the
software elements that correspond to the levels of a specific variability. First,
possible reflections of the resolved variability are searched in the process model.
Any configuration conducted, as a result, may also induce further configuration
steps in the process model, or the other assets, as the result of propagating the
constraints. Later, corresponding modifications need to be conducted in the component and finally, in the connector sets. The software assets come with configuration interfaces accepting guidance from variability specifications.

The rest of the chapter includes a background section that provides terminology
related to cloud computing and other required topics. Problem definition is provided
next along with the related work. Then suggested development paradigm is presented in detail and a case study is provided, which shows how to develop a small
application with the proposed notion. The chapter concludes after the authors'
remarks in the discussion section.

5.2 Background

Various paradigms and technologies are presented in this section that this study is built on. These include cloud computing, SOA, business process management (BPM), as well as software ecosystems, variability and service composition on cloud computing.

5.2.1 Cloud Computing

The rapid development of technology has enabled computing resources to become cheaper and more powerful. This trend has led to the emergence of a paradigm called cloud computing. Cloud computing is using services provided by suppliers over the internet connection. These services can be various; however, commonly they accept input from the user, process the input and provide a result to the user [2]. Users only need an internet connection to use desired services deployed on clouds, probably distributed on different geographic locations. Using cloud lightens the burden of the user for modification and maintenance of the software. Also, cloud technologies also allow users to easily collaborate on their project and provide mobility [3]. Generally, cloud computing services are reviewed in three categories: Software as a Service (SaaS), Platform as a Service (PaaS) and Infrastructure as a Service (IaaS).

Software as a Service (SaaS)

SaaS provides access to applications of the provider that run on a cloud infrastructure. This allows multiple users to access an application over the Internet. Customers can access the applications from their own devices using a web browser, a client interface or a program [4]. The customer is not able to manage or control the cloud infrastructure features, such as network, servers, operating systems or storage, except for some configuration settings.

Platform as a Service (PaaS)

PaaS allows the development of software applications on clouds where infrastructure (such as operating systems, application programming interfaces and programming languages) is arranged by the PaaS provider [5]. Web-based infrastructure empowers collaborative development. However, there are technologies allowing developers to work offline. The customer has control over the deployed applications and some configuration settings of the application or hosting environment [4].

Infrastructure as a Service (IaaS)

With the widespread use of cloud computing, the tendency of software and infrastructure to exploit the cloud has begun to increase significantly [6]. The reason for this trend is that especially small businesses can obtain more reliable, flexible and cheaper systems than deploying their own infrastructures. When choosing a cloud provider or technology, it is necessary to determine what features are needed first. The classification framework for IaaS developed by Repschlaeger et al. [7] can be used to identify the needs. They focus on six target dimensions for cloud computing in terms of characteristics of cloud and the IaaS provider market. The specified target dimensions correspond to flexibility, costs, scope and performance, IT security and compliance, reliability, trustworthiness and service and cloud management.

5.2.2 Service-Oriented Architecture (SOA)

SOA is the technical enabler of the open systems which enables reusability of services and standardization efforts to interactions. SOA brings service providers, consumers and brokers together, which is a means of facilitating inner and inter-organizational computing by reusing services and service descriptions while relying on architecture [8]. With the use of SOA, adaptability and flexibility in design and runtime have become emerging research issues. The basic goal of SOA is to design a system that provides a set of services to end-users and other services for fulfilling business needs. Services are orchestrated to realize business processes helping business-IT alignment with a well-understood architecture. As every application has a different architecture outlined by orchestration and/or choreography, service composition plays an important role to achieve flexibility to respond to rapid changes [9–11].

Experiences learned from SOA for more than a decade and prevalent usage of containerization techniques paved the way for microservices. Although there are different definitions of microservice architectures, their characteristics can be identified: building up the system via independently deployable services. The idea of microservice diverges from SOA in ownership of data and functionalities. Microservices enable a rapid deployment environment but bring decentralized data management and governance onto the table. There are still debates over microservice granularity or the distribution of the developer teams over microservices. To this end, domain-driven design techniques aid system stakeholders for decomposing the system into microservices. Besides, independent deployability of microservices supports flexibility and adaptability: the system can change its behaviour at runtime by employing different microservices.

5.2.3 Business Process Modelling (BPM)

Business Process Modelling represents the flow of operations of an enterprise. It helps to understand, change, improve or run the modelled processes. Thus, while efficiency and quality increase in enterprise processes, the cost is reduced [12, 13].

Business Process Modelling Notation has been leading the modelling notations in terms of wide usage. Both business engineering concerns and software development usages have been exploiting the rich features of this easy-to-use graphical language. New versions of this notation are being created and its capabilities are increased. It has been promoted by SOA practices and also more recently, it is being used as a software specification notation.

5.2.4   Variability

Software Product Lines (SPL) provide the environment to construct a set of software systems that are developed from common assets in order to satisfy specific requirements and to manage features of a final product. Management of core assets and handling of variability are some of the main concerns in SPLs. In [14], a requirement for a framework with two distinct processes is specified: domain engineering for the characterization and realization of necessary assets; application engineering for the reproduction of distinctive applications with the utilization of variability. A product is defined by the resolution of variability throughout the phases. With every decision on the constraints and the features to be included in the final product, the number of possible systems decreases. Each design decision causes the system to differentiate, thus constitutes a variation point. Late-binding techniques yield more features and options to be utilized in the configuration of a product. This capability also supports development through freeing the creational minds from detailed fixations at the early stages. That is why delayed design decisions improve the effectiveness of the SPLs by allowing core assets to be used more resiliently with respect to changing requirements [15].

Variability resolution is also supported in some early approaches that are using domain feature models. However, the management of variability is discovered to be very complicated due to variability being superimposed on a feature model. There are variability models such as OVM [14] and Covamof [16] that have been used in software development successfully.

OVM is the pioneering variability modelling approach. The fundamental motivation has been the independent representation of variability, especially from the feature model. In practical cases, a feature model quickly grows out of manageable size and complexity. Addition of feature models information further complicates variability management. For both OVM and Covamof, the decoupled variability representation is central. Besides, some support for the hierarchical organization of variability is provided. This property supported with implementation-specific links to the configurability of the software assets becomes critically important in the automated construction of software applications.

OVM, besides offering the basic variability in a domain model, specifically addresses the following notions:

- Variability in Time: different versions of an artefact that can be valid at different times.
- Variability in Space: different shapes of an artefact that can exist at the same time.
- External Variability: variability of domain artefacts that is visible to customers.
- Internal Variability: variability of domain artefacts that is hidden from customers.

Internal variability is instrumental in interpreting external variability to propagate constraints to various abstraction levels in the application.

Covamof, introduced with the motivations that are mostly common to that of OVM, specifically addresses the following issues that support the holistic modelling with variability as the driving engine for configurations:

- Variability is the core notion in development, applied at all abstraction levels.

  - Variability should be organized hierarchically.
  - Different views are needed.

- Tools are required for both product and domain engineering.

  - Traceability should be supported.
  - Dependencies are also first-class entities.

- Explicit modelling of interactions between dependencies.
- Tightly coupled model and artefacts for integrity during evolution.

It can be observed that Covam of mentions more detailed capabilities, as a result of its appearing later at the arena.

XCOSEML

Component-Oriented Software Engineering (COSE) is a development paradigm where components are at the centre of the development activity [17]. Different from the component-based approaches, development of the inner component functionality is not in focus. Instead, only integration is aimed. Modification of existing components and the development of new components are done based on need. The COSE approach does not suggest an approach for component development; the developer can select among existing methods. This approach requires component-aware development stages from requirements to application development.

COSE Modelling Language (COSEML) is the graphical modelling tool of the COSE approach [18]. COSEML allows the developer to decompose the system hierarchically. It allows the representation of abstract entities (e.g. a package) and physical entities (e.g. a component) of the system. Later, the language was extended with a process model and a variability model which is inspired by Covamof and OVM [19, 20]. This version of the language is called XCOSEML and it is text-based. The process model of XCOSEML is directly affected by the configuration of the variability model. Process blocks are included or excluded from the process itself based on the configuration. Other assets of the system, namely components, connectors and interfaces, are configured and added to the system if they are referenced in the process model.

5.2.5 Software Ecosystems

As systems and organizations have opened up their technologies and capabilities to others to be used or extended in the creation of a collaboration environment, inter-organizational computing came into play which is a way of developing distributed and autonomous systems. This change is the indicator of open systems theory [21] which refers simply to the concept that organizations are strongly influenced by their environment. The environment also provides key resources that sustain the organization and lead to change and survival. In other words, an organization's survival is dependent upon its relationship with the environment. Likewise, open innovation and open business systems [22] increase the significance of collaboration with inter-organizational entities. Moreover, Global Software Engineering [23] efforts have gained importance and have been adopted by several organizations.

Once the organization decides to make its platform available outside the organizational boundary, the organization transitions to a software ecosystem. The increasing value of the core assets through offering to existing users, accelerating innovation through open innovation capability provided by the ecosystem, collaborating with partners in the ecosystems to share the cost of innovation are some of the reasons for transitioning to software ecosystems [24].

Software ecosystem is defined as "the interaction of a set of actors on top of a common technological platform that results in a number of software solutions or services" [25]. Software ecosystems have three main aspects: organization, business and software. Software ecosystems can be put into three categories: operating system-centric, application-centric and end-user programming [24]. Operating system-based software ecosystems are domain-independent and assume that third-party developers build applications that offer value for customers. A specific category under end-user programming category explicitly focuses on application developers that have good domain understanding, but no computer science or engineering degree. Such a capability can be offered for instance, by providing a very intuitive configuration and composition environment (in other words modelled in terms of the concepts of the end-user) that the end-user can create the required applications himself/herself. Application-centric category is domain-specific and often starts from an application that achieves success in the marketplace without the support of an ecosystem around it.

5.2.6   Cloud Computing and Service Composition

Increased use of cloud computing across the world leads to publishing more cloud services in the service pool [26]. In many real-world applications, it is not possible for a single service to meet existing functional requirements. This fact exists despite the presence of complex and discrete services. That's why it's important to have a group of simple services that work with each other in a complementary manner towards achieving complex systems. Moreover, a service composition mechanism is required to be embedded in cloud computing.

Many simple services provided by different service providers and a variety of effective parameters in the cloud pool have made service composition an NP-hard problem [27]. The existing methods for tackling this problem spread into five categories: classic and graph-based algorithms, combinatorial algorithms, machine-based approaches, structures and frameworks. Despite the existence of a variety of methods, only a limited number of these potential solutions can offer near-optimal solutions to specific problems.

5.3   Motivation and Related Work

The ever-increasing demand in the software field has attracted new approaches to meet this demand through non-conventional methods. Naturally, automation and reuse have been the mechanisms behind the new supply attempts.

Component and later web service technologies offered the elementary support for reuse-based development. Later, SOA and SPL arrived in the arena to organize this struggle. On the automation end, model-driven development (MDD) offered successful solutions that are currently part of modern software engineering practices. However, these developments can still benefit by offering more intuitive and simple use. Also, a holistic view of the modelling of complex systems would prove helpful.

End-User Development appeared as a related concept, although not necessarily targeting "no-code development" capabilities. However, this concept is so supportive of its expected meaning for such capabilities. A market also developed for no-code development, with the ambitions stated here. However, the tools offered that are classified as no-code are evaluated to be very simple. With their current abilities, they are not suitable for the construction of general purpose and complex software products.

It is an ambitious goal to construct complex software easier. Current approaches therefore naturally confronted difficulties. The next section exposes some of those difficulties.

5.3.1 Challenges for Improvement

An interesting avenue, especially related to MDD, is Domain-Specific Languages (DSL). Such languages prove an attractive tool once domain-specific models gain emphasis. Domain-driven development becomes necessary because both developers are more efficient in defined domains, rather than working on any field and automation is more successful in software construction for limited domains. Models are basically graphical; however, their textual counterparts are handy in machine processing and in various development modes, even excluding MDD. First, challenges related to DSLs will be introduced.

Developing a system or an application necessitates both domain and technical skills, including programming. Domain-driven design is around for years to extract domain knowledge and create a set of DSLs to be used by domain experts. By doing so, domain experts lay down their knowledge to several system models which in turn are mapped to components and connectors. However, there are some challenges related to incorporating DSLs in software construction:

- Domain experts still need to learn one or more DSLs to operate. But for some, the model may seem too complex and learning requires time investment that is not always available.
- Sometimes, many new DSLs get created during one project and soon there are too many new languages in an organization to be learned, remembered and managed. This is referred to as the tower of Babel phenomenon.
- It is not too easy to develop a tool comprising all steps of the development which also maps DSL artefacts to an execution environment. It takes a considerable amount of time, even if there are efforts to guide developers to create such a tool from scratch.
- End users of the system are generally treated more as the target community, excluded from the development process. There are attempts to incorporate them such as in agile methodologies that have been applied for several years. However, it has not been practically possible to observe end-user preference and reaction on developed assets as a community. An immediate feedback mechanism is required to further develop complete and high-quality assets.
- Specification and management of variable parts of the DSL artefacts need a considerable amount of effort that is incorporated into both DSLs and components and connectors in the technical domain.

Other kinds of assets in a domain-driven approach also present their specific challenges. Providing motivation for the creation of assets, their reusability and quality can be supported through sharing and existence of powerful frameworks. Such factors will also support cost and sustainability:

- An open market developer can earn money by deploying new assets and some assets can be provided free. Economical aspects of asset development should be addressed as we are moving towards more independently deployable assets over time.

- A sustainable environment is needed to enable the growth of domain models for a better understanding of domain capabilities which then incites new end users to enrol and grow the existing community.

5.3.2 Related Work

Service composition techniques have been used in many studies for cloud computing systems since 2009. Jula et al. [26] examined and compared the studies in this context in detail with the systematic literature survey.

Many PaaS alternatives with different features are available on the market. It would be useful to narrow the alternative space according to some criteria. Lucassen et al. have created a list of PaaS providers taking into account two criteria [28]:

- Easy deployment should be supported in at least one development framework among PHP, .Net or Ruby.
- The technology should be mature and actively used.

The list contains a number of solutions such as Microsoft Azure, Google App Engine, CloudFoundry, DotCloud, Engine Yard, Heroku, Nodejitsu and OpenShift. This section covers these solutions briefly.

Microsoft Azure is an IaaS since it provides a Windows Server operating system. Azure, though renowned as the IaaS provider, has a wide range of features as services [29]. It is therefore known as both IaaS and PaaS. PaaS offerings of Azure contain Azure Websites, Azure SQL Database, Azure Mobile Services, Azure BizTalk Services and Azure Content Delivery Network.

Google App Engine is another well-known example of PaaS. It is possible to build and host web or mobile applications using the scalable infrastructure of Google [30]. Therefore, there is no need for a server to be maintained. In addition, the need for an administrator has ended. App Engine's runtime environment uses Python programming language.

One of the first cloud platforms, Heroku, allows to build and deploy web applications and supports many programming languages and runtime environments. Nodejitsu is another cloud platform for PaaS and developed based on Node.js. This solution helps developers to deploy data-intensive and real-time systems.

CloudFoundry and OpenShift are two open-source cloud platforms [31]. Both platforms support many popular programming languages and platforms. They also can be hosted on popular IaaS such as AWS and OpenStack. DotCloud is a PaaS and uses an open-source engine called Docker to pack, ship and run the application as a lightweight container [32]. Engine Yard is an open-source PaaS platform that runs over Amazon EC2 [33]. It supports PHP, Ruby on Rails and Node.js.

5.4   Suggested Development Paradigm

Suggested development capabilities to be offered to the non-programmer developer
are fundamentally compositional. Here, existing components are located and integrated into the solution. Further, the architecture frees the developer from designing
the overall structure of the solution. The products are shaped around a two-level
hierarchy. Such integration suggests a two-level hierarchy in the product architecture. Top-level modelling corresponds to a process model that defines the control
flow of the application: This part is responsible for the ordering of the function
calls. Once the program starts to execute, what function should be called first and
then which one is next is determined here. The ordering of the functions can be
sequential, conditional (such as using if statements) or repetitive as in loops.
Functions correspond to component methods that are implemented as different
executable fields in the forms provided by the developed tool.

Similar to the capabilities offered by SOA, this environment offers a graphical
flow specification that can be mapped to services (or other software assets). The
functional units are provided by the existing assets. However, the not so common
application of this approach is the configuration of the whole constituents by
variability modelling. The users will be offered a friendly interface to define the
differences of their application from the rest of the family of products. Mostly, this
definition is conducted through the selection of predetermined capabilities.

Figure 5.2 shows how variability affects involved models. Developers make
decisions on the variation points first. These choices propagate to other models and
assets, namely the process model, components and connectors. For example, the
effects of the user choices reflect on the process model as inclusion and exclusion of
sub-processes. Components and connectors are selected in this way directly or after
configuration. This relationship is shown with the three arrows at the left side of
Fig. 5.2. Moreover, the configuration of the process model can affect the component and connector selections, which is how the variability model has an indirect
effect on components and connectors. This type of configuration is represented by
the longer arrow at the right side of Fig. 5.2.

Variability is assumed to take place in the highest level of abstraction. Changes
made in this model can affect two other models directly: the process model and the
component model. Also, any change imposed on the process model can further
affect the components; therefore, the variability can affect the component configurations indirectly over processes as well as directly.

The environment incorporates different kinds of actions and different kinds of
users. The domain engineering and product engineering layers of the SPL process
are also adapted here. Therefore, it will be beneficial to continue the descriptions
based on what kinds of users interact with what kinds of capabilities. The following
sections are organized based on this view.

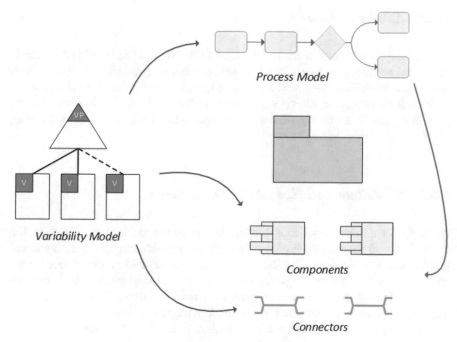

Fig. 5.2 Variability affecting involved models

5.4.1   User Roles

There are three different kinds of users for the environment. Administrators allocate access rights, conduct policies for offer and demand and administer the tests of any quality control procedures. The second group of users is domain developers and product developers. These users develop and publish software assets (components and associated development artefacts such as design models)—they could be free or paid items. The third kind of users is software users or buyers. They can acquire a software application.

There are two basic kinds of users on the development side: domain developers and product developers. Domain developers add new assets to the existing framework, clearly destined for a specific domain environment. A domain environment offers three main graphical models. These are a variability model in conjunction with the other models: a feature model, a process model and a component model. Domain developers can define such models partially or fully for a domain. They can also start a new domain with its specific models.

The product developers use these models to configure their applications. A pure product developer only selects among predefined assets. If the absence of some required capabilities is noticed, such deficiencies can be supplied through the domain developer role.

5.4.2 Administration

As a principle, any constituent to be added to the framework is subject to verification and security screening. Based on the business model, different organizations can utilize the platform for open/free or paid resources. It is possible to charge users with some percentage of the sales, for assets or for whole applications or even for publishing new assets through domain developer roles. There is an expected testing effort on the administration side.

5.4.3 Modelling and Execution Environment

Attempts for a commercial tool supporting the proposed ideas date back to the 2010s. Original versions evolved from database-centric architectures that supported graphical screen elements. Similar attempts have been widely experienced elsewhere, especially offering drag-and-drop development of graphical user interfaces that are windows based and connected to a database table. Different from the majority of these developments, the specific tool targeted no-code development and was partially successful; however, with limited capabilities. The idea was funded through national agencies and products have been completed. Few licences were issued and the customers were able to create applications for small enterprise business functions. Two separate projects were funded for different capabilities that were incorporated into the tool. The no-code support came in forms of expression editors, validation controls and finally, a process modeller. The final version accommodates a graphical editor for the core capabilities of the BPMN whose outcome is executable.

Overview

As a solution towards the suggested development paradigm, the Geneu tool [34] is utilized for demonstration purposes. The current implementation of the tool guides the developers to a form-based development. Currently, process models can be defined as internal to forms; fortunately, they have global access to all the assets defined for the project. To utilize the tool complying with the methodology, one form should be dedicated to the global flow control mechanism. The process defined in this form will be the main process for the application. It is possible for this process to invoke other executions as the tasks of this process. Such invocations will target the other processes or equations in other forms. The set of components corresponds to forms and the methods of the components correspond to the fields of these forms such as processes, equations and value assignments for the simple fields. This usage of the tool both corresponds to the suggested view comprising the specifics of models and at the same time allows easy development by non-programmer users. An overview of the tool is provided in Fig. 5.3.

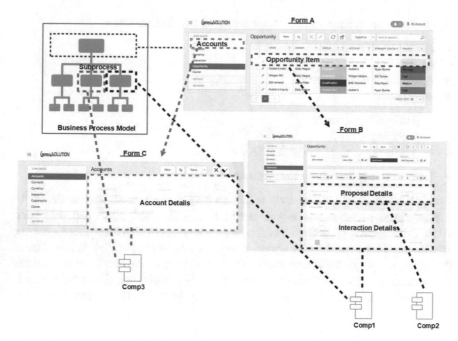

Fig. 5.3 An overview of the tool

Figure 5.3 presents an example development, where forms A, B and C correspond to real screen views, whereas the Business Process Model and the Comp1, Comp2 and Comp3 elements correspond to their logical definition according to the suggested paradigm. The process defined in Form A provides the global flow model which has tasks or sub-processes that will be implemented by processes in the other forms or other fields.

The tool also supports validation rules for error elimination in the filling or calculation of the fields on these forms. One advantage of implementing the assets in a form-based structure is the conception induced in the developers' minds. The no-code developer-user will associate component-level units as forms that are visual structures easy to understand. However, the complex scheme of hierarchical variability management and related constraint propagation is not implemented in the current tool—they take place in experimental implementations.

Geneu mainly targets application development in the Enterprise Resource Planning (ERP) domain in which users generally develop form-based solutions by using spreadsheet applications. However, these applications are usually error-prone and cannot be used in the long term because of the absence of non-functional properties such as user-friendliness. The tool aims to provide fast and cost-effective solutions to the end-users who are non-programmers but domain experts.

Project development with Geneu contains the following stages:

- Requirement analysis
- Application design
- Application development
- Application test

    - Test data preparation
    - Defining roles
    - Verifying defined authorization

- Delivering application to the customer.

Application development starts with including desired packages into the application. These packages are report, notification, dashboard, SMS and scheduler. After the selection of desired packages, the developer can create menus for the application. Menu entities correspond to forms, as shown in Fig. 5.4. Widgets are at the left-hand side of the screen that can be added to the form shown in the middle. Features related to an entity are allocated at the right-hand side. When the design is finished, the application is created and a link is provided to the user to the created application.

Fig. 5.4 The designer view of the tool

5.4.4   Case Study

In this section, the development of a Manuscript Evaluation System (MES) by using Geneu is demonstrated. The system is designed for scientific journals. MES allows journals to collect and evaluate submissions with ease.

One of the main concerns in the evaluation process is anonymity. Reviewers and editors should not know the authors of a manuscript in order to prevent a bias that may occur in the process. Therefore, the authors will be asked to black out the fields which contain their personal information such as their names, e-mails and affiliations. Then the manuscript will be recorded by the system and the author will be given a unique key for tracking purposes. This key will be stored in the MES database along with the user's e-mail address and it will only be accessible by system administrators. Once the submission is completed, it will be forwarded to an editor.

The MES will keep records for registered referees with their research interests/ fields. The appointed editor will assign referees to the submitted manuscript from a pool. The number of referees that are going to be reviewing a submission will be determined by the editor.

When a registered referee receives an evaluation request, he/she will be notified by an e-mail. If the refcree chooses to accept it, the submission will be forwarded to his/her account. Then the referee will be able to download and evaluate the submission.

Figure 5.5 illustrates the general approach for designing forms for the MES project. It displays the screen view corresponding to the design environment for the reviewer panel. The icons allocated on the left-hand side are used to create necessary fields. A simple drag-and-drop mechanism allows users to include the elements in the application. Entity and Property Panels allocated on the right-hand side of the screen allow designers to edit the properties of the inserted elements. The

Fig. 5.5 Design environment for the reviewer panel

selected elements can be renamed, assigned values or can be associated with other elements. The area in the middle shows the designed form to be used by the referees. Assigned referees will be able to select a submission from the drop-down menu to the right of the "select submission" area, download it and see its due date. They will provide feedback via a text field (reviewer's comments). Finally, they will evaluate the submission by selecting a value from the "overall evaluation" field that will present values in the range (strong reject—strong accept) via a drop-down menu.

The MES uses a majority voting algorithm in order to decide if a submission is accepted or rejected. Basically, the average of the overall evaluation fields from all reviewers will be calculated and the result of the assessment will be forwarded to the authors along with the reviewers' comments. The functionality of the designed system is provided with the inclusion of a process model that will be running on the implemented forms. A process model can be created for each form with the "event management" button located in the entity panel shown at the right-hand side of Fig. 5.5.

Figure 5.6 shows the window used for creating a process model. The Geneu uses BPMN 2.0 rendering toolkit "bpmn-js" [35] to model the forms. Tools section shown on the left-hand side includes the process design elements that are based on BPMN 2.0. These elements can be dragged and dropped into the upper middle part of the window to create a workflow. New variables can be created and existing variables can be inspected/updated from the bottom middle part of this screen. The right-hand side of the window includes the properties panel and the activities panel. Properties panel allows users to name/rename the workflow elements such as events, gateways and choose the time of execution for the workflow as: when saved, loaded or constructed. Activities panel is used to bind created variables and

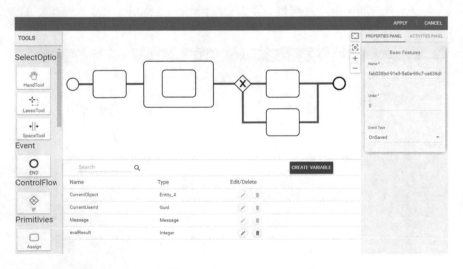

Fig. 5.6 An example process model

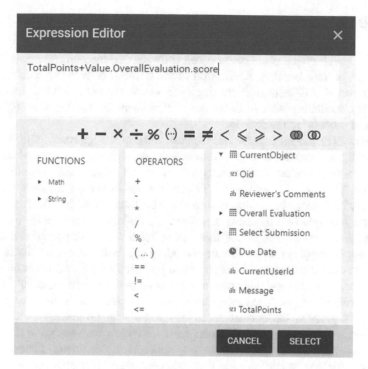

Fig. 5.7 The expression editor window

tables to the design entities and run mathematical expressions through the implemented workflow with its "expression editor" window shown in Fig. 5.7. Usable operators, form entities and all of the created variables are accessible through this window.

An example process model is presented in Fig. 5.6. It is responsible for finding the average of the reviewers' overall evaluations and presenting the results to the editor. The workflow starts with the entity "start". Then an event is defined to initialize a variable "total" with the initial value of zero. The second event is contained in a "for each" entity. In this part, the evaluation score of each reviewer is added to previously defined "total" for all of the submitted evaluations from the reviewers. After that, an exclusive gateway is used in order to check if all of the reviewers have submitted their evaluations. If not, the system will print an error message stating that evaluation cannot be completed because of missing reviews from the referees. If the condition for the gateway holds, the system will be calculating the average with the subsequently defined event. Finally, the result of the evaluation will be sent to the editor and the author) along with the detailed feedback from the referees. The definitions for the gateway and actions are not directly readable on the screen. An item must be selected to see its details on the properties panel at the right-hand side.

5.5 Discussion

The conceptual model of the cloud-based software development framework brings challenges besides benefits. Realization of the framework without variability support surfaces the difficulties in creating such an environment. We will incorporate variability modelling and management environment in the light of previously developed models by the authors.

Current experience demonstrates the feasibility of the approach, with tool support and limited industrial usage. Although the capabilities that were addressed in this chapter as "targeted" have already been achieved, there is still more that can be improved. There have been many different approaches for easing software development; with Geneu, being one tool in that direction. It can be observed that the developer's insight into database structures is a definite advantage for the current environment. This is due to the tools allowing lower-level access due to their philosophy to support users with various coding capability levels, luckily including also the "none" alternative. The current configuration of the tool suggests that the process model is the top-level construct. No-code development expectation be met through guided usage. Therefore, the current achievement is fine for proof of concept. In general, the current status of the tool can be analogous to the programming languages that include undesirable constructs such as "go to statements" and leaves it to the programmer to use appropriate constructs. Geneu has the required modelling capabilities and their interconnection; it is up to the developer to organize the priorities of the different models.

With the amount of experimentation achieved so far, it is possible to say that a slightly more organized usage of the tool can benefit more than a little. The next enhancement to the tool should align the development activities, starting with variability and corresponding dependent configuration actions rippling down to process and functional models. This process should rather be enforced to a certain extent. Also, to be supported with matured domain infrastructures, such usage will provide the biggest amount of automation to development in the current setup.

5.6 Conclusion

Recent prevalent use of cloud, independently deployable services, containerization techniques and their support from bigger companies such as Amazon and Google are the prominent enablers of cloud-based software development framework. Besides, the idea of ecosystems is around for years but its application to software development needed to wait for advancements in this area. The proposed framework employs both people who have domain knowledge, but no programming skills and people from a technical background. We also bring end-users into the equation as a community to create an open market where paid and free services can be provided. The framework enables more domain knowledge enrolment by focusing on integration rather than inner development. Besides, it supports the flexibility of the domain by providing variability to application developers while managing dependencies between variable parts under the hood. To demonstrate a part of our approach, the Geneu tool is presented which provides a modelling environment for a form-based and process-driven software development on the cloud. As future studies, assessment and suggestive ideas on better use will be developed to guide the reader about benefiting more from such environments. Incorporation of variability in the tool environment is an immediate next step.

Acknowledgements We are thankful to Yalin Software who opened up their tool, Geneu and provided expertise that greatly assisted the research. We especially thank Ozcan Manavoglu for his tutorial for the tool. Also, we appreciate the support from TÜBİTAK (Scientific and Technological Research Council of Turkey) for partially funding two projects (3150612 and 3110392) related to the development of GENEU, within the TEYDEB project program.

References

1. Togay C, Dogru AH, Tanik JU (2008) Systematic component-oriented development with axiomatic design. J Syst Softw 81(11):1803–1815
2. Prajapati AG, Sharma SJ, Badgujar VS (2018) All about cloud: a systematic survey. In: International conference on smart city and emerging technology (ICSCET). Mumbai, India, pp 1–6. https://doi.org/10.1109/ICSCET.2018.8537277
3. Hayes B (2008) Cloud computing. Commun ACM 51(7):9–11
4. Mell P, Grance T (2011) The NIST definition of cloud computing
5. Lawton G (2008) Developing software online with platform-as-a-service technology. Computer 41(6):13–15. https://doi.org/10.1109/MC.2008.185
6. Serrano N, Gallardo G, Hernantes J (2015) Infrastructure as a service and cloud technologies. IEEE Softw 32(2):30–36
7. Repschlaeger J, Wind S, Zarnekow R, Turowski K. (2012) A reference guide to cloud computing dimensions: infrastructure as a service classification framework. In 2012 45th Hawaii international conference on system sciences. IEEE, pp 2178–2188
8. Erl T (2005) Service-oriented architecture: concepts, technology, and design. Prentice Hall, Upper Saddle River
9. Papazoglou MP, Traverso P, Dustdar S, Leymann F (2007) Service-oriented computing: State of the art and research challenges. Computer 40(11):38–45
10. Sommerville I (2011) Software engineering, 9th edn. Addison-Wesley, Boston
11. Stojanovic Z, Dahanayake A (2005) Service-oriented software system engineering challenges and practices. IGI Publishing, Hershey
12. Havey M (2005) Essential business process modeling. O'Reilly Media Inc., Sebastopol
13. Weske M (2007) Business process management—concepts, languages and architectures. Springer, Berlin
14. Pohl K, Böckle G, van Der Linden FJ (2005) Software product line engineering: foundations, principles and techniques. Springer Science & Business Media, Berlin
15. Van Gurp J, Bosch J, Svahnberg M (2001) On the notion of variability in software product lines. In: Proceedings working IEEE/IFIP conference on software architecture. Amsterdam, The Netherlands, 28–31 August, pp 45–54
16. Sinnema M, Deelstra S, Nijhuis J, Bosch J (2004) Covamof: a framework for modeling variability in software product families. In: International conference on software product lines. Springer, Berlin, pp 197–213

17. Dogru AH, Tanik MM (2003) A process model for component-oriented software engineering. IEEE Softw 20(2):34–41
18. Dogru AH (1999) Component oriented software engineering modeling language: COSEML. Computer Engineering Department, Middle East Technical University, Turkey
19. Kaya MC, Suloglu S, Dogru AH (2014) Variability modeling in component oriented software engineering. In Proceedings of the society for design and process science. Kuching Sarawak, Malaysia
20. Cetinkaya A, Kaya MC, Dogru AH (2016) Enhancing XCOSEML with connector variability for component oriented development. In Proceedings of SDPS 21st international conference on emerging trends and technologies in designing healthcare systems, Orlando, FL, USA, pp 120–125
21. Scott WR (2002) Organizations: rational, natural, and open systems. Prentice Hall, Upper Saddle River
22. Chesbrough H (2003) Open innovation: the new imperative for creating and profiting from technology. Harvard Business Review Press
23. Herbsleb JD (2007) Global software engineering: the future of socio-technical coordination. In: Proceedings of FOSE '07 future of software engineering, 23–25 May Minneapolis, Minnesota, USA, pp 188–198
24. Bosch J (2009) From software product lines to software ecosystems. In: Proceedings of SPLC '09 The 13th international software product line conference, San Francisco, California, USA, August 24–28, pp 111–119
25. Manikas K, Hansen KM (2013) Software ecosystems—a systematic literature review. J Syst Softw 86(5):1294–1306
26. Jula A, Sundararajan E, Othman Z (2014) Cloud computing service composition: a systematic literature review. Expert Syst Appl 41(8):3809–3824
27. Tao F, Zhao D, Hu Y, Zhou Z (2008) Resource service composition and its optimal-selection based on particle swarm optimization in manufacturing grid system. IEEE Trans Industr Inf 4 (4):315–327
28. Lucassen G, Van Rooij K, Jansen S (2013) Ecosystem health of cloud PaaS providers. International conference of software business. Springer, Berlin, pp 183–194
29. Copeland M, Soh J, Puca A, Manning M, Gollob D (2015) Microsoft Azure: planning, deploying, and managing your data center in the cloud. Apress, Berkely
30. Zahariev A (2009) Google app engine. Helsinki University of Technology, Espoo, pp 1–5
31. Lomov A (2014) OpenShift and cloud foundry PaaS: high-level overview of features and architectures. White paper, Altoros
32. Fingar P (2009) Dot cloud: the 21st century business platform built on cloud computing. Meghan-Kiffer Press
33. Teixeira C, Pinto JS, Azevedo R, Batista T, Monteiro A (2014) The building blocks of a PaaS. J Netw Syst Manage 22(1):75–99
34. Yalin Software (2019) Geneu tool. https://geneu.app/. Accessed 1 Apr 2019
35. BPMN-JS (2019) BPMN 2.0 rendering toolkit and web modeler, https://bpmn.io/toolkit/bpmn-js/. Accessed 1 Apr 2019

Chapter 6
Security Challenges in Software Engineering for the Cloud: A Systematic Review

Mohamed Alloghani and Mohammed M. Alani

Abstract Cloud computing is among the fastest growing technologies, and it has brought noticeable growth in security concerns. Despite the security challenges, cloud computing has proven pivotal in the development and success of distributed systems. This comes from certain features such as rapid elasticity, on-demand service deployment, and support for self-service. All these features are associated with security challenges such as data breaches, network security, data access, denial of service attacks, hijacking of accounts, and exploitable system vulncrabilities. Regardless of the cloud model, the cloud software development process and the consideration of integrated security features are critical for securing cloud computing. As such, software engineering is required to play an essential role in combating cloud security issues in the future applications. In this paper, we introduce a systematic review of articles in the area of software engineering security challenges on the cloud. The review examines articles that were published between 2014 and 2019. The procedure for article qualification relied on the elucidation of Preferred Reporting Items for Systematic Reviews and Meta-Analyses premises. Meta-analysis checklist was employed to explore the analytical quality of the reviewed papers. Some of the issues considered were included, but were not limited to, cloud models of service delivery, access control, harm detection, and integrity. All these elements are discussed from the perspective of software engineering and its prospect in improving cloud security.

Keywords Security · Software engineering security · Service security · Security survey

M. Alloghani
Liverpool John-Moores University, Liverpool, UK
e-mail: M.AlLawghani@2014.ljmu.ac.uk

M. M. Alani (✉)
Senior Member of the ACM, Abu Dhabi, UAE
e-mail: m@alani.me

© Springer Nature Switzerland AG 2020
M. Ramachandran and Z. Mahmood (eds.), *Software Engineering in the Era of Cloud Computing*, Computer Communications and Networks,
https://doi.org/10.1007/978-3-030-33624-0_6

6.1 Introduction

Cloud computing, being an emerging and fast-growing trend, has brought global attention in pursuit of cloud data management and storage systems. However, it is imperative to note that cloud systems require data manipulation algorithms that are not necessarily integrated within the system. That is, third-party data manipulation systems are usually used in the cloud regardless of whether it is private, public, or otherwise. As such, many cloud systems are susceptible to breaches that are related to cross-boundary usage of data manipulation programs [1, 2]. From a nuanced perspective, cloud systems attract threat actors based on issues such as network security, access to data, locality of the data, and vulnerability of the cloud system itself.

Arguably, cloud systems store a vast amount of data, including sensitive personal information, and as such have become attractive and profitable for malicious actors. Regardless of the severity of the attack, the damage caused by confidentiality exposure from the perspective of the data owner can be devastating. It is also worth to note that service models such as software as a service (SaaS) bring new attack surface [3, 4]. In most SaaS deployments, data is stored in localized servers and as such leakage of confidential information is more probable because many of these systems rely on poor user verification processes.

The unique nature of the cloud makes common non-cloud attacks even more damaging. An attack like distributed denial of service (DDoS) can cause websites and service to go down for prolonged periods of time. However, the damage can be much higher when this attack is applied to a cloud-based system. As cloud services are charged per usage, a DDoS attack can cost the cloud client extremely high cost due to the exhaustion of cloud resource [5]. In addition to DDoS attacks, cloud services can be susceptible to exploitation of bugs within core operating systems leading to serious compromise in security.

For several decades, when organizations needed to expand or upgrade their computing capacity, they had one of the two options: either purchase additional hard drives, memory, or other hardware component, or try to streamline and fine-tune all information technology operations to become more efficient [6, 7]. Regardless of choice, such organizations were forced to embark on sophisticated engineering or re-engineering tasks, some of which included replication of databases, scaling the capability of processes, and expansion tasks that would support a significant increase in the number of users as well as concurrent processes. However, it is also imperative to note that these options brought in a noticeable increase in hardware, software, and maintenance costs. From these costs and undesired characteristics of these standard options, companies resorted to cloud computing. The basic feature of cloud computing is the delivery of data services based on a lease of storage and processing capacity.

Security challenges that cloud computing users face are clearly seen in software as a service (SaaS) deployment model [8]. Despite cloud benefits, including agility, availability, cost-effectiveness, and elasticity, some of the recent developments have exposed cloud systems to multiple issues. For instance, the architecture and design of cloud services have exposed many organizations to many attacks because of the vulnerabilities within their systems. Another concern is the method of deploying some of the cloud service software as well as the technique used to manipulate the data stored within the cloud [9, 10]. From a deployment perspective, the cloud environment combines comprehensive and interdisciplinary deployment strategies and platforms. However, each of these platforms and strategies brings their weaknesses, and most of them do not focus on a conventional design as well as a typical architecture. The lack of uniformity exposes these systems to different attacks and hence the importance of considering software engineering approaches in handling cloud security issues.

This chapter starts with an introduction to the topic and then moves on to explain the motivation behind the work. The third section discusses related works and introduces a thorough literature review of works within the area of software engineering security for the cloud. The fourth section dives deeply into the methodology used to produce the results. In the methodology section, the rationale behind the selection of PRISMA along with the literature input, quality search assessment, search strategy, and inclusion and exclusion criteria are explained. The fifth section shows the results of research conducted. The results section includes different distributions of papers along with the security challenges in software engineering. The last section includes the conclusions derived from the research with some pointers to future directions.

6.2   Motivation

Of the many plausible solutions to security issues in the cloud, software engineering remains of key value because of its role in cloud system development cycle. That is, software engineering provides rich tools and techniques for modeling cloud software requirements and testing besides other software design needs but with a focus on security issues. It is pertinent to note that security issues are specific to given attributes of software and operating systems, and as such, software engineering is the best option of dealing with such security issues. Regardless of the opportunities that software engineering brings with it, security in the context cloud computing is becoming more precarious given its rapidly growing adoption.

Although not known as cloud computing back then, in 1961, a well-known computer scientist named John McCarthy stated

> computers of the kind I have advocated become the computers of the future, then computing may someday be organized as a public utility just as the telephone system is a public utility… The computer utility could become the basis of a new and important industry.

From that statement, the term utility computing was identified as a computer-on-demand service that can be used by the public with a pay-for-what-you-use financial model. The term kept evolving with more maturity till the end of the 1990s when Sales-force.com introduced the first remotely provisioned service to organizations. Near the end of the 1990s, the concepts started to focus on an abstraction layer used to facilitate data delivery methods in packet-switched heterogeneous networks. In 2002, Amazon.com introduced Amazon Web Services (AWS) platform. The platform, back then, provided remote rapidly provisioned computing and storage. Commercially, the term cloud computing emerged in 2006 when Amazon introduced service named Elastic Compute Cloud (EC2). The service model was based on "leasing" elastic computing processing power and storage where organizations can run their apps. Later that year, Google also started providing Google apps [11].

Cloud computing was identified by NIST in [12] as

> a model for enabling ubiquitous, convenient, and on-demand network access to a shared pool of configurable computing resources (e.g., networks, servers, storage, applications, and services) that can be rapidly provisioned and released with minimal management effort or service provider interaction.

As such, it is imperative to profile software engineering models that developers integrate into cloud systems to combat increasing malicious attempts from threat actors. In principle, cloud computing faces fast-growing security threats such as abstraction layer threats, corporate espionage, spamming, and identity theft among other Internet-based crimes that are fast shifting to cloud-based applications. Hence, it is essential to understand and explore the historical approach of dealing with security issues right from the identification of software requirements, designs, and testing as means of pursuing build-in security features as opposed to the conventional security batching approaches.

6.3   Related Works

Some of the notable reviewed articles regarding either software engineering and its role on cloud security issues or just cloud computing issues include an article by Kaur and Singh [13], which is a review of cloud computing security issues. The authors assert that third-party inclusion on cloud services is one of the significant security concerns although it is important to note that the authors only address security issues without discussing the possible solutions, especially the role of software engineering [13]. Of concern is the revelation that cloud servers and data storage systems are vulnerable to unauthorized access; primarily because one data is transferred to the cloud, cloud providers have unfettered access to the information.

Our previous work in [14] also introduced a survey of the most common threats, attacks, and mitigation mechanisms. The survey focused on threats and attacks on different layers of the cloud and concluded with general security improvement directions.

One of the systematic review papers that addressed elements of cloud computing focused on software engineering but from the perspective of integration, delivery, and deployment [15]. The other systematic review focused on source code analysis from a multilingual perspective [16]. Both papers focus on specific codes including recommendations on maintenance of source codes and applications to improve security. However, it was noted that the articles focused on programs and codes from a general perspective.

6.4   Methodology

The systematic review was based on Preferred Reporting Items for Systematic Reviews and Meta-Analyses (PRISMA) methodology, and as such explicit steps and checklist were used to determine the articles that qualified for the review.

The PRISMA framework was developed by the EQUATOR Network and was derived from the methodological framework for guideline development [17]. The rationale behind selecting PRISMA is that it has exclusively been developed to support the completion of meta-analyses and systematic reviews and facilitates the researchers in improving their overall research findings. It offers a structured approach and a minimum set of items that streamline the completion of systematic reviews. It provides the researchers with a 27-item checklist accompanied by a flow diagram comprising of four phases to help guide in effective reporting of findings from systematic reviews [17]. Despite using PRISMA techniques and protocols, the articles were reviewed using a double-blind approach and as such the reports generated regarding each of the articles were anonymous, but the assessment formed a critical part of this review article. The checklist used in the review was based on the 2009 list that was adopted for KIN 4400.

The primary sections or topics of the checklist include but not limited to title, abstract, introduction, methods, results, discussion, and conclusion. In specific, each of the subjects explores different aspects of the qualified article. For example, the abstract focuses on the structure of the summary with emphasis on whether the author specified the objectives, background, sources of data, and eligibility of the study among others. As for the introduction, the consideration for the rationale and objectives of the articles as per the guidelines required modification. That is, PRISMA rationale evaluation requires recognition of participants, interventions, comparisons, outcomes, and study design (PICOS), yet some of these elements may not be an application in the field of research interest. Hence, instead of rationale, the review focused on the novelty of the article, and the eligibility criteria of the research methods also focused on the same. However, some elements such as search and study selection were also ignored from the evaluation because ideal systematic reviews focus on literature review papers on the subject of interest.

6.4.1   Literature Input

The articles used in the review were retrieved from IEEE Xplore Digital Library, ProQuest Central database, and search arXiv.org. The queries implemented in each of the databases were dependent on the requirement of the specific search engine although the keywords were the same. It is also imperative to note that the search results meet year of publication (2014–2019), access or retrieval capabilities (open access), and content of the document (full text). It is imperative to note that IEEE Xplore Digital Library returns both free and subscription-based articles, and the interest was on the open access articles. Conversely, ProQuest Central provides abstracts of articles that require purchase or active subscriptions.

6.4.2   Quality Assessment and Processing Steps

The quality assessment and appraisal of the documents were based on the PRISMA checklist as well as the flow diagram. Notably, the PRISMA diagram consists of identification, screening, eligibility, and inclusion steps. Each of these steps has distinct quality assessment and article appraisal protocols regardless of the search strategy and string that was implemented to retrieve the articles. Firstly, the identification process only recognized documents that were open access and available in any of the databases. As previously noted, the articles were restricted to those available in IEEE Xplore Digital Library, ProQuest Central database, and search arXiv.org although IEEE Xplore Digital Library was the primary database given its focus computer science-related issues including having several journals and conferences on cloud and Internet of Things (IoT) related issues.

The screening process proceeded systematically, and it started with abstract screening, full-text screening, study allocation, appraisal for criteria suitability and data contribution, and summary of the study. Both abstract and full-text screening focused on the relevance of the papers on addressing software engineering for cloud security. The appraisal of the methods and subsequent analytical quality of the papers focused on the data and the contribution of the study to further development and applications of software engineering in cloud security. Hence, it is imperative to note that the quality of the papers was based on the above elements but with a specific emphasis on how each of the articles addressed software engineering and its role on improving security on the cloud.

6.4.3   Search Strategy and Search Strings

The search string was based on a combination of Boolean operators and thematic areas that includes software engineering, cloud computing, and cloud security.

The standard structure for the search string including the full title Security challenges in software engineering for the cloud AND software engineering AND Cloud Computing AND cloud security. Table 6.1 summarizes the research strings implemented in the different databases (Fig. 6.1).

6.4.4   Inclusion and Exclusion Criteria

The inclusion and hence exclusion criteria did not precisely abide by the standard rules and regulations because most of the research articles did not abide by the standard research processes. However, the common exclusion/inclusion criteria included data (2014–2019), exposure of interest (cloud systems that have resorted to software engineering as a means for improving security), and reported outcomes. Furthermore, factors such as peer review, type f publication, and reported outcomes also influenced the inclusion or exclusion of the articles. Regarding the date of the review, attempts were undertaken to ensure that some of the existing reviews on the same issues were not submitted within the duration of the current review; otherwise, if a review existed, then the dimensions and approach used were different and did not conflict with the current review.

Besides the date, the review process focused on articles that addressed software engineering as a means of addressing security issues in cloud environments. Further, the review excluded non-peer-reviewed articles, and as such ignored all grey literature including conference papers as well as technical reports on given research questions. As for reported outcomes, each paper was appraised for the consistency between the reported outcomes relative to its objective and cited literature. All articles that had elements of self-reporting were excluded, and suitable articles were searched and reviewed accordingly. Finally, the systematic review focused on original studies based on the type of publication and based on the metrics; only papers published in reviewed journals were considered. Therefore, the review rejected publications categorized as reviews, editorials, reports, and letters. The exclusion was meant to eliminate paper documents that had been used in small-scale documents because their contribution to the course of the analysis would not make much difference given their lack of authority from a scholarly perspective.

Table 6.1 Summary of the search strategy and keywords

Subject area	Search phrase	Boolean operator	
Computer science	Security challenges in software engineering for the cloud	AND	Title (TI)
Software engineering	Cloud computing	AND	Abstract and full text
Cloud computing	Cloud security	AND	Title, abstract, and full text

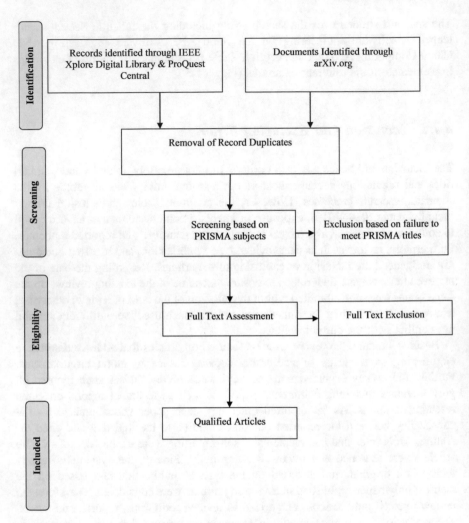

Fig. 6.1 Steps were taken in sourcing and appraising the articles to identify the qualified ones

6.5 Results

It is imperative to note that the implementation of the search strings in different databases yielded different results. Specifically, the IEEE Xplore Digital Library results returned 216 articles although only 12 meet the inclusion/exclusion criteria [18–29]. From arXiv.org such search results, a total of 315 articles matched the full-text title search although attempts to narrow the search using any combination of the Boolean operations failed to yield any results. Of the 315 articles, only seven meet all the appraisal requirements while the rest violated either one or a combination of the PRISMA topics and specifications [30–36].

Furthermore, focusing on full-text peer-reviewed scholarly journals published between 2014 and 2019, ProQuest Central search returned 4837 results. However, given that the focus was on cloud computing, additional filters were added, so that 771 articles addressing cloud computing security issues were addressed. Excluding literature review, feature, reports, and general information articles reduced the search results to 761 articles. For further processing and publication relevance, some of the publication excluded from the search results included Multimedia Tools and Applications, PLoS One, Sensors, Journal of Medical Systems, The Scientific World Journal, and Future Internet among many others. The exclusion of irrelevant publications reduced the number of articles to 247, and all these articles underwent screening and subsequent quality assessment based on the PRISMA model. The summary of the documents retrieval and processing is as presented in Fig. 6.2. From Fig. 6.2, it is apparent that about 51 articles were included in the quality assessment, although it is paramount to state that all the 266 articles meet the initial search criteria, but most of them were excluded after further assessment and considerations.

6.5.1 Quality of Methodology of the Analytical Papers

Of the 266 papers, it suffices to deduce that met the PRISMA themes and subtopics. That is, all the papers had a title, an introduction, a methodology, results, discussion, and conclusion sections. However, with regard to the eligibility criteria, 215 of the articles lacked explicitly stated characteristics of the period of the study, the characteristics of the research, and the credible data sources for replicating the studies. It is imperative to note that over 76% of the reviewed articles were qualitative and the authors resorted to exploring qualitative aspects of the software engineering applications in resolving security issues. The remaining 24% employed both qualitative and quantitative approaches, and it suffices to deduce that both approaches were suitable for the research articles. It was also observed that about 88% of the articles addressed issues related to security improvements based on software engineering approaches. Hence, it suffices to deduce that the reviewed articles had considerably high accuracy and scholarly concern regarding the use of software engineering in addressing.

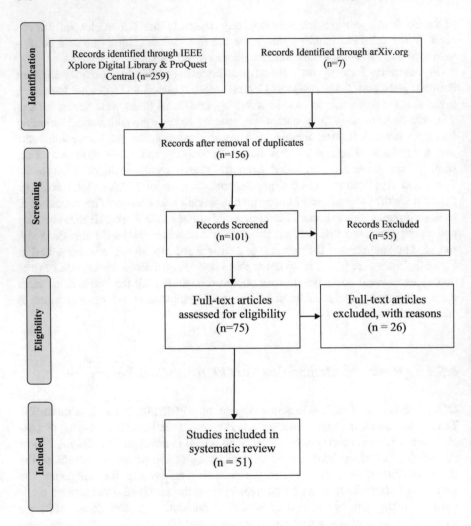

Fig. 6.2 Summary of the articles searched and used in the review process

6.5.2 Distribution of Papers Based on Year of Publication

One of the noted differences while using the three databases was the classification of documents. For instance, the search results from ProQuest despite filtering using scholarly journals as a key search parameter, the outcome included articles that have the feature, journal article, literature review and a combination of both features. Table 6.3 summarizes the different classification of articles by document type despite restricting the search to scholarly articles (Table 6.2).

Table 6.2 Summary of the number of articles based on document type

Document type	Count
Feature	12
Journal article	118
Journal article; feature	1
Journal article; literature review	2
Literature review	1
Grand total	134

IEEE Xplore Digital Library, unlike ProQuest Central, returned exclusively the number of documents published as scholarly articles. The distribution of these articles is shown in Fig. 6.3.

From Fig. 6.3, it is apparent that the number of open access document published in the IEEE Xplore Digital Library increased between 2015 and 2016. However, it is imperative to note that articles addressing the same are published in other IEEE non-open access journals. As such, the data was more sensible when used in the context of search results from ProQuest Central (Fig. 6.4).

A similar trend is evident in the annual distribution of the number of articles published between 2014 and 2018. Between 2017 and 2018, the number of articles tends to decrease although more research should be leaning toward improving security in the cloud. It was also prudent to explore the distribution of the articles based on publishers.

6.5.3   Distribution of Papers Based on Publishing

The following disposition did not focus on the impact factors of the respective journals but rather prioritized the title and requirements of the publishing journal with regard to the title of the literature review. As such, journals related to software engineering, systems engineering, cloud computing, and software or software developments were prioritized.

Fig. 6.3 Distribution of articles retrieved from IEEE Xplore Digital Library by year

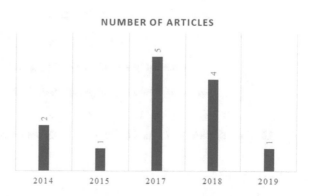

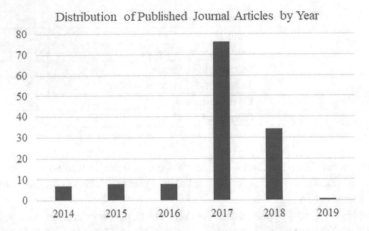

Fig. 6.4 Distribution of journal articles based on ProQuest Central search results

From Fig. 6.5, it is evident that most of the articles were from Journal of Systems Engineering and Electronics. The other two major journals included IEEE Internet of Things Journal and Computing in Science and Engineering. A similar distribution chart for the journals that served as the source of the articles retrieved from ProQuest Central is presented in Fig. 6.6.

Interestingly, the search data from ProQuest Central drew data from the Journal of Cloud Computing and International Journal of Advanced Research in Computer Science. A majority of the articles were retrieved from the latter journal.

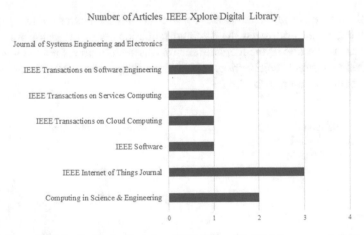

Fig. 6.5 Journal article distribution based on IEEE Xplore Digital Library search data

Distribution of Articles by Journal

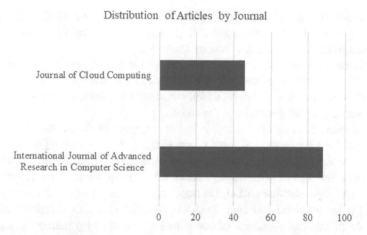

Fig. 6.6 Journal article distribution based on ProQuest central search data

6.5.4   Analytic Based on the Literature Search Results

Even though the search results based on year of publication and journals tend to be diverse, the word cloud shown in Fig. 6.7 suggests a pattern based on the key search strings as well as the implemented search strategy.

Fig. 6.7 Word cloud based on the subjects and keywords from the search results data

As shown in Fig. 6.7, computing was the dominant word (232 instances) followed by cloud (207), information (81) security (68), computation (53), and management (29). This can be seen in Table 6.3.

As shown in Table 6.3, it is evident that a total of 670 words—based on relevance to the systematic review—were extracted. These words were found in various combinations, such as cloud computing, cloud security, information security, cloud management, and computation in the cloud.

Despite the focus on computing in the cloud, none of the articles had software engineering as a key although most of the discussions about cloud computing and emerging security threats were related to software engineering. Further, it was noted that discussion regarding software and cloud security focused on cryptography, privacy, and data transfer within, between, and across private and cloud systems. Hence, both word cloud and topic modeling established that most articles addressed cloud computing and elements of software engineering as a means of improving security.

6.5.5 Security Challenges in Software Engineering

Supposing that security issues in the cloud are related to software engineering challenges then some of the articles addressed both directly and indirectly. As most of the authors acknowledge, security issues in public computing clouds are attributed mostly to external threat although insider threats are also increasingly becoming menacing in cases where mitigation measures and restrictions do not exist or are poor [37–40]. The security issues are categorized, and threat actors infiltrate the systems based on the layers of a cloud computing platform and the countermeasures placed against unauthorized access [41–45]. The three layers that are susceptible to attack are the infrastructure with about multiple virtual machines, the platform layer, and the application layer [34, 46, 47]. Of these layers, the software layer that serves as the platform for the core of the cloud service as well as the stack for hosting customer application tends to be a pathway for most of the attacks.

Table 6.3 Keywords and subjects extracted from the search results data

Keywords	Instances count
Computing	233
Cloud	207
Information	81
Security	68
Computation	53
Management	29
Total (based on relevance)	670

The debate has been around how to identify attackers masquerading as attackers; it is difficult to tell apart the actions of the two groups of users. In most cases, network eavesdroppers successfully position themselves in between perpetrating man-in-the-middle attacks regardless of the existence of firewalls as well as aggressive intrusion detection systems strategically installed within all the layers of the cloud system [48, 49]. Some of the common attacks include authentication, virtual machines attacks, denial of service attacks, and insider abuse.

Most of the authentication attacks occur between the end users and the cloud platform, and in most cases, public cloud environments tend to more vulnerable compared to private ones. The vulnerability arises from the fact that most systems rely on public and private keys to authenticate users, and the requirement for username–password authentication allows a man in the middle to harvest credentials and abuse them [50–52]. As for virtual machine attacks, the challenge can be attributed to the inevitable need for multiple cloud tenants to share the same physical computer but using virtually separated drives. In case an adversary gains access and assumes a legal identity, attacks based on calls to a virtual network device would lead to an intrusion into the physical device. Such acts support the insertion of malicious codes and subsequent devastating attacks on the entire physical system [53–55]. Furthermore, once an adversary has access to virtualized machines, such access can be used to exploit open access control as well as intercept inter-virtual machine communication, especially for virtual machine spaces on the same physical computers. Most researchers noted that denial of service attacks is unlikely in cloud computing environments due to a large number of servers involved in supporting the transfer of data and communication [56, 57]. However, from the perspective of users, adversaries can target specific servers and spoof them to retrieve specific information or in some case sought collusion with a legitimate user to expose sensitive data [58]. Finally, computers in cloud environments tend to have active connections with cloud service vendors who have access to sensitive information, and one of the challenges is regulating and controlling the information that the vendors can access. Third-party vendors have detailed information about the location of the physical machines, and any collusion or compromise from the side of the vendor can prove to be fatal.

6.5.6   Securing Applications on the Cloud

Securing applications and data on the cloud have its unique challenges because once both data and applications are moved to the cloud, cloud service providers gain access to all the details increasing the risk of misuse of the stored data or hosted applications [59]. Some of the ideal techniques for securing the cloud investing in file distribution systems and securing multiple storage to ensure confidentiality and availability of the data as well as the sources of the applications

Table 6.4 Security attacks and recommended mitigation

Security attacks	Recommended mitigation
Insider threats	Rigorous access control
Eavesdropping	Encryption, intrusion detection and prevention systems (IDS/IPS)
Denial of service (DoS)	Intrusion detection and prevention systems (IDS/IPS)
Man-in-the-middle attack	Firewall, authentication, and IDS
Unauthorized access	Multi-factor authentication and updated public-key cryptography
Virtual machines attacks	Distributing application to multiple instances within the cloud and integration of private virtual cloud environment

[60, 61]. Some of the primary methods for protecting data and applications include a regular update of encryption key pairs alongside enforcement of multi-factor authentication, distributing the application to multiple instances within the cloud environment, segregating the virtual machines, and migrating virtual machines of the public cloud to private virtual cloud environments [62, 63]. Most importantly, all important source files and data transferred to the cloud must be encrypted to deter man-in-the-middle interception and misuse. Table 6.4 lists the major security attacks and possible recommended mitigation techniques based on the current findings.

6.5.7 Recommended Best Practices

Some industry best practices with respect to cloud security are highlighted in the list below.

1. User Access Management: Implementation of access control policies, including role-based access control (RBAC), mandatory access control (MAC), and discretionary access control (DAC).
2. Data Protection: Implementation of the web applications firewall, the on-premise firewall, and hardware-based multi-factor authentication, along with strong encryption through hardware security modules (HSMs).
3. Monitoring and Control: Integration of strong intrusion detection and prevention systems over the cloud to prevent intrusions.
4. Updates: Ensure that the cloud, servers, and applications have been provided with the latest updates to security patches, updating user IDs and passwords on a frequent basis.

Further general cloud security recommendations can be found in [64].

6.6   Conclusion and Future Work

Even though a majority of the reviewed articles addressed security in cloud computing environments, very few considered software engineering as a technique for mitigating such issues. As such, it is pertinent to consider security features during all phases of developing a cloud-based platform, application, or the third app for data transfer and manipulation. Based on the concerns that arise after transferring data and applications to the cloud, it is not promising when most of the articles do not address security after the effective transfer. Granting data and applications to cloud providers exposes both users and organizations to unknown adversaries. Even though it is possible and prudent to encrypt data before transferring to the cloud, further modalities are necessary to ensure that the privacy of the data remains with the owner. That is, the further prospect should focus on algorithms for compressing and encrypting data so that manipulation programs in the cloud cannot have access to the content.

The systematic review conducted in this chapter contributes to the existing literature by specifically highlighting the major security issues and challenges underlying software engineering in the cloud. The review identifies insider threats, eavesdropping, man-in-the-middle attacks, virtual machine attacks, and denial of service attacks to be the most common issues in cloud computing security management. Implementation of firewalls, intrusion detection and prevention systems, rigorous multi-factor authentication, and strong access control policies was identified as key recommendations to help overcome these security challenges. However, one key limitation of the review is that it entailed a rather generic scope and did not consider the specific cloud computing service delivery models, such as software as a service or platform as a service. It is arguable that more severe security challenges may be involved in specific cloud computing environments. This calls for future research to be conducted in a more extensive way that identifies security issues for each of the three cloud delivery models of SaaS, PaaS, and IaaS. Attention to the CIA triad comprising of confidentiality, integrity, and availability as core components is also a key direction for future research.

References

1. Armbrust M, Fox A, Griffith R, Joseph AD, Katz RH, Konwinski A, Lee G, Patterson DA, Rabkin A, Stoica I, Zaharia M (2009) Above the clouds: a Berkeley view of cloud computing. EECS Department, University of California, Berkeley. https://doi.org/10.1145/1721654.1721672
2. Hunt SE, Mooney JG, Williams ML (2014) Cloud computing. In: Computing handbook, 3rd edn. Information systems and information technology. https://doi.org/10.1201/b16768
3. Foster I, Zhao Y, Raicu I, Lu S (2008) Cloud computing and grid computing 360-degree compared. In: Grid computing environments workshop, GCE 2008. https://doi.org/10.1109/gce.2008.4738445

4. Marston S, Li Z, Bandyopadhyay S, Zhang J, Ghalsasi A (2011) Cloud computing—the business perspective. Decis Support Syst. https://doi.org/10.1016/j.dss.2010.12.006

5. Alani MM (2016) Security attacks in cloud computing. In: Elements of cloud computing security. Springer, Berlin, pp 41–50

6. Hashem IAT, Yaqoob I, Anuar NB, Mokhtar S, Gani A, Ullah Khan S (2015) The rise of "big data" on cloud computing: Review and open research issues. Inf Syst. https://doi.org/10.1016/j.is.2014.07.006

7. Kalapatapu A, Sarkar M (2017) Cloud computing: an overview. In: Cloud computing: methodology, systems, and applications. https://doi.org/10.1201/b11149

8. Fernando N, Loke SW, Rahayu W (2013) Mobile cloud computing: a survey. Future Gener Comput Syst 29(1):84–106, Elsevier

9. Grobauer B, Walloschek T, Stöcker E (2011) Understanding cloud computing vulnerabilities. IEEE Secur Priv. https://doi.org/10.1109/msp.2010.115

10. Xu X (2012) From cloud computing to cloud manufacturing. Robot Comput-Integr Manuf. https://doi.org/10.1016/j.rcim.2011.07.002

11. Alani MM (2016) What is the cloud? In: Elements of cloud computing security. Springer, Berlin, pp 1–14

12. Mell P, Grance T et al (2011) The NIST definition of cloud computing. Computer Security Division, Information Technology Laboratory, National Institute of Standards and Technology

13. Kaur M, Singh H (2015) A review of cloud computing security issues. Intl J Adv Eng Technol 8(3):397

14. Alani MM (2016) Elements of cloud computing security: a survey of key practicalities. Springer, Berlin

15. Shahin M, Babar MA, Zhu L (2017) Continuous integration, delivery and deployment: a systematic review on approaches, tools, challenges and practices. IEEE Access 5:3909–3943. https://doi.org/10.1109/ACCESS.2017.2685629

16. Mushtaq Z, Rasool G, Shehzad B (2017) Multilingual source code analysis: a systematic literature review. IEEE Access 5:11307–11336. https://doi.org/10.1109/ACCESS.2017.2710421

17. Moher D, Liberati A, Tetzlaff J, Altman DG (2009) Preferred reporting items for systematic reviews and meta-analyses: the prisma statement. Ann Intern Med 151(4):264–269

18. Al-Kaseem BR, Al-Dunainawi Y, Al-Raweshidy HS (2019) End-to-end delay enhancement in 6LoWPAN testbed using programmable network concepts. IEEE Internet Things J 1. https://doi.org/10.1109/jiot.2018.2879111

19. Al-Kaseem BR, Al-Raweshidyhamed HS (2017) SD-NFV as an energy efficient approach for M2M networks using cloud-based 6LoWPAN testbed. IEEE Internet Things J 4(5):1787–1797. https://doi.org/10.1109/JIOT.2017.2704921

20. Chen T, Bahsoon R (2017) Self-adaptive and online QoS modeling for cloud-based software services. IEEE Trans Softw Eng 43(5):453–475. https://doi.org/10.1109/TSE.2016.2608826

21. Goodacre J (2017) Innovating the delivery of server technology with Kaleao KMAX. Comput Sci Eng 19(5):77–81. https://doi.org/10.1109/MCSE.2017.3421544

22. Hu G, Sun X, Liang D, Sun Y (2014) Cloud removal of remote sensing image based on multi-output support vector regression. J Syst Eng Electr 25(6):1082–1088. https://doi.org/10.1109/JSEE.2014.00124

23. Kantarci B, Mouftah HT (2014) Trustworthy sensing for public safety in cloud-centric internet of things. IEEE Internet Things J 1(4):360–368. https://doi.org/10.1109/JIOT.2014.2337886

24. Mocskos EH, C.J.B., Castro H, Ramírez DC, Nesmachnow S, Mayo-García R (2018) Boosting advanced computational applications and resources in latin america through collaboration and sharing. Comput Sci Eng 20(3), 39–48 (2018). https://doi.org/10.1109/mcse.2018.03202633

25. Wang Y, Wang J, Liao H, Chen H (2017) Unsupervised feature selection based on Markov blanket and particle swarm optimization. J Syst Eng Electr 28(1):151–161. https://doi.org/10.21629/JSEE.2017.01.17
26. Wu Y, He F, Zhang D, Li X (2018) Service-oriented feature-based data exchange for cloud-based design and manufacturing. IEEE Trans Serv Comput 11(2):341–353. https://doi.org/10.1109/TSC.2015.2501981
27. Xiaolong X, Qitong Z, Yiqi M, Xinyuan L (2018) Server load prediction algorithm based on CM-MC for cloud systems. J Syst Eng Electr 29(5):1069–1078. https://doi.org/10.21629/JSEE.2018.05.17
28. Yuan H, Bi J, Li B (2015) Workload-aware request routing in cloud data center using software-defined networking. J Syst Eng Electr 26(1):151–160. https://doi.org/10.1109/JSEE.2015.00020
29. Zhang W, Xie H, Hsu C (2017) Automatic memory control of multiple virtual machines on a consolidated server. IEEE Trans Cloud Comput 5(1):2–14. https://doi.org/10.1109/TCC.2014.2378794
30. Alnasser A, Sun H, Jiang J. Cyber security challenges and solutions for V2X communications: a survey. Comput Netw. doi S1389128618306157
31. Brenier JL (1967) The role of the Halsted operation in treatment of breast cancer. Int Surg 47 (3):288–290. https://doi.org/arXiv:1609.01107
32. Cruz L, Abreu R, Lo D (2019) To the attention of mobile software developers: guess what, test your app!. Empirical Softw Eng, 1–31, Springer
33. Ibrahim AS, Hamlyn J, Grundy J (2010) Emerging security challenges of cloud virtual infrastructure. In: Proceedings of APSEC 2010 cloud workshop. doi 10.1.1.185.603
34. Li ZH (2014) Research on data security in cloud computing. Adv Mater Res 930(5):2811–2814. doi 10.4028/www.scientific.net/AMR.926-930.2811. http://www.scientific.net/AMR.926-930.2811
35. Hu P, Dhelim S, Ning H, Qiu T (2017). Survey on fog computing: architecture, key technologies, applications and open issues. J Netw Comput Appl 98:27–42, Elsevier
36. Tian Z, Su S, Li M, Du X, Guizani M et al (2019) Automated attack and defense framework for 5G security on physical and logical layers. https://doi.org/arXiv:1902.04009
37. Geng R, Wang X, Liu J (2018) A software defined networking-oriented security scheme for vehicle networks. IEEE Access 6:58195–58203. https://doi.org/10.1109/ACCESS.2018.2875104
38. Heartfield R, Loukas G, Gan D (2017) An eye for deception: a case study in utilizing the human-as-a-security-sensor paradigm to detect zero-day semantic social engineering attacks. In: 2017 IEEE 15th international conference on software engineering research, management and applications (SERA), 371–378. https://doi.org/10.1109/sera.2017.7965754
39. Martin W, Sarro F, Jia Y, Zhang Y, Harman M (2017) A survey of app store analysis for software engineering. IEEE Trans Software Eng 43(9):817–847. https://doi.org/10.1109/tse.2016.2630689
40. Siboni S, Sachidananda V, Meidan Y, Bohadana M, Mathov Y, Bhairav S, Shabtai A, Elovici Y (2018) Security testbed for internet-of-things devices. IEEE Trans Reliab 1–22. https://doi.org/10.1109/tr.2018.2864536
41. Luo M, Zhou X, Li L, Choo KR, He D (2017) Security analysis of two password-authenticated multi-key exchange protocols. IEEE Access 5:8017–8024. https://doi.org/10.1109/ACCESS.2017.2698390
42. Mingfu X, Aiqun H, Guyue L (2014) Detecting hardware trojan through heuristic partition and activity driven test pattern generation. In: 2014 communications security conference (CSC 2014), pp 1–6. https://doi.org/10.1049/cp.2014.0728
43. Su Q, He F, Wu N, Lin Z (2018) A method for construction of software protection technology application sequence based on petri net with inhibitor arcs. IEEE Access 6:11988–12000. https://doi.org/10.1109/ACCESS.2018.2812764

44. Wang B, Chen Y, Zhang S, Wu H (2019) Updating model of software component trustworthiness based on users feedback. IEEE Access 1. https://doi.org/10.1109/access.2019. 2892518
45. Wang S, Wu J, Zhang S, Wang K (2018) SSDS: a smart software-defined security mechanism for vehicle-to-grid using transfer learning. IEEE Access 6:63967–63975. https://doi.org/10. 1109/ACCESS.2018.2870955
46. Cox JH, Chung J, Donovan S, Ivey J, Clark RJ, Riley G, Owen HL (2017) Advancing software-defined networks: a survey. IEEE Access 5:25487–25526. https://doi.org/10.1109/ ACCESS.2017.2762291
47. Zahra S, Alam M, Javaid Q, Wahid A, Javaid N, Malik SUR, Khan MK (2017) Fog computing over IoT: a secure deployment and formal verification. IEEE Access 5:27132–27144. https://doi.org/10.1109/ACCESS.2017.2766180
48. Sharma PK, Chen M, Park JH (2018) A software defined fog node based distributed blockchain cloud architecture for IoT. IEEE Access 6:115–124. https://doi.org/10.1109/ ACCESS.2017.2757955
49. Wang D, Jiang Y, Song H, He F, Gu M, Sun J (2017) Verification of implementations of cryptographic hash functions. IEEE Access 5:7816–7825. https://doi.org/10.1109/ACCESS. 2017.2697918
50. Ashraf MA, Jamal H, Khan SA, Ahmed Z, Baig MI (2016) A heterogeneous service-oriented deep packet inspection and analysis framework for traffic-aware network management and security systems. IEEE Access 4:5918–5936. https://doi.org/10.1109/ACCESS.2016.2609398
51. Bangash YA, Rana T, Abbas H, Imran MA, Khan AA (2019) Incast mitigation in a data center storage cluster through a dynamic fair-share buffer policy. IEEE Access 7:10718–10733. https://doi.org/10.1109/ACCESS.2019.2891264
52. Zou D, Huang Z, Yuan B, Chen H, Jin H (2018) Solving anomalies in NFV-SDN based service function chaining composition for IoT network. IEEE Access 6:62286–62295. https:// doi.org/10.1109/ACCESS.2018.2876314
53. Dehling T, Sunyaev A (2014) Information security and privacy of patient-centered health IT services: what needs to be done? In: 2014 47th Hawaii international conference on system sciences, pp. 2984–2993. https://doi.org/10.1109/hicss.2014.371
54. Li X, Wang Q, Lan X, Chen X, Zhang N, Chen D (2019) Enhancing cloud-based IoT security through trustworthy cloud service: an integration of security and reputation approach. IEEE Access 7:9368–9383. https://doi.org/10.1109/ACCESS.2018.2890432
55. Shu X, Yao D, Bertino E (2015) Privacy-preserving detection of sensitive data exposure. IEEE Trans Inf Forens Secur 10(5):1092–1103. https://doi.org/10.1109/TIFS.2015.2398363
56. Sheikh NA, Malik AA, Mahboob A, Nisa K (2014) Implementing voice over Internet protocol in mobile ad hoc network—analysing its features regarding efficiency, reliability and security. J Eng 2014(5):184–192. https://doi.org/10.1049/joe.2014.0035
57. Ullah R, Ahmed SH, Kim B (2018) Information-centric networking with edge computing for IoT: research challenges and future directions. IEEE Access 6:73465–73488. https://doi.org/ 10.1109/ACCESS.2018.2884536
58. Chin T, Xiong K, Hu C (2018) Phishlimiter: a phishing detection and mitigation approach using software-defined networking. IEEE Access 6:42516–42531. https://doi.org/10.1109/ ACCESS.2018.2837889
59. Sun J, Long X, Zhao Y (2018) A verified capability-based model for information flow security with dynamic policies. IEEE Access 6:16395–16407. https://doi.org/10.1109/ ACCESS.2018.2815766
60. Dorey P (2017) Securing the internet of things. In: Smart cards, tokens, security and applications, 2nd edn. https://doi.org/10.1007/978-3-319-50500-8_16
61. Jarraya Y, Zanetti G, PietikÄInen A, Obi C, Ylitalo J, Nanda S, Jorgensen MB, Pourzandi M (2017) Securing the cloud. Ericsson review (English edn). https://doi.org/10.1016/c2009-0-30544-9

62. Biswas K, Muthukkumarasamy V (2017) Securing smart cities using blockchain technology. In: Proceedings—18th IEEE international conference on high performance computing and communications, 14th IEEE international conference on smart city and 2nd IEEE international conference on data science and systems, HPCC/SmartCity/DSS 2016. https://doi.org/10.1109/hpcc-smartcity-dss.2016.0198
63. Yi S, Li C, Li Q (2015) A survey of fog computing: concepts, applications and issues (#16). In: Proceedings of the 2015 workshop on mobile big data—Mobidata'15. https://doi.org/10.1145/2757384.2757397
64. Alani MM (2016) General cloud security recommendations. In: Elements of cloud computing security, pp 51–54. Springer, Berlin

Part II
Cloud Design and Software Engineering Analytics with Machine Learning Approaches

Chapter 7
Software Engineering Framework for Software Defect Management Using Machine Learning Techniques with Azure

Uma Subbiah, Muthu Ramachandran and Zaigham Mahmood

Abstract The presence of bugs in a software release has become inevitable. The loss incurred by a company due to the presence of bugs in a software release is phenomenal. Modern methods of testing and debugging have shifted focus from 'detecting' to 'predicting' bugs in the code. The existing models of bug prediction have not been optimized for commercial use. Moreover, the scalability of these models has not been discussed in depth yet. Taking into account the varying costs of fixing bugs, depending on which stage of the software development cycle the bug is detected in, this chapter uses two approaches—one model which can be employed when the 'cost of changing code' curve is exponential and the other model can be used otherwise. The cases where each model is best suited are discussed. This chapter proposes a model that can be deployed on a cloud platform for software development companies to use. The model in this chapter aims to predict the presence or absence of a bug in the code, using machine learning classification models. Using Microsoft Azure's machine learning platform, this model can be distributed as a web service worldwide, thus providing bug prediction as a service (BPaaS).

Keywords Machine learning · Machine learning as a service · Bug prediction as a service · Microsoft Azure

U. Subbiah (✉)
Computer Science and Engineering, Amrita School of Engineering, Amritanagar,
Ettimadai, Coimbatore 641112, Tamil Nadu, India
e-mail: umasubbiah19@gmail.com

M. Ramachandran
School of Computing, Creative Technologies and Engineering, Leeds Beckett University,
Headingley Campus, Churchwood Ave, Leeds LS6 3QS, UK
e-mail: m.ramachandran@leedsbeckett.ac.uk

Z. Mahmood
Northampton University, Northampton NN2 7AL, UK
e-mail: dr.z.mahmood@hotmail.co.uk

© Springer Nature Switzerland AG 2020
M. Ramachandran and Z. Mahmood (eds.), *Software Engineering in the Era of Cloud Computing*, Computer Communications and Networks,
https://doi.org/10.1007/978-3-030-33624-0_7

7.1 Introduction

The presence of bugs in any written software is inevitable. However, the cost of fixing these bugs varies significantly, depending on when the bug is detected. If software developers are able to detect bugs at an earlier stage in the life cycle, the cost incurred in fixing the bug will be significantly lower.

Recent trends revolve around the fact that bugs can now be predicted, much before they are detected. Large collections of previous bug data are vital to be able to predict bugs with reasonable accuracy. Software analytics has opened up endless possibilities for using data analytics and reasoning to improve the quality of software. Actionable analytics uses the results of the software analysis as real-time data, to make useful predictions.

By determining the presence or absence of a bug in a software version, developers can predict the success of the software version even before it is released, based on a few features (characteristics) of the release version. If this prediction is performed at an earlier stage in the software development cycle, it will reduce the cost of fixing the bug. Moreover, by incorporating various software analytic techniques, we might be able to develop a bug prediction model that is both agile and efficient enough to be used commercially by the software development industry.

Machine learning has been successfully applied to make predictions in various datasets. Given the huge number of bug datasets available today, predicting the presence of bugs too can be done using various machine learning techniques. This chapter uses Microsoft's popular machine learning as a service (MLaaS) tool Azure to build machine learning models and deploy them on the cloud. By employing a cloud-based machine learning tool, this chapter facilitates the easy deployment of a model on the cloud as a web service for companies to use. Various metrics are used to evaluate the models, and their results are shown in this chapter, for software developers to use. We also offer suggestions on which model is best suited for a given scenario.

This chapter proposes the use of machine learning as a service (MLaaS) to provide a viable solution to software developers for predicting the presence of bugs in written software, thereby providing bug prediction as a service (BPaaS).

This chapter has been organized as follows: Sect. 7.2 contains a literature review of research results that have been used and reviewed in this chapter and similar ongoing work. While Sect. 7.3 discusses a machine learning approach to software engineering, Sect. 7.4 discusses the application of big data techniques to software engineering analytics. Section 7.5 discusses software defects in detail. Section 7.6 cover software defect detection techniques used, along with related tools. Section 7.7 discusses bug prediction in software development, from various perspectives. A neural network approach to bug prediction to resolve software costs and to feed new requirements is discussed in Sect. 7.8. A service-oriented approach to providing bug prediction is seen in Sect. 7.9. Cloud software engineering for machine learning applications is detailed in Sect. 7.10. Section 7.11 discusses the experiment performed, and the results obtained. Section 7.12 contains a critical evaluation of neural network approaches and their application in software engineering analytics. Section 7.13 concludes this chapter and describes the scope for this model.

7.2   Machine Learning Application to Software Engineering Analytics: Literature Review

Software companies around the world use predictive analysis to determine how many bugs will appear in the code or which part of the code is more prone to bugs. This analysis has helped cut down the loss caused by commercial failure of software releases. Software analytics and the application of big data and neural network learning algorithms on this analysis are also being dealt with in software companies.

However, the extent to which these measures reduce the cost of changing the code is yet to be explored. By looking at the cost of change curve [1, 2] for various software development methods, it is evident that the earlier a bug is fixed, the less it will cost a company to rectify the bug. More recently, service-oriented computing allows for software to be composed of reusable services, from various providers. These service components can be obtained from a variety of service providers. Bug prediction methods can thus be provided as a reusable service component with the help of machine learning on the cloud.

This section will look at software engineering approaches to machine learning as well as the new area of software analytics which focuses on the application of machine learning and data analytics to improve software development practices. The reciprocal use of software development for machine learning and bug prediction is discussed. This section ends with a literature review of cloud computing, the integration of which this chapter aims to achieve.

This chapter introduces three distinct processes:

1. Machine learning for software engineering.
2. Software engineering for machine learning.
3. Software engineering analytics: This is a new area of software engineering which combines machine learning, data analytics, and software engineering repositories which provide nearly 50 years of software development experiences.

This section also covers a literature review of big data and neural network approaches to software engineering analytics.

7.2.1 Early Use of Machine/Deep Learning in Software Analytics

The use of machine learning to create an entirely automated method of deciding the action to be taken by a company when a bug is reported was first proposed by [3]. The method adopted uses text categorization to predict bug severity. This method works correctly on 30% of the bugs reported to developers. Feature selection can be used to improve the accuracy of the bug prediction model [4]. In this paper, information gain and chi-square selection methods are used to extract the best features to train a naive Bayes multinomial algorithm and a k-nearest neighbor's algorithm.

Neural networks are undoubtedly the go-to solution for prediction. An early research conducted by [5] shows that neural networks can be used in analytic models. The model in the paper uses neural networks to predict software failure. The same paper concludes that this approach is far better that other analytic models. More recently, [6] use SVMs, ensemble models, and deep neural networks to make software analytic predictions. By analyzing the structure of the neural network used, the paper [6] is able to overcome the disadvantage of using a neural network as a black box. There are numerous neural network techniques that have been successful. However, there is a need for a systematic process to be applied when collecting data and classifying data for efficient use of neural networks.

7.2.2 Software Engineering for Machine Learning

Software engineering for machine learning refers to the application of software engineering principles and techniques to the machine learning domain. The 2019 conference 'Software Engineering for Machine Learning Applications' (SEMLA) effectively bridged the gap between the two domains, enabling experts to collaborate [7]. Of the observable advantages, the most important is possibly the ability of software engineering practices to improve the overall efficacy of machine learning [8].

The process of software engineering for machine learning is shown in Fig. 7.1 which consists of a nine-stage workflow as discussed in [9]. This includes single and multistage feedback loops, depicted by the closed and open loops, respectively.

While the use of machine learning in software engineering has been extensively researched in recent years (e.g., [10]), the reciprocal use of software engineering in machine learning has not been studied as extensively [8]. Similarly, the use of machine learning in software engineering and vice versa is strongly advocated by [7], which states that while AI helps automate manual tasks, it also introduces a high level of complexity in the software systems concerned. Figure 7.1 (adapted from [9]) shows a nine-stage process of how machine learning can be incorporated in software development in companies. As the datasets used become increasingly larger, it is essential to employ special analysis techniques that come under the broad term 'software engineering analytics.'

Fig. 7.1 Simplified process diagram for the integration of machine learning into software development

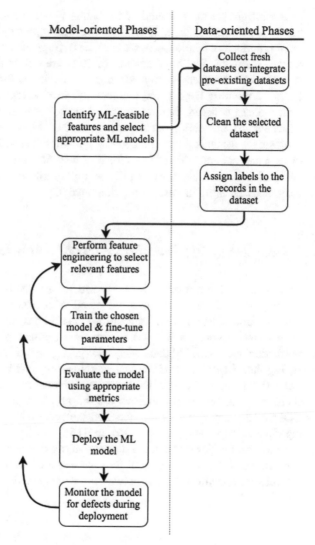

7.2.3 Software Engineering Analytics

Software engineering analytics (SEA) involves processing and inspecting data pertaining to the functional and structural aspects of software. Big data is an efficient, popular method of computational analysis that can be used in SEA. Similarly, machine and deep learning are proven methods of making predictions from given data. The use of these two technologies on SEA datasets can be extremely beneficial, as described in this chapter.

An insight on the potential of analyzing and utilizing data [11] obtained from previous software releases helps improve the efficiency of the software development process as a whole (software analytics). Bugs are an unavoidable part of every written software. The huge amount of software defect data available—both open source and otherwise—serves as an excellent input for predictive software analytics. Combining the existing methods of machine learning and big data analytics with modern software engineering can have a tremendous impact on the cost and efficiency of the entire development process. More importantly, providing the analysis results in real time (actionable analytics) keeps bug prediction systems up to date and accurate. This is an effective way of increasing the 'agility' and 'success' of software development [12]. A highly efficient way of dealing with such large datasets is by the use of big data analytics.

7.2.4   Use of Big Data for Software Analytics

The amount of information in the world reportedly increases tenfold every five years [13], and the world has witnessed a corresponding improvement in the storage, computation, and processing capabilities of computer systems. This information revolution has led to the development of many novel methods of predicting and analyzing data. Maryville [14] give a detailed description of the uses of big data in the field of software engineering. With the amount of data being generated from bug reports and statistics collected from previous software releases, there exist numerous possibilities for the application of big data throughout the SDLC. The importance of datasets and the processing steps applied to them is of significant importance, as detailed in [15]. Finally, DeLine [16] describes the various possibilities that exist in the current big data era of software engineering. In the next section, the background for a potential candidate for the predictive aspect of software analytics—neural networks—is discussed.

7.2.5   Neural Network Approach to Bug Prediction and Cost Estimation

A novel approach to addressing previously unresolved problems in using learning models for software engineering analytics is found in [16]. Initially, [17] introduced the concept of neural networks in the software cost estimation domain. This was further refined by modeling it with a fuzzy system to provide a cost estimate. Optimized evolutionary neural networks can also be used to predict bugs [18]. The use of deep learning on the cloud is an increasingly popular trend. Li et al. [19] present a way to use deep learning on the cloud, while ensuring privacy is maintained, thereby overcoming one of the drawbacks of using the cloud.

AI approaches to software analytics, especially for software improvement, are not limited to machine/deep learning, but extend to various fields of computational intelligence like fuzzy logic systems [20]. The use of case-based reasoning (CBR) is discussed in this chapter, an example of which is found in [20]. Moreover, the application of fuzzy logic to software fault prediction under semi-supervised conditions has been empirically studied in [21]. We shall now look at the various machine and deep learning approaches to software engineering in detail.

7.3 Machine/Deep Learning Approaches to Software Engineering

Software engineering analytics involves huge amounts of data being processed for prediction. The most effective way of analyzing and predicting on such datasets is using deep learning. Mahapatra [22] provides an insight as to why data scientists prefer deep learning to machine learning even though both methods can be employed to take intelligent decisions. To summarize, deep learning has greater power and flexibility, achieved by using a nested hierarchy and abstraction. The performance of deep learning is far superior when compared to larger datasets, shown in Fig. 7.3 adapted from [22]. Moreover, deep learning does not require a prior feature extraction to have been performed [23] as depicted in Fig. 7.2, though it does require state-of-the-art infrastructure. For these reasons, deep learning is preferred over machine learning. This chapter performs a comparative analysis of both machine learning and deep learning algorithms, with the understanding that there are scenarios where we understand the software domain enough to extract features by hand.

One of the earliest applications of neural networks to software analysis was in 1996, by [24], which uses principal component analysis to train two neural networks to predict the number of software faults in the testing phase. They achieved an average RMS error of <0.10. Given the large number of bug reports, metrics

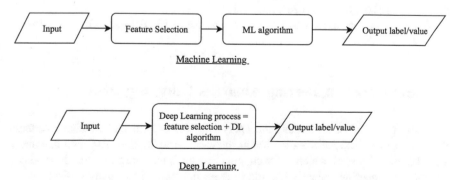

Fig. 7.2 Difference in preprocessing required by machine and deep learning

Fig. 7.3 Increase in
performance of deep learning
algorithms over ML
algorithms as the size of the
data increases

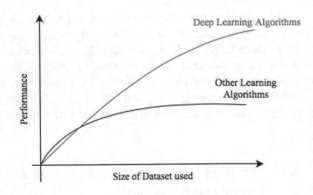

Table 7.1 Applications of
deep learning in various
phases of the SDLC

Phase	Application of deep learning
Requirement analysis	Effort estimation
Design and coding	Software metrics
Software testing	Software testing activities
Operation and maintenance	Other software metrics

used, and feedback collected from software customers, there is enough descriptive analytic data to train a neural network to make predictions based on analysis of past software trends. Given a scenario, a neural network will be able to foresee the presence or absence of a bug using predictive analysis. Future work will include using these predictions to take preventive steps using prescriptive analytics, as discussed in the previous section of this chapter. Li et al. [25] provide a bibliographical analysis as to which phases of the software development life cycle will benefit from deep learning approaches to analysis. Considering the six phases of software development mentioned in [25], we have requirement analysis, software design, development, testing, maintenance, and project management. While all phases can be benefitted by deep learning, studies such as [25] show that the development phase is benefitted the most, followed by the maintenance phase then testing, management, and finally requirements and design.

The various applications of deep learning in various phases of the SDLC are described in [26], as shown in Table 7.1.

The next section takes a closer look at the use of big data analytics in software engineering analytics.

7.4   Software Engineering Analytics Using Big Data

Big data involves the computational analysis of datasets to find trends or patterns that reveal insightful information about the data being studied. Big data has many varied applications, of which software engineering is an important one. Nowadays, software engineering datasets include (but are not limited to) datasets like:

- Cost estimation datasets, which hold details of the project like team experience, manager experience, duration of the project, effort (in person-hours), transactions, entities, and the type of language used to code [27].
- Repository datasets, which track the number of commits, authors, comments, messages, the contents of files, languages used, licenses, etc. [28].
- Bug tracking datasets, which provide information about the various attributes and metrics of the software code, and how many bugs of different types were detected in the past [29].

These datasets are highly invaluable, when it comes to analyzing past trends and predicting future outcomes. As the cliché saying goes, 'Garbage in, garbage out.' Data quality is an extremely important feature of any dataset; its importance increases with the size of the dataset. Sparse datasets with many missing components are known to have caused significant losses and even failure of the predictive model trained using them; an example is found in [15].

Big data plays a huge role in the analysis of software, by providing a mechanism to perform this computational analysis. Companies have started turning to processes like data analytics, machine learning, deep learning, AI, and business intelligence to answer their questions and provide data required to prevent future mishaps [30]. Not only do these methods help 'see the future,' they also help developers and testers understand the impact that a change made in one stage will have across the software development life cycle [31].

In any software engineering analysis, there are three main processes that play important roles, in the attempt to develop low-cost, high-quality software [31]. They are:

1. Descriptive analytics
2. Predictive analytics
3. Prescriptive analytics.

Descriptive analytics is a process mainly concerned with past events. It analyzes huge datasets of past data, searching for a relation between the causes and effects of a particular outcome (positive or negative outcomes). In the field of software engineering, this phase would include a thorough analysis of the software systems available, to document attributes and features of the software. The factors influencing a particular outcome of software and the reasons behind certain outcomes are studied. This phase generates huge datasets, which will be used later in the analytic process.

Next, predictive analytics comes into picture. The predictive analytics phase employs a predictive model, which may be either a machine learning model or a deep learning model. Based on this difference, there is a slight difference in the steps taken after the first phase. Machine learning algorithms require a prior feature extraction to be performed. In this case, inferences are drawn from the dataset and fed as input to the next phase. On the other hand, deep learning algorithms are capable of extracting features and drawing inferences from a given dataset. Here, the dataset produced during the descriptive analytics phase is directly given to the predictive model. Incorporating big data and ML with SE is largely a cyclic process that begins with large datasets obtained from data mining. These are analyzed, and predictions are made upon them. The analysis leads to a course of action that is in turn recorded. Huge records are mined for bug data, and issues that arise, which feed big datasets again, as shown in Fig. 7.4.

Overall, the second phase gives companies a prediction or 'forecast' of what may occur, given a scenario. It takes into account the outcomes of previous scenarios, derived from the descriptive analysis phase. This phase depends heavily on machine learning and deep learning algorithms.

Figure 7.4 is a visual depiction of the process of incorporating big data and learning algorithms into software engineering and related analysis. Data pertaining to software engineering is collected from large datasets and analyzed. Analysis takes place in the three-step process described above, i.e., descriptive analysis followed by predictive analysis and finally prescriptive analysis. The predictive analysis stage entails predictive modeling, which may be machine or deep learning. After the suggested course of action (the output of prescriptive analytics) is taken,

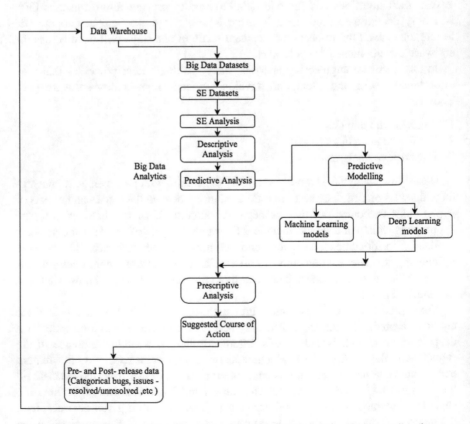

Fig. 7.4 Process of incorporating big data and machine learning with software engineering

all the data (pre- and post-release data) collected is gathered and stored in the data warehouse to be used in future datasets.

Finally, prescriptive analytics helps visualize the various courses of action available and determine the optimal solution to the problem/situation at hand. This is a process-intensive task [32], since it requires a heuristic understanding of which decisions might be taken, the weight of impact of each decision, and how they may influence/be influenced by the outcomes of other decisions. The techniques employed by prescriptive analytics to obtain an optimal solution will involve a combination of optimization, game theory, artificial intelligence, and decision taking algorithms. Since the entire process can be viewed as a series of components that are executed in sequence, there is scope for a service-oriented approach to this problem. In the next section, software defects, their types, and classification schemes are discussed in detail.

7.5 Software Defects

A software bug can be defined as a defect in the code or program that leads to a deviation from the expected behavior of the system. The bug life cycle [33] is an essential component to understand the defect/bug. Bugs may occur at any stage of the software development life cycle (SDLC).

There may be many different reasons for the presence of software bugs, all of which can be summarized in the following four categories, as detailed in [34]:

1. Defects due to incorrect and incomplete methods
2. Defects due to the use of faulty and unreliable tools
3. Defects due to the intervention of inexperienced people in software development
4. Defects due to incomplete or defective requirement specifications.

Many methods of software bug classification exist. A popular method is the MR classification, detailed in [35] which classifies bugs based on the defect modification requests (MRs) that the software developer receives. These are stored in a database, and root cause analysis (RCA) is performed, studying the defects based on various properties, such as which phase the defect was detected in, where the defect occurred and defect location. In this phase, severity of the bug is also considered.

The two main characteristics that this chapter and [36] consider are bug severity and the presence of high priority. While severity is determined by the loss that will be incurred if the bug persists, priority is defined by how soon the defect needs to be fixed in order to avoid further complications. Priority and severity are often, but not always, highly correlated [37]. Moreover, they must be assigned cautiously, since they determine the amount of time and resources that will be invested in the rectification of a defect.

In the Eclipse bug dataset [29], seven types of bugs have been specified, based on severity. They are detailed below, as obtained from [38]:

- Blocker—a term used to describe a defect that prohibits further development. There exist no known solutions to these defects
- Critical—defects causing the system to enter a 'frozen' state, data/memory leak
- Major—major defects are those that cause the software to malfunction
- Normal—defects that are characterized by circumstantial loss of function
- Minor—defects that exhibit a small loss of function, where a solution is known, and easy to implement
- Trivial—defects that do not affect the function of the software at all, but may impede ease of use/comprehension for the user
- Enhancement—MRs to enhance the function of the software.

Bug priority is a measure used to determine the sequence of rectifying bugs in a software. Bugs with higher priority need to be resolved earlier. Priority of a bug is assessed on a scale of 1–5 in the Eclipse bug dataset. Level 1 priority defects are of highest priority, while level 5 are the lowest. High-priority bugs are identified by partial loss of functionality of the system—they do not render the system unusable, but cause the software to be 'undeployable' [39].

Given the classification of bugs and characteristics like priority and severity, it is essential to consider how they are *detected*, before we look at how bugs can be *predicted* instead.

7.6   Software Defect Detection Techniques and Tools

As the complexity of software increases, so does the number of bugs in the software. This section covers the various methods of detecting defects and the metrics used to measure software defects. Defect detection is essentially the process of finding bugs in a system. As the cost of fixing bugs increases with time, regardless of the method of software development, it is desirable to detect and fix the bug as early as possible. While bugs are certain to occur in software, finding bugs is an elusive process. There are various defect detection methods in use, each with its own advantages and disadvantages. Moreover, several measures have been proposed to quantify the bugs in a software release.

In practice, there are three main approaches to bug detection [40], as described below. It has been observed [41] that the effort required for bug detection is almost the same for all three methods.

- **Code reading/review**:

In this method of software testing, the software tester identifies the specifications of subprograms of the whole program. This is repeated for larger subprograms and the program to find any deviations from the expected specifications.

This method is largely immune to hidden defects [41]. Moreover, it is a comparatively time-consuming process.

- **Functional testing**:

In functional testing, the software tester executes the code for a set of test cases, based on the specifications of the program, and records the obtained output. The obtained and the expected outputs are compared. Functional testing reveals more observable faults in the code [41]. The cost of functional testing is relatively higher.

- **Structural testing**:

The internal structure of the software is assessed in this type of testing. The tester examines the method of implementation, rather than the functional details of the software. Structural testing takes equal or less time for bug detection, with a lower computational cost than that of functional testing [41].

Some formulae that help quantify defects in a software release are:

Defect Detection Efficiency (DDE) is defined as the ratio of the number of injected defects detected in a phase to the number of injected defects in the same phase.

$$\text{Defect Detection Efficiency} = \frac{\text{No.of injected defects detected in a phase}}{\text{No.of defects injected in the phase}} \times 100$$

Defect Detection Percentage (DDP) is defined as the ratio of the number of defects discovered during the SDLC to the number of defects found after the software was released.

$$\text{Defect Detection Percentage} = \frac{\text{No.of defects detected prior to the software release}}{\text{No.of defects detected after the software release}} \times 100$$

Defect leakage rate (DLR) is a useful indicator of the number of defects that the tester was able to discover, as opposed to the bug being discovered by a customer or end user.

$$\text{Defect Leakage Rate} = \frac{\text{No.of defects detected by the tester before release}}{\text{No.of defects detected by the customer after release}} \times 100$$

Bug tracking tools are extremely useful tools that assist in reporting, analyzing, and monitoring the life cycle of bugs in a software. There is a wide variety of bug tracking tools available in the market today. Bug tracking tools can also help testers view trends in the software defects, report defects, and track customer bug reports, while ensuring that duplicate defects do not appear too often. The importance of a feature for the bug tracking system depends on the type of software being developed, the software development team, and many other factors. An intricate web of dependencies is formed, while deciding which bug tracking tool is to be employed.

An insightful comparison of five popular tools—Flyspray, JTrac, Mantis, phpBugTracker, and WebIssues—is found in [42]. Another paper [43], also published in 2015, studies Bugzilla, JIRA, Trac, Mantis, BugTracker.Net, Gnats, and Fossil and adds that Track + bug tracking tool is one of the most effective available. More recently, [44] compared ten bug tracking tools—Bugzilla, Eventum, Fossil, Mantis, OTRS, Redmine, Request tracker, The Bug Genie, Trac, and WebIssues, concluding that Bugzilla has the highest number of essential features.

Despite the existence of many methods for defect detection, it is often easier to avoid defects altogether. This can be done by a thorough analysis and review of the requirements with the customer and root cause analysis [45]. While this process of tracking and fixing bugs is efficient, prevention is better than cure. Predicting bugs even before they arise can save developers enormous amounts of time, and software companies can save millions of dollars every year, as seen in the next section.

7.7   Bug Prediction in Software Development

A software defect may either be an error in the code causing abnormal functionality of the software or a feature that does not conform to the requirements. Either way, the presence of a bug is undesirable in the commercial release of a software or a version thereof. This section discusses the vitality of bug *prediction* over *detection*.

The fifth edition of the Software Fail Watch report [46] by a software company called Tricentis claimed that 606 reported software bugs had caused a loss of $1.7 trillion worldwide, in 2017. The cost of fixing bugs rises throughout the software development life cycle. Further, the overall cost of fixing software defects rises year over year. It is evident that an efficient means of predicting software defects will help cut down the loss due to software production globally.

The waterfall model of software development suggests testing for defects after integrating all of the components in the system. However, testing each unit or component after it has been developed increases the probability of finding a defect. The iterative model incorporates a testing phase for each smaller iteration of the complete software system. This leads to a greater chance of finding the bugs earlier in the development cycle. The V-model has intense testing and validation phases. Functional defects are hard to modify in this model—it is hard to go back once a component is in the testing phase. The agile model also uses smaller iterations and a testing phase in each iteration. The various prototyping models also have testing methods for each prototype that is created. The testing phase is always done later on in the development cycle. This will inevitably lead to larger costs of fixing the defect.

If a bug can be predicted, it may be possible to prevent it altogether. Given the advantages and reduction in cost that will result from preventing bugs, this novel method may replace traditional methods of software testing like black box, white box, gray box, agile, and ad hoc testing in the future. The cost of change curve is an important consideration, while considering the point in the software development life cycle that the bug prediction algorithms have to be run.

Boehm's [1] cost of change curve is an exponential curve, implying that the cost of fixing a bug at a given stage will always be greater than the cost of fixing it at an earlier stage. Paper [36] describes two models—one that predicts the presence of a bug based only on the types of bugs found in versions before this release and another that uses a dataset of CK metrics [47] and code attributes to predict a defect.

This chapter proposes two models—one based solely on previous version data and a second based on attributes of the class in the current version. If Ambler's cost of change curve is followed (for the agile software development cycle), the first model is preferred, since it can predict buggy code at an earlier stage. However, Kent Beck's cost of change curve [2] for eXtreme Programming (XP) tends to flatten out. Here, the second model's higher AUC score (though only available at a later stage) might be more desirable, since the cost does not grow exponentially. Figure 7.5 from [36] shows the feedback of data after the gain of defect knowledge of previous versions to the requirements stage of the next version, to facilitate bug prediction. This process of bug prediction can be provided as a service on the cloud, as proposed by the same paper.

As shown in the schematic flowchart of the process of bug prediction (Fig. 7.6) using machine learning, the bug reports from various development environments along with various software metrics are stored in a bug database. From this database, a dataset is taken. This dataset is used to train a suitable machine learning model. By deploying the machine learning model on the cloud, bug prediction can be provided as a cloud-based service to software development companies across the world. In addition to bug prediction, neural networks can also be used to resolve extraneous software costs and to aid in the requirements analysis phase.

Fig. 7.5 Feedback of defect knowledge gained can facilitate bug prediction

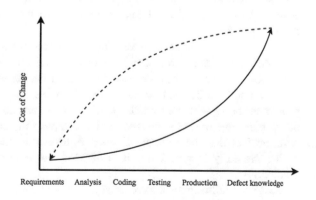

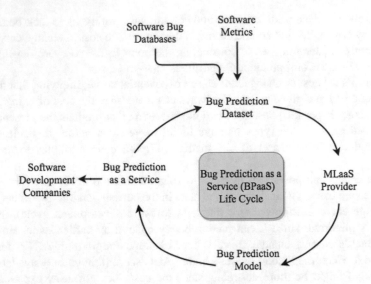

Fig. 7.6 Bug prediction as a service (BPaaS) life cycle diagram

7.8 Neural Network Approach for Bug Prediction to Estimate Software Costs and to Feed New Requirements

Software engineering is a complicated procedure. The main cause for its complexity lies in the fact that the desired requirements are hard to define. Moreover, every project is unique. Though off-the-shelf components do exist, it is often very hard to replicate code, even in projects that appear to be similar. Bugs appear from every line of the code, and in an attempt to fix them, developers often resort to quick patch jobs, leading to messy code and hence a higher cost of debugging. This section discusses two main ideas—the application of neural networks to resolve software costs and the application of neural networks to feed new requirements into the analysis phase.

Ambler's cost of change curve (Fig. 7.5) shows that the optimal time to estimate cost and take steps to avoid any bugs that may arise is as early as possible. Hence, prediction models are best deployed in the earlier stages of the SDLC. An early estimate needs to be made to determine the sales price of the software, to plan a project properly with reasonable timeline goals, and to determine if the software developers have the resources required to complete the project [48]. This is coupled with the fact that the software developers have less knowledge about the project during the early stages. Thus, it is difficult, but essential to make accurate cost predictions.

Cost estimation models have been in use, and [49] details the three major types of cost estimation models as:

- **Expert's prediction**: Where a software engineer with experience predicts the effort that will be required to complete the project
- **Algorithmic models**: Which include the COCOMO model (discussed below). These models work on prediction algorithms, taking size of the software into account. Size of the software can be input into the model in various measures, such as lines of code (LOC) and function points (FPs).
- **Machine learning models**: These models are usually used for predicting cost along with algorithmic models, or combined with fuzzy logic concepts and evolutionary computing methods. Neural network models are gaining increasing popularity, especially multi-layer perceptrons (MLPs). They work on the basis of the backpropagation algorithm.

The COCOMO II model (or the Constructive Cost Model) [50] uses a model trained on the number of lines of code. It is primarily a regression model. There are three levels of the COCOMO model, based on the level of detail considered. They are: basic, intermediate, and detailed COCOMO models. It is one of the most widely used cost estimation models.

The COCOMO model has been enhanced using neural networks as detailed in [51]. The neural network described in [51] uses 17 inputs to estimate the effort required to complete the project in person months. They have trained the model using the perceptron learning algorithm and the identity activation function. They use Magnitude of Relative Error (MRE) [51] as a metric to compare the accuracies of the predicted effort and the actual effort.

$$\text{MRE} = \frac{|\text{Actual effort} - \text{Estimated effort}|}{\text{Actual Effort}} \times 100$$

Figure 7.7 shows the general process of a case-based reasoning (an artificially intelligent method) which consists of four major stages—retrieve, reuse, revise, and retain. Overall, any given description of a situation is matched against known cases (stored in a 'case-base') to find the closest match. The action taken for the matching case along with its outcome (positive or negative) is noted, and the course of action to be followed this time is decided. The outcome of this case is then recorded along with the outcome, for future use.

Since most problems that arise in the analysis phase are repetitions of previous problems faced by developers worldwide, this is a good way to enhance the requirements analysis process. By applying neural networks to predict what may go wrong, combined with the CBR's knowledge of what will go wrong under certain circumstances, we can effectively produce an error-free environment for software development.

Since the entire process of providing bug prediction as a service on the cloud can be viewed as a series of components that are executed in sequence, there is scope for a service-oriented approach to solve this problem.

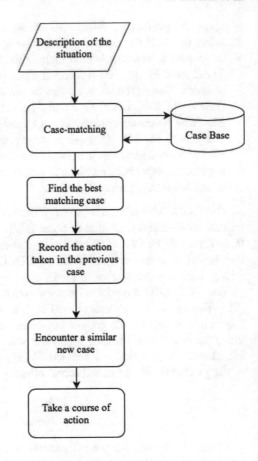

Fig. 7.7 General flow of process for case-based reasoning

7.9 Service-Oriented Approach to Providing Bug Prediction

Service-oriented software engineering can be used to provide bug prediction as a service component to software developers. It adopts a service-oriented architectural approach to the problem at hand. This means that it looks at software development as the use and development of reusable components that act as services, usually provided by a third-party developer. The software system dynamically connects to the service required, when it is required. This introduces a whole new perspective to the reusability quotient of a software. Moreover, since the services used are usually invoked as a black box by the software, this approach to software design decreases the necessity for expert software developers in various domains to exist in every software development team. This helps improve the cost of software development from the human resource perspective. Additionally, services can further invoke other services, resulting in a hierarchy of services that can be utilized by the software. Services may also be rendered from completely different networks in the world, hence making it a distributed computing system. The process and framework of the service-oriented software engineering methodology are described in [52], while its reliance on cloud computing is discussed at length by [53].

As mentioned above, service-oriented software engineering heavily depends on black box testing of the services requested. Since software developers often sell their services as propriety code, debugging by examining the source code is no longer possible. There arises a need to generate and test as many test cases as possible, before relying on the service to benefit the software. Furthermore, there is often a lack of time and resource to perform the black box testing for as many test cases as desired. In this situation, prioritization of test cases becomes inevitable. Neural networks are promising means of deterring the priority of test cases in tight situations where the need arises [54].

Machine learning can be used to predict the performance of a design instance of service-oriented architecture. This can even be done in the early stage of the software development life cycle, as described in [55], which uses a procedure that involves converting an annotated UML diagram into a Queuing Network Model (QNM) to obtain various parameters of the model. Since the above are highly resource-intensive processes, there arises a need to follow an effective resource-utilization scheme such as cloud computing.

7.10 Cloud Software Engineering for Machine Learning Applications

Cloud computing involves the utilization of resources over an Internet connection, wherein the resources reside on a computer at a geographically different location. Cloud software is an umbrella term for the various types of applications and cloud service providers that allow users to access and utilize the cloud resources in different ways. This software may involve storage, data manipulation, or even security providers on the cloud. Cloud Software Engineering is the application of various engineering concepts to the development and maintenance of cloud software.

The term 'cloud' has been used to describe various services that are offered by service providers over the Internet [56]. Recently, the application of software engineering principles and techniques has proven useful in the cloud domain. Conferences like the SE Cloud 2018 commune to discuss novel methods of cloud engineering. Most importantly, software engineering can help systematically and methodically approach cloud services to prevent service failure. Security of the cloud has also been improved by software engineering techniques. The methods of using cloud computing for services like continuous integration and continuous deployment ultimately aiding the software development life cycle are detailed by [57]. A detailed view of the evolution of cloud computing, its role in software deployment, and newer features are discussed at length in the chapter by [58].

Section 7.3 takes a closer look at the use of big data analytics in software engineering analytics.

As specified, the cloud provides services from resources at distant locations over the Internet. Machine learning and deep learning algorithms are often processor intensive and do not run efficiently on processors commonly found in personal use laptops. With the recent surge in the amount of data available and the re-emergence of machine learning, there has arisen a necessity for powerful processors like graphic processing units (GPUs). These services can be hosted on the cloud. Moreover, application-specific integrated circuits (ASICs), like Google's TPUs (tensor processing units), are also made available on their cloud, for extremely processor-intensive applications. Neural networks can also be distributed over the cloud, as distributed neural networks (DNNs) proposed by [59], which shows that this approach can lead to lower costs of communication and an improved accuracy. Thus, cloud services and software are irreplaceable in machine/deep learning scenarios.

Machine learning and artificial intelligence have also proven useful to cloud software. One example is the use of machine learning to improve digital advertisements of companies [60]. In a novel approach, the use of neural networks has also been applied to the prediction of server load and server down time in a cloud environment. This knowledge allows clouds service providers to switch off extraneous servers and thus save energy. [61] reports a power saving in the range of 46.3% to 46.7%. Neural networks can also be employed for cloud security, by predicting whether a specific request to the cloud is genuine or not. They can be used to predict server crashes, resource demands (for efficient scaling) [62], and fault tolerance of the cloud [63].

The following section covers an experiment that integrates the technologies discussed so far. It uses machine learning to predict bugs and provides the prediction as a service to software development companies worldwide on the cloud. Essentially, the service provision follows service-oriented architecture methods.

7.11   Experiment with Microsoft Azure Machine Learning

An experiment was conducted [36] to test the utility of a bug prediction service based on machine learning as a service provided by cloud providers. MLaaS is an increasingly popular area of interest today. We can now run processor-intensive applications on the cloud, using powerful hardware located virtually anywhere in the world. The paper discusses a classification-based machine learning model that uses Microsoft Azure's ML Studio (sample flowchart shown in Fig. 7.8).

A schematic flowchart [36] for the process executed with sample data is shown in Fig. 7.8. The bug dataset is imported, and features are extracted. The dataset is preprocessed and split into training and testing datasets. Then, the machine learning algorithm (or model) is trained. Finally, the model is scored and evaluated, according to the desired metrics.

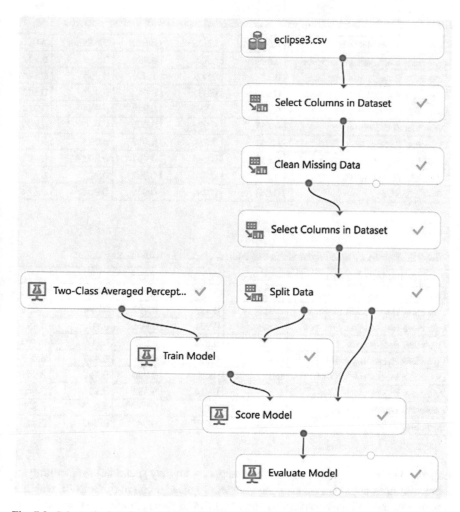

Fig. 7.8 Schematic flowchart of the machine learning experiment in Azure

There are nine models offered by Azure ML Studio for binomial classification. They are logistic regression, decision forest, decision jungle, boosted decision tree, neural network, averaged perceptron, support vector machine, locally deep support vector machine, and Bayes' point machine. The results from training each of the nine models on two datasets ('Change metrics (15) plus categorized (with severity and priority) post-release defects' dataset for model 1 and the 'Churn of CK and other 11 object-oriented metrics over 91 versions of the system' dataset for model 2—both of which are freely available at http://bug.inf.usi.ch)—are shown in Table 7.2 and Table 7.3, respectively.

The threshold is a measure of trade-off between false positives and false negatives. Here, a false positive would be a clean software version being classified as

Table 7.2 Results obtained from various classification models with training dataset 1

Classification model used	Accuracy	Precision	Recall	F1 Score	AUC
Two-class averaged perceptron	**0.855**	**0.858**	**0.980**	**0.915**	**0.821**
Two-class Bayes point machine	0.811	0.808	1.000	0.894	0.544
Two-class boosted decision tree	0.839	0.867	0.942	0.903	0.803
Two-class decision forest	0.843	0.869	0.944	0.905	0.714
Two-class decision jungle	0.851	0.874	0.949	0.910	0.805
Two-class locally deep SVM	0.841	0.859	0.957	0.905	0.635
Two-class logistic regression	0.819	0.816	0.997	0.897	0.796
Two-class neural network	0.837	0.840	0.982	0.905	0.826
Two-class SVM	0.841	0.864	0.949	0.905	0.824

Table 7.3 Results obtained from various classification models with training dataset 2

Classification model used	Accuracy	Precision	Recall	F1 Score	AUC
Two-class averaged perceptron	0.837	0.837	0.977	0.902	0.840
Two-class Bayes point machine	0.813	0.813	0.992	0.894	0.559
Two-class boosted decision tree	0.843	0.901	0.901	0.901	0.853
Two-class decision fores	0.845	0.873	0.942	0.906	0.799
Two-class decision jungle	**0.843**	**0.855**	**0.967**	**0.907**	**0.865**
Two-class locally deep SVM	0.823	0829	0.980	0.898	0.841
Two-class logistic regression	0.815	0.814	0.995	0.895	0.840
Two-class neural network	0.827	0.834	0.977	0.900	0.840
Two-class SVM	0.843	0.858	0.962	0.907	0.817

buggy. This is of great burden on the developer who may spend hours searching for a bug that does not exist. A false negative would mean a bug in the release, which is a bother to the end user. Assuming the cost of both these situations to be the same, the threshold was set to 0.5.

From Tables 7.2 and 7.3, we conclude that a two-class averaged perceptron model for the first dataset and a two-class decision jungle for the second dataset are the best suited.

The ROC curves for both the datasets are plotted in Figs. 7.9 and 7.11. The high area under the ROC curve indicates a high chance that a positive prediction chosen at random will be ranked higher than a negative prediction chosen at random.

The precision–recall curves are plotted in Figs. 7.10 and 7.12. The area under the precision–recall graph is very high in Fig. 7.10 and reasonably high in Fig. 7.12, denoting a high precision and a high recall. Since high precision corresponds to a low FP rate and high recall corresponds to a low FN rate; this denotes that this model is very accurate.

Fig. 7.9 ROC curve for model 1

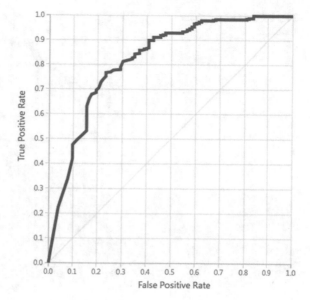

Fig. 7.10 Precision–recall curve for model 1

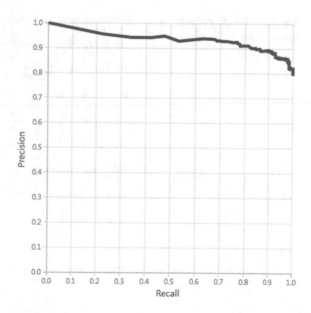

While equal weightage has been given to all five evaluation metrics, the software tester may feel that a different evaluation metric describes his needs better, for instance when a false positive costs more than a false negative or vice versa. In such cases, the machine learning model can easily be switched for a more suitable

Fig. 7.11 ROC curve for model 2

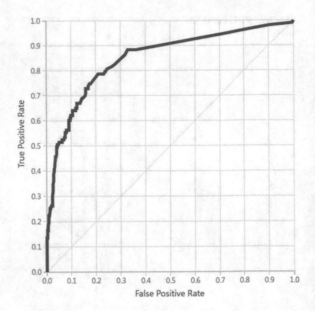

Fig. 7.12 Precision–recall curve for model 2

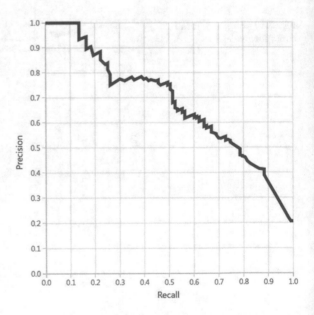

machine learning model. This is an advantage of using machine learning as a service (MLaaS) on the cloud for bug prediction. Despite such advantages, there has been a fair amount of criticism for neural networks and their use in software engineering analytics.

7.12 Critical Evaluations of Neural Network Approaches and Their Application in Software Engineering Analytics

While the use of neural networks for prediction in software engineering analytics is extremely beneficial and has proven to be a feasible solution to a number of problems encountered frequently in the software engineering domain, it still has its disadvantages. A problem common among all neural networks is the tendency of the network to overfit the data very quickly. As the number of layers in the neural network increases and becomes denser, the learning capacity of the network increases. This coupled with a small dataset which leads to overfitting of the data. This can lead to unpredictably high errors during real-time deployment.

The constant necessity for datasets large enough to satisfactorily train the neural network is yet another drawback of employing neural networks in software engineering analytics. Obtaining such large datasets is time consuming and subjective to proprietary rights violation. The increasing number of data privacy restrictions further complicates the process of obtaining datasets. Even within a dataset, the validation data that we provide the network may not be an accurate predictor of the outcome of the system. There will often be features that arise in future software systems that we are currently unaware of.

Another disadvantage is that neural networks require powerful processors for their effective computation. Though cloud computing has facilitated the on-demand availability of GPUs and similar powerful processors, it has also introduced the need for the system to be constantly connected to the Internet. If the remote server that a neural network is running on goes down for a while, this can affect the entire system.

A major difficulty faced by neural network programmers is the determination of a neural network structure that can optimally solve the task at hand. This seems simple—if the developer is aware of the different types of neural networks and their strengths, then improvising on a model or developing a hybrid of previously known models should be easy. However, developers are often in for a shock; models often behave erratically—simple models may yield more accurate predictions than well-developed models are common.

Further factors include the cost of hardware/cloud costs incurred to run the model, costs incurred to obtain the dataset, time consumed to train and re-train the same model to perfection and debugging difficulties.

While all of the above general factors also apply to software engineering analytics, there are unique difficulties in training a neural network for software engineering analytical purposes. A few a listed below:

There still exists a certain degree of uncertainty in determining the safety buffer required to pad the neural network's predictions. For example, in the application of neural networks for effort estimation, how many story points does the team leave as leeway for the estimated value? The most effective known method of solving this problem would be to employ a fuzzy system like the Mamdani or Sugeno fuzzy interpretation system to model the scenario. This, however, adds on to the overall cost of the software development process.

Software development teams often consist of cross-functional members, who have extensive experience in fields other than deep learning. Further, there exists a tendency of software development companies to try to cut costs wherever possible. So, instead of having a dedicated deep learning team, software development teams may end up using neural networks as black boxes, to analyze their data and come up with conclusions. This is a tolerable practice as long as the impact is kept to a minimum. However, problems arise when this black box usage of neural networks is adopted for large-scale analytics.

In conclusion, AI systems are far from 100% accuracy, but so are humans [7]. Though AI systems merely aim to automate processes, they require significantly more training data than humans. Moreover, there is a high level of complexity introduced into the system by the AI algorithm. This article [7] also points out that AI shifts the process of constructing software systems from a traditional deductive process to an inferred inductive one.

7.13   Conclusion and Future Work

The models proposed by [36] have an F1 score of 91.5% for model 1, which works with data from previous releases and an F1 score of 90.7% for model 2 that works with details known at the design and coding phase of the current release. The accuracy and precision of the models are high enough for these models to be commercially and profitably used in software development companies. Moreover, the memory footprint of the two-class decision jungle used by model 2 is lower than any other model. Future work may include increasing the accuracy of these models with commercial datasets (as opposed to the open-sourced datasets used in the experiment discussed). The use of MLaaS in this chapter allows the bug prediction models to be deployed on the cloud, as a web service, using the service-oriented architecture methods discussed. We believe that this work has proved that it is important to apply systematic approach to machine learning with 50 years of successful software engineering practices for mutual benefits.

References

1. Boehm B (1976) Software engineering and knowledge engineering. In: Proceedings of IEEE transactions on computers. IEEE, pp 1226–1241
2. Beck K (1999) Extreme programming explained: embrace change. Addison-Wesley Longman, Boston
3. Čubranić D, Murphy GC (2004) Automatic bug triage using text classification. In: Proceedings of software engineering and knowledge engineering, pp 92–97

4. Sharma G, Sharma S, Gujral S (2015) A novel way of assessing software bug severity using dictionary of critical terms. Procedia Comput Sci 70:632–639
5. Karunanithi N, Malaiya Y, Whitley D (1991) Prediction of software reliability using neural networks
6. Khanh Dam H, Tran T, Ghose A (2018) Explainable software analytics. https://doi.org/10.1145/3183399.3183424
7. Khomh F, Adams B, Cheng J, Fokaefs M, Antoniol G (2018) Software engineering for machine-learning applications: the road ahead. IEEE Softw 81–84
8. Famelis M (2018) Applying software engineering principles to a machine learning algorithm. In: Poster presented at software engineering for machine learning applications (SEMLA), Montreal, Quebec, Canada, 12–13 Jun 2018
9. Amershi S, Begel A, Bird C, DeLine R, Gall H, Kamar E, Nagappan N, Nushi B, Zimmermann T (2019) Software engineering for machine learning: a case study. In: International conference on software engineering (ICSE 2019)—software engineering in practice track
10. Zhang Du, Tsai JJP (2002) Machine learning and software engineering. Softw Q J SQJ 11:22–29
11. Menzies T, Zimmermann T (2013) Software analytics: so what? IEEE Softw 30(4):31–37
12. Yang Y, Falessi D, Menzies T, Hihn J (2018) Actionable analytics for software engineering. IEEE Softw 35(1):51–53
13. Hendrickson S (2010) Getting started with Hadoop with Amazon's Elastic MapReduce, EMR
14. Maryville (2019) Where big data and software development collide. In: Maryville Online. https://online.maryville.edu/blog/where-big-data-and-software-development-collide/. Accessed 13 Apr 2019
15. Shepperd M, Song Q, Sun Z, Mair C (2013) Data quality: some comments on the NASA software defect datasets. In: IEEE Transactions on software engineering, pp 1208–1215
16. DeLine R (2015) Research opportunities for the big data era of software engineering. 2015 IEEE/ACM 1st international workshop on big data software engineering, Florence, pp 26–29
17. Venkatachalam R (1993) Software cost estimation using artificial neural networks. In: Proceedings of 1993 international conference on neural networks (IJCNN-93-Nagoya, Japan), Nagoya, Japan, 1993, pp 987–990
18. Yu L, Tsai W-T, Zhao W, Wu F (2010) Predicting defect priority based on neural networks, pp 356–367
19. Li P, Li J, Huang Z, Li T, Gao C-Z, Yiu S-M, Chen K (2017) Multi-key privacy preserving deep learning in cloud computing. Future Gener Comput Syst 74:76–85
20. Rashid E, Patnayak S, Bhattacharjee V (2012) A survey in the area of machine learning and its application for software quality prediction. ACM SIGSOFT Softw Eng Notes 37. https://doi.org/10.1145/2347696.2347709
21. Arshad A, Riaz S, Jiao L, Murthy A (2018) The empirical study of semi-supervised deep fuzzy C-mean clustering for software fault prediction. IEEE Access 6:47047–47061. https://doi.org/10.1109/ACCESS.2018.2866082
22. Mahapatra S (2019) Why deep learning over traditional machine learning? In: Towards data science. https://towardsdatascience.com/why-deep-learning-is-needed-over-traditional-machine-learning-1b6a99177063. Accessed 13 Apr 2019
23. Alex D. Difference between machine learning and deep learning. https://artificialintelligencehow.com/2017/10/18/difference-machine-learning-deep-learning/
24. Khoshgoftaar TM, Szabo RM (1996) Using neural networks to predict software faults during testing. IEEE Trans Reliab 45(3):456–462
25. Li X, He J, Ren Z, Li G, Zhang J (2018) Deep learning in software engineering
26. Singh Y, Bhatia P, Kaur A, Sangwan O (2009) Application of neural networks in software engineering: a review. Commun Comput Inf Sci 31:128–137. https://doi.org/10.1007/978-3-642-00405-6_17

27. Sayyad Shirabad J, Menzies TJ (2005) The PROMISE repository of software engineering databases. School of I.T. and Engineering, University of Ottawa, Canada
28. GitHub Activity Data. In: github.com. https://console.cloud.google.com/marketplace/details/github/github-repos?pli=1
29. D'Ambros M, Lanza M, Robbes R (2010) An extensive comparison of bug prediction approaches. In: Proceedings of the 7th IEEE working conference on mining software repositories (MSR)
30. Maryville. Where big data and software development collide. In: Maryville Online. https://online.maryville.edu/blog/where-big-data-and-software-development-collide/
31. TechBeacon. How predictive analytics will speed software development, improve quality. In: TechBeacon. https://techbeacon.com/app-dev-testing/how-predictive-analytics-will-disrupt-software-development
32. SearchCIO. What is prescriptive analytics?—Definition from WhatIs.com. In: SearchCIO. https://searchcio.techtarget.com/definition/Prescriptive-analytics
33. Software Testing Class. Bug life cycle in software testing. https://www.softwaretestingclass.com/bug-life-cycle-in-software-testing/. Accessed 13 Apr 2019
34. Card DN (1998) Learning from our mistakes with defect causal analysis. In: IEEE Software, vol 15, no 1, pp 56–63, Jan–Feb 1998
35. Leszak M, Perry D, Stoll D (2002) Classification and evaluation of defect in a project retrospective. J Syst Softw
36. Subbiah U, Ramachandran M, Mahmood Z (2018) Software engineering approach to bug prediction models using machine learning as a service (MLaaS)
37. Huynh T (2019) Difference between severity and priority in defect report—finally revealed. In: AskTester. https://www.asktester.com/severity-vs-priority/
38. Eclipse/Bug Tracking—Eclipsepedia. In: Wiki.eclipse.org. https://wiki.eclipse.org/Eclipse/Bug_Tracking
39. Softwaretestinghelp.com. Severity and priority in testing with examples and difference. https://www.softwaretestinghelp.com/how-to-set-defect-priority-and-severity-with-defect-triage-process/
40. Roper M, Wood M, Miller J (1997) An empirical evaluation of defect detection techniques. Inf Softw Technol 39(11)
41. Juristo N, Vegas S (2003) Functional testing, structural testing and code reading: what fault type do they each detect? In: Conradi R, Wang AI (eds) Empirical methods and studies in software engineering. Lecture notes in computer science, vol 2765. Springer, Berlin
42. Sharma Y, Kumar Sharma A (2015) Comparative study of the bug tracking tools. Int J Adv Res Comput Sci Softw Eng
43. Soner S, Soner S, Yadav M (2015) A survey on software bug evaluation. Int J Comput Appl
44. Bhattacharya S, Radha Suja, Jat DS (2016) Comparative analysis of bug tracking tools. Int J Pharm Technol 8:4989–4998
45. Huynh T (2019) A beginner's guide to software defect detection and prevention. In: AskTester. https://www.asktester.com/defect-detection-and-prevention/
46. Tricentis (2018) Software fail watch, 5th edn. Tricentis. Available at: https://www.tricentis.com/software-fail-watch
47. Chidamber SR, Kemerer CF (1994) A metrics suite for object oriented design. Proc IEEE Trans Softw Eng 20(6):476–493
48. Effort Estimation for Software Development. In: Open-works.org. http://open-works.org/?e=effort-estimation-for-software-development
49. Nassif A, Azzeh M, Capretz L, Ho D (2016) Neural network models for software development effort estimation: a comparative study
50. Boehm B (1981) Software engineering economics. Prentice-Hall, Upper Saddle River
51. Kaushik A, Chauhan A, Mittal D, Gupta S (2012) COCOMO estimates using neural networks. Int J Intell Syst Appl

52. Karhunen H, Jantti M, Eerola A (2005) Service-oriented software engineering (SOSE) framework. In: Proceedings of ICSSSM'05. 2005 international conference on services systems and services management, 2005, Chongqing, China, pp 1199–1204

53. Yau S, An H (2011) Software engineering meets services and cloud computing. Computer 44 (10):47–53. https://doi.org/10.1109/MC.2011.267

54. Harsh B, Esha K (2014) Neural network based black box testing. SIGSOFT Softw Eng Notes 39(2):1–6

55. Moniem H, Ammar H (2015) A framework for performance prediction of service-oriented architecture. Int J Comput Appl Technol Res 4:865–870. https://doi.org/10.7753/IJCATR0411.1013

56. Kiswani J, Dascalu S, Harris Jr. F (2018) Cloud-RA: a reference architecture for cloud based information systems. SE Cloud

57. Yara P, Ramachandran R, Balasubramanian G, Muthuswamy K, Chandrasekar D (2009) Global software development with cloud platforms. In Gotel O, Joseph M, Meyer B (eds) Software engineering approaches for offshore and outsourced development. SEAFOOD 2009. Lecture notes in business information processing, vol 35. Springer, Berlin

58. Dashofy EM (2019) Software engineering in the cloud. In: Cha S, Taylor R, Kang K (eds) Handbook of software engineering. Springer, Cham

59. Teerapittayanon S, McDanel B, Kung HT (2017) Distributed deep neural networks over the cloud, the edge and end devices. 2017 IEEE 37th international conference on distributed computing systems (ICDCS), Atlanta, GA, pp 328–339

60. Dialani P (2018) The fusion of artificial intelligence and cloud computing | analytics insight. In: Analytics Insight. https://www.analyticsinsight.net/the-fusion-of-ai-and-cloud-computing/. Accessed 13 Apr 2019

61. Duy TVT, Sato Y, Inoguchi Y (2010) Performance evaluation of a green scheduling algorithm for energy savings in cloud computing. 2010 IEEE international symposium on parallel & distributed processing, workshops and PhD Forum (IPDPSW), Atlanta, GA, 2010, pp 1–8

62. Islam S, Keung J, Lee K, Liu A (2012) Empirical prediction models for adaptive resource provisioning in the cloud. Future Gener Comput Syst 28(1):155–162

63. Amin Z, Singh H, Ahluwalia N (2015) Review on fault tolerance techniques in cloud computing. Int J Comput Appl

Chapter 8
Sentiment Analysis of Twitter Data Through Machine Learning Techniques

Asdrúbal López-Chau, David Valle-Cruz and Rodrigo Sandoval-Almazán

Abstract Cloud computing is a revolutionary technology for businesses, governments, and citizens. Some examples of Software-as-a-Services (SaaS) of cloud computing are banking apps, e-mail, blog, online news, and social networks. In this chapter, we analyze data sets generated by trending topics on Twitter that emerged from Mexican citizens that interacted during the earthquake of September 19, 2017, using sentiment analysis and supervised learning, based on the Ekman's six emotional model. We built three classifiers to determine the emotions of tweets that belong to the same topic. The classifiers with the best accuracy for predicting emotions were Naive Bayes and support vector machine. We found that the most frequent predicted emotions were happiness, anger, and sadness; also, that 6.5% of predicted tweets were irrelevant. We provide some recommendations about the use of machine learning techniques in sentiment analysis. Our contribution is the expansion of the emotions range, from three (negative, neutral, positive) to six in order to provide more elements to understand how users interact with social media platforms. Future research will include validation of the method with different data sets and emotions, and the addition of new artificial intelligence techniques to improve accuracy.

Keywords Cloud computing · Sentiment analysis · Machine learning · ML · Twitter · Naive Bayes · Ekman's model

A. López-Chau
Autonomous University of the State of Mexico, 55600 Valle Hermoso, Zumpango, Estado de México, Mexico
e-mail: alchau@uaemex.mx

D. Valle-Cruz (✉) · R. Sandoval-Almazán
Autonomous University of the State of Mexico, Instituto Literario # 100, Toluca, Mexico
e-mail: davacr@uaemex.mx

R. Sandoval-Almazán
e-mail: rsandovala@uaemex.mx

© Springer Nature Switzerland AG 2020
M. Ramachandran and Z. Mahmood (eds.), *Software Engineering in the Era of Cloud Computing*, Computer Communications and Networks,
https://doi.org/10.1007/978-3-030-33624-0_8

8.1 Introduction

Cloud computing is a revolutionary technology for businesses, governments, and citizens [1], providing different types of services based on consumer Internet services. It represents a model for enabling ubiquitous, convenience, on-demand network access to a shared pool of configurable computing resources that can be rapidly provisioned and released with minimal management effort or service provider interaction [2]. There are three cloud computing delivery models: Software-as-a-Service (SaaS) where the consumer uses an application, but does not control the computer system; Platform-as-a-Services (PaaS), the consumer uses a hosting for his/her applications; and Infrastructure-as-a-Service (IaaS), where the consumer uses fundamental computing resources such as processing power.

Some examples of cloud computing are banking apps, e-mail, blog, online news, and social networks. Social networks represent an ideal platform for establishing relationships with customers or users, and for understanding the interaction between them, but in an unstructured form. One way to understand how users behave on social networks is through sentiment analysis.

Furthermore, e-governance uses information and communication technologies (ICT) to provide government services and information exchange for developing government to citizen interaction. However, the traditional e-governance solutions are incapable of fulfilling the current need because of its increasing demand, application complexity, infrastructure management, cost overhead, and other technical challenges. Emerging technologies such as cloud computing, big data, and machine learning could overcome these challenges using the modern approach for computing, storage, and data processing, because they provide unique features to e-governance: lower cost, scalability, easy management, disaster recovery, accountability, resource provisioning, distributed storage, data analytics, mobility, etc. [3].

Social networks represent a communication media useful to express feelings and thoughts, meet other people, do business, or speak about someone or something. Among these social networks, the most used ones are Facebook, Twitter, YouTube, and Instagram [4]. Each of these social networks offers different types of content generation on text, images, and video. These data nurture big data, which contains a large amount of information, valuable for decision making of political actors, businesses, and government organizations [5].

There are artificial intelligence techniques and tools for the automatic analysis of data generated in social media in the areas of machine learning, natural language processing, and text mining which allow finding patterns in data or relationships [6, 7]. Sentiment analysis of Twitter posts has attracted the attention of many scholars [8, 9] since the unstructured data generated every day, in this kind of social media, contain valuable information for decision making in organizations and how to react to certain types of events.

This chapter aims to present our experience in analyzing Twitter data sets with sentiment analysis techniques. Unlike the vast previous research, which only considers three categories: negative, positive, or neutral to identify emotions, our proposal, using the Ekman's emotion model, classifies six types of emotions: joy or happiness, anger, sadness, surprise, disgust, and fear [10]. The traditional use of this model is the detection of emotions in facial expressions, but in this chapter, we used it to identify sentiment analysis from Twitter posts. As part of the framework, we provide some recommendations on how to deal with the problems found on implementing systems for sentiment analysis based on our experience analyzing several Mexican data sets in recent local elections, Mexico's 2017 earthquake, and the presidential campaign [11–13].

This chapter consists of five parts: the first section explains the introduction related to sentiment analysis on Twitter, as well as the purpose of the study. The second section consists of three parts: a literature review related to social networks, Ekman's model, and sentiment analysis literature. The third section describes our proposal of a method based on classification methods to predict sentiments using Ekman's model. The fourth section describes the results of classifiers systems. The final section describes conclusions, experiences, and future research.

8.2   Literature Review

This literature review section is divided into three parts: social networks; Ekman's model; and sentiment analysis. These are explored in the following sections.

8.2.1   Social Networks

Social networks changed communication and represent a technological tool, useful to disseminate any kind of information through the world. With the development of 5G, wireless and Internet connections, it will enable different kinds of new applications, more personalized, connected and interactive services become available with resource-limited mobile terminals [14], and it is easier to express feelings and thoughts and communicate ideas to other people. More and more data will be generated on Internet which represent ground gold for all kind of organizations. For this reason, it is important to study how to analyze big data through techniques such as sentiment analysis.

In this context, different data sources feed the big data every day. Devices such as wearables, smartphones, tablets, and personal computers allow access to programmed applications that maintain us immersed in a hyperconnected world. Internet of Things, sensors, data clouds, Google searches, Amazon and eBay shopping, and the use of social media are some examples that generate much information that cannot be analyzed with traditional tools.

In particular, social networks generate large amounts of data about each person, where they express their tastes, feelings, and moods; social networks have become a sensor in real time. With the advent of social networks, users create records of their lives by daily posting details of activities they perform, events they attend or live, places they visit, pictures they take, and things they enjoy, want, and feel. Social network is a platform where millions of people interact, share opinions, and express their feelings.

The most used social networks worldwide are Facebook, YouTube, WhatsApp, Instagram, and Twitter. Twitter allows us to publish content in a microblogging format, quickly, compactly, and in real time. Differently, Facebook allows us to use six basic impressions (like, love, haha, wow, sad, and angry). Twitter has no mechanism to express impressions, but several researchers have developed studies on the sentiment analysis on Twitter, to understand what the users of this social media express.

The rise of social networks has generated today a tremendous interest among Internet users, organizations, and researchers. Data from these social networking sites can be used for several purposes, like prediction, marketing, or sentiment analysis [15]; some other researchers have developed personalized recommendation systems based on learning automata and sentiment analysis [16]. The millions of tweets received every day could be subjected to sentiment analysis, but handling such a huge amount of unstructured data is a tedious task to take up [15]. Since data generated in social networks are in an unstructured way, analytics solutions that mine structured and unstructured data are important as they can help organizations gain insights not only from their privately acquired data but also from large amounts of data publicly available on the Web [17].

The development of computational intelligence techniques enables these platforms to become modern large-scale laboratories in which the development of intelligent emotion-aware applications can be incubated to maximize the quality of computerized solutions [18]. The optimum solution for this kind of problem is analyzing the information available on social network platforms and performing sentiment analysis [19]. The study of social networks using sentiment analysis allows identifying patterns in large data sets.

8.2.2 Ekman's Model

People express emotions provoked by the events they live in, the environment in which they are immersed, their personality, and the experiences they have lived. An emotion is an alteration by a shock or impulse in the brain caused by impressions of senses, ideas, or memories. Emotions are shown through facial and body expressions, the tone of voice, among other characteristics. Nowadays, social networks invite us to express our emotions and feelings through texts, images, videos, or any multimedia element. In order to understand and classify emotions, different researchers and psychologists have provided answers to questions such as: How do we have emotions and what does it cause to have these emotions [20]? Some answers are simplistic but very concise in the classification of emotions [21] and some others consider different factors to identify a wide variety of emotions [22].

Sreeja and Mahalakshmi [20] classify theories of emotions into three main categories: physiological: where the response within the human body is responsible for the generation of emotions; neurological: where brain action leads to emotional reactions; and cognitive: where thoughts and other mental activities form emotions [20]. Affective computing scientist has developed computational solutions to identify and react to user emotional states, in this sense, the representation of emotions is designed in two main ways, categorically: the generated emotion is selected from a set of emotions and labeled; dimensional: the representation of emotions is based on a set of quantitative measures using multidimensional scales.

In the categorical models, there are important representations such as: Ekman's [18] basic emotions and Navarasa models; on the other hand, in the dimensional models have been designed representations such as circumplex, Plutchik, Pad, and Thayer [20]. In categorical models, emotions are identified by a class label such as anger, disgust, fear, joy, sadness, and surprise; and they are easy to understand. In contrast, in dimensional models, it is necessary to quantitatively define the value of each emotion, in addition to defining combined emotional states of different numerical levels.

Several studies have carried out sentiment analysis with the help of a frame of reference or model of emotions [22–24]. These emotional models make it possible to identify different kinds and numbers of emotions. However, one of the most commonly used in the area of affective computing, intelligent agents, and sentiment analysis is the Ekman's model which is based on emotions generated in facial expressions, emotions that have been identified in humans and inherited from ancestral times.

Ekman classifies emotions into six types: anger, fear, disgust, surprise, joy, and sadness (see Fig. 8.1). These six emotions are basic and universal for facial expressions since they are defined as adaptations selected by biological mechanisms with evolutionary value and, in a general way, since when expressing any of the six emotions, the same facial features are found [21]. These emotions are presented in social media posts because social media users express themselves depending on the event they are living, and the event generates an impression or reaction that some people display on their social media.

The six emotions proposed by Ekman are classified into positive and negative, depending on the reaction expected, and the event they are living. Positive emotions are produced by reacting to pleasant events or people's liking, such as joy and surprise. Negative emotions are the result of an event that people do not like, such as anger, fear, sadness, and disgust. Humans present a combination of these emotions by reacting to events that happen in their daily lives. For example, if someone receives a birthday present that he or she has wanted for some time, the emotions of surprise and joy will be present, but if a person is frightened, he or she may show fear, anger, and disgust, but also surprise.

Although Ekman's model is used to identify emotions in people's expressions and for modeling intelligent agents, there is some research in the area of sentiment

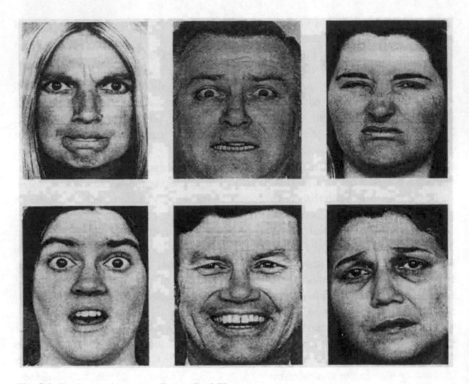

Fig. 8.1 Basic emotions according to Paul Ekman

analysis that has adopted this model to classify emotions in texts generated in blogs, social media, and Web pages [25–27]. For the purpose of this research, we adapted the Ekman's model to classify emotions in large data sets generated on Twitter, during the earthquake of September 19, 2017, in Mexico.

8.2.3 Sentiment Analysis

Traditionally, sentiment analysis studies have been based on classifying or identifying the polarity of Twitter posts, classifying sentiment as positive, negative, or neutral, getting very good results in the precision of the data analysis, and predictions carried out. However, applying a classification based on a most varied number of emotions is a difficult task.

Sentiment analysis is a technique which involves natural language processing, text analysis, and data mining [28, 29]. The sentiment content of the text is characterized by using techniques such as natural language processing (NLP), statistics or any of the machine learning methods. Sentiment analysis can also be proceeded by based on rule-based classifier or supervised learning [30]. Sentiment analysis of short texts and reviews available on different social networking sites is challenging because of the limited contextual information. Based on the sentiments and available opinions, developing a recommendation system is an interesting concept, which includes strategies that combine the small text content with prior knowledge [31].

The increase of smartphone and tablet applications allows users to interact on different service platforms at any time through mobile Internet, social media, cloud computing, and others. However, there are very few studies of classification methods applied to this area [28]. Nowadays, large volumes of data are in an unstructured manner, and it is very difficult to perform operations in unstructured data. So, the data need to be structured and organized before any analysis. The sentiment analysis technique is used to analyze the sentiments of a user based on text analysis [32].

The technology within text analytics comes from fundamental fields including linguistics, statistics, and machine learning. In general, modern text analytics uses statistical models, coupled with linguistics and emotional theories, to capture patterns in human languages in such a way that machines can understand the meaning of texts and perform various text analytics tasks. Text mining in the area of sentiment analysis helps organizations to uncover sentiments and improve their customer relationship management [33, 34]. This is useful to identify patterns on data, and for decision making.

Some of the techniques used to develop sentiment analysis on Twitter have been artificial neural networks, vector support machines, logistic regression models, Bayes' theorem, decision trees, and fuzzy logic. In this chapter, we analyze data sets generated by trending topics on Twitter that emerged from the Mexican citizens that interact during the earthquake of September 19, 2017, using sentiment analysis, supervised learning, and based on the Ekman's six emotional model.

8.3   Methodology

In this section, we introduce the proposed methodology to perform sentiment analysis of Twitter posts. This methodology is summarized in Fig. 8.2.

Each one of the steps presented in Fig. 8.2 is explained in the following subsections.

8.3.1   Data Collection

The large number of posts that are made at any time in social media, such as Twitter and Facebook, have attracted the attention of researchers to apply methods for sentiment analysis. Although there are public data sets to test and compare these methods (see, for example, https://www.kaggle.com/kazanova/sentiment140,

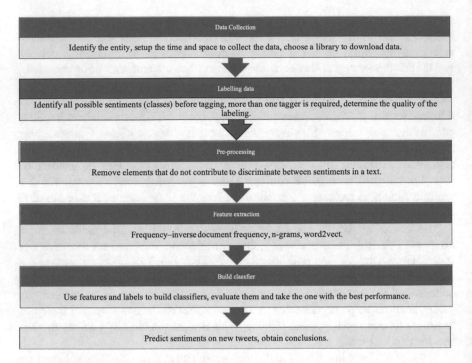

Fig. 8.2 Summary of the methodology

www.kaggle.com/eliasdabbas/5000-justdoit-tweets-dataset, and https://data.world/datasets/twitter), it is now possible to download data directly from social media to analyze data. To achieve the above goal, it is necessary to register as a developer in the social media from which you wish to download data and use special libraries. As part of the methodology, we recommend to take into account the following recommendations for this phase:

- Identify the entity. This can be an event, topic, service, institution, person, or thing that users of social media post about. These users use the symbol# before a keyword or a series of words to refer to a specific topic in a message. On the other hand, the symbol @ is usually used before the username to refer someone or something (a brand, for example). These two symbols can help to group publications related to an entity.
- Set up the time and space to collect data. Data downloaded from social media (in this case: microblogging sites) can correspond to an interval, or they can be collected in real time. According to the literature, the most common approach is to apply sentiment analyses methods on data of a period. Also, location can be important in some cases, for example, in the case of earthquakes, tsunamis, etc.
- Choose a library to download the data. Several alternatives are available, depending on the chosen programming language. For the R programming language, the packages twitterR or retweet are good options. For the Python

programming language, there are many libraries; some of them are the following: python-twitter, tweepy, TweetPony, Python Twitter Tools, Twitter Search, TwitterAPI, and Birdy. It is recommended to make sure that the coding (UTF-8 for example) is correct to avoid getting data corrupted or with incorrect symbols.

8.3.2 Labeling Data

One of the crucial steps to apply sentiment analysis successfully is the labeling of data. In order to label data, it is necessary that texts are read carefully and then assigned to each of the labels that correspond to a specific sentiment. Therefore, this step is one of the most difficult ones, as well as time-consuming.

Before starting the labeling, it is necessary that all the labels that will be used have been defined, including one for cases where it is not possible to accurately determine the sentiment in the analyzed text (for example, "ambiguous" label). Each tagger (it is recommended to have more than one) must first determine if the analyzed text is relevant. If so, the entity and the aspects or attributes of the entity in the text must be identified. Then, the labels (sentiments) that best represent the full text are assigned. It is important to point out that one tweet can have different sentiments.

Once the data have been labeled, it is necessary to carry out a type of evaluation to determine the quality of the labeling. One of the metrics is the Kappa index, which is used to know the degree of agreement between the taggers.

The number of labels of each type should be balanced to avoid a poor performance of machine learning methods. It has been suggested that there are around 3000 texts labeled.

8.3.3 Preprocessing of Texts

The texts post in social media can contain emoticons, numbers, exclamation and question marks, and other symbols. In some languages, such as Spanish, some words have accented vowels. It has been studied that most of these elements do not contribute to discriminate between sentiments in a text. Therefore, in the preprocessing step, these elements are removed or transformed from texts. Basic preprocessing for sentiment analysis includes the following tasks:

- Remove numbers, exclamation marks, question marks, and punctuation marks.
- Identify mentions (@UserName), these can be substituted by the keyword USERNAME.
- Identify topics (#topic), these can be substituted by the keyword TOPIC.
- Remove URLs, these begin with the string "http".

- Remove html tags, these begin and end with "<" and ">", respectively.
- Remove other symbols, such as $, %, ^, *.
- Change accented vowels by the same vowels without accents.
- Apply a stemming algorithm, such as snow-ball.
- Most of these previous tasks can be implemented using a programming language, and therefore done automatically.

A more detailed preprocessing can be applied to the texts, for example:

- Identify misspelled words and correct them using a dictionary. Identify words that contain extra letters [for example: "Fueraaaa" (get out), "Goool," (goal)], and remove these extra letters.
- Words transformation, these include many variants, for example:

  - In some texts in Spanish, letters in words are substituted by numbers or symbols, (for example "H014" instead of "HOLA," Hello in English). Detecting these symbols and transforming the string into a known word is one task of the preprocessing phase.
  - It is common to interchange one word instead of another; quite similar to synonyms. Some authors present a Spanish specific lexicon of social networks. It is a list of words in Spanish that is used in social networks, and that is understood with a completely different meaning to the common one. Identifying these words and substituting them with other ones makes the text clearer to understand.

8.3.4 Feature Extraction

Machine learning methods produce better results when they are fed with characteristics extracted from a text, instead of providing them with the raw text. The features extracted can be very simple such as identifying the presence of terms (bag-of-word or BoW), or more complex such as lexical and syntactic features.

For this research, we used a lexicon-based feature which consists of counting sentiment terms in each document. From these frequencies, derived features can be obtained, such as a ratio of term frequency on the document, the ratio of term frequency on the whole corpus, and the absolute value of the difference between both previous ratios.

Word2vect calculates the distribution probability of terms in a document. This technique can discover semantic relations among terms in the corpus. It is computed by training a neural network, and the vectors that represent each word are the synaptic weights.

One of the most common features used for sentiment analysis is the term frequency–inverse document frequency. It is a matrix that contains the inverse of the frequency of terms that occur in a set of documents. Therefore, each entry of the matrix has a low value for terms that occur very frequently in the document set, and a high value for those that occur rarely.

Regardless of the feature extracted, each document (or tweet in our case) is transformed into a vector. We have assigned a label (sentiment) to some of these vectors that are set manually. This way, we obtain a labeled data set, and we use it as the input of machine learning methods.

8.3.5   Classification Methods for Sentiment Analysis

Classification methods are supervised learning methods of machine learning. This means that they need labeled data to build a model that is called a classifier. For sentiment analysis, each sentiment (positive, negative, and neutral in most and simple cases, or joy, anger, sadness, surprise, displeasure, and fear in Ekman's model) is considered a category or class. The purpose of classifiers is to predict the class of previously unseen data. Therefore, they are used to determine the polarity or sentiment of opinions post on social networks as Twitter. Most common classifiers for sentiment analysis are the following:

- Support vector machine. It is a classifier that computes the optimal separating hyperplane. It solves a quadratic programming problem to compute the hyperplane with maximal margin.
- Naive Bayes. It is a probabilistic model that considers that each variable is independent of the rest.
- Decision tree. Decision trees are classifiers whose structure resembles a flowchart. A classifier of this type is induced by partitioning the input space recursively, up to a level of purity of each partition is satisfied.
- Logistic regression. It is a statistical procedure that estimates relationships between attributes and classes. Logistic regression is quite similar to linear regression, but is oriented for categorical outputs.
- Neural network. It is a classification method inspired by the human brain. The most popular training method for neural networks is backpropagation.

Currently, some libraries facilitate the generation of classifiers without the need to implement them from scratch. Some of these libraries for the Python programming language are scikit-learn, NLTK, and SciPy.

8.3.6   Evaluation of Classifiers

One of the techniques most commonly used for evaluating classification methods is 10-cross-validation. In this technique, the data set is partitioned into two subsets; one of them is the training set, used to build the prediction model (train a classifier).

Table 8.1 Confusion matrix for two classes

		Actual class		
		Positive	Negative	Total
Prediction	Positive	TP	FP	Number instances predicted as positive
	Negative	FN	TN	Number instances predicted as negative
	Total	Number of actual positive instances	Number of actual negative instances	N: Number of instances

The other subset is the test set, used to observe the prediction of a classifier and to compare it against the true value. This process is repeated 10 times to obtain an average of the performance of a classifier.

It is useful to build a confusion matrix to assess the performance of classifiers. The entries of this matrix contain the counts of the actual categories or classes in a data set and the number of correct and incorrect predictions made by a classifier. Table 8.1 shows the confusion matrix for the simplest case of two classes that are called the positive and the negative class.

In Table 8.1, TP is the number of instances (tweets, in our case) of the positive class that is predicted as positive by a classifier; TN is the number of instances of the negative class that is predicted as negative by a classifier; and FP/FN is the number of instances of positive/negative class that is predicted as negative/positive by a classifier. For these two entries, the classifier commits an error.

Based on the values of a confusion matrix, the following measures can be computed:

- Classification accuracy is the percentage of correct predictions. Classification accuracy = (TP + TN)/(N).
- Precision measures how accurate the classifier to predict positive instances is. Precision = TP/(TP + FP).
- Recall measures how accurate the classifier to predict negative instances is. Recall = TP(TN + FN).
- F1 score combines precision and recall in one formula. F1 score = 2 * (Precision * Recall/(Precision + Recall)).

8.3.7 Using Classification Methods for Sentiment Analysis

Once a classifier has been evaluated, it can be applied to predict new instances. It is recommended to build more than one classifier, and then choose the one with the best performance. Depending on the problem being solved with machine learning methods, the data can have two or more classes. The first case is a binary classification problem; the second case is a multiclass problem.

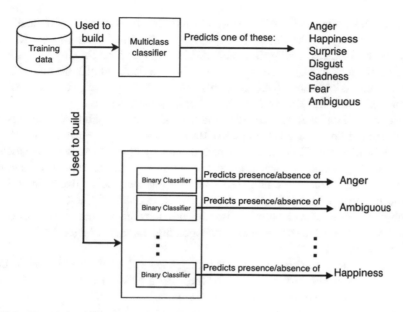

Fig. 8.3 Classifier architecture for sentiment analysis

If there are more than two classes (sentiments) in the analyzed data, then there are at least two possibilities for applying classification methods. The first option is to merge all classes in one categorical attribute. In our case, the possible values of it are the seven classes (six from Ekman's emotions and the class ambiguous). As stated before, it is a problem of multiclass classification. It is well-known that this type of problem is more complex than a binary class classification problem, and that the performance of classifiers is usually better for the latter. The second option—the one used in this chapter—is to consider each sentiment independently of the others. This is a binary classification problem. Therefore, instead of predicting one of the sentiments for a new document (tweet), the presence/absence of only one of them is predicted with each classifier. Figure 8.3 shows the two explained approaches.

8.4   Results

As an example of the application of the methodology explained above, we used data about Mexico's earthquake occurred on September 19th, 2017. For our experiments, we downloaded a data set (corpus) that correspond to each trend identified during this earthquake, these trends are #TvAztecaMiente (TV Azteca Lies), #TodosSomosMexico (WeAreAllMexico), #Terremoto (Earthquake), #TemblorMx (Earthquake Mx), #SomosMexico (WeAreMexico), #skyAlertMx (SkyAlertMx), #SismoMX (EarthquakeMX), #SismoMexico2017 (EarthquakeMexico2017), #Sismo (Earthquake), #RoboComoGraco (StealLikeGraco), #PueblaSigueDePie

(PueblaStillStanding), #PrayForMexico, #PartidosDenSuDinero (PoliticalParties
GiveTheirMoney), #PartidosDenNuestroDinero (PoliticalPartiesGiveOurMoney),
#Millenials, #MexicoEstaDepie (PueblaStillStanding), #Jojutla (Jojutla), #Fuerza
Mexico (ForceMexico), #FuerteMexico (FortMexico), #FinDelMundo (EndOf
World), #AyudaMexico (HelpMexico), #AyudaCdMx (HelpMxCity), #Alerta
Sismica (SeismicAlert), and #SismoMX (EarthqueakeMX).

We focused on labeling manually a portion of tweets, avoiding to label retweets
of the corpora downloaded. In our case, for each corpus, we selected about 10% of
the tweets randomly. Each tweet was read independently by three people to identify
the presence/absence of each one of the following: anger, happiness, surprise,
disgust, sadness, fear, and ambiguous. We chose as the label of tweets the opinion
of the majority of taggers. The column "only tweets" of Table 8.2 shows the
number of tweets for each corpus. After labeling the data, the number of times each
feeling was detected is shown in the corresponding column ("Anger," "Happiness,"
etc.) of Table 8.2.

It can be seen in Table 8.2 that the frequency values for sentiments in these
corpora are very low. For example, for the corpus #TVAztecaMiente, 119 tweets
were labeled; the sentiment surprise was present in 85.7% (102 out of 119) of them.
However, disgust and fear are only 0.8% (1 out of 119) of tweets read and labeled.
According to the labeled corpora, the predominated emotions were sadness, surprise, and happiness. There was also much ambiguous information, as some Twitter
users posted information that was not relevant for the event (13.9%).

We applied preprocessing to text removing numbers, exclamation marks,
question marks, and punctuation marks from tweets, as mentioned earlier. On the
other hand, although many features can be extracted from the analyzed tweets, it is
recommended to begin with the simplest features. We computed the term
"frequency-inverse document frequency matrix" with the preprocessed texts. This
matrix was used to feed three classification methods. The purpose was to build
classifiers to identify automatically the sentiments presented in not labeled tweets.

We built three classifiers from data, Naive Bayes (NB), decision tree (DT), and
support vector machine (SVM). The parameters of DT and SVM were tuned using
the grid search technique; NB classifier does not have any parameter to optimize.
Table 8.3 shows the performances of three classifiers built with each corpus. The
method to obtain these performances was 10-cross-validation.

In Table 8.3, the best performances among the three classifiers are marked in
bold; the white spaces mean that the corresponding classifier could not produce a
response. This is usually the case for data with only one class. It is important to
clarify that the classification accuracy is not the only evaluation that needs to be
applied to classifiers. Especially, for cases of imbalanced data sets, more measures
such as precision, recall, and F1 score can be necessary. Naive Bayes and support
vector machine were the classifiers with the best accuracy.

The classifiers achieve the following performance on average for each emotion,
for anger: 91.4%, joy: 78.8%, surprise: 81.3%, disgust: 89.7%, sadness: 76.8%,
fear: 66.1%, and for ambiguous: 50.3% (see Table 8.3).

Table 8.2 Sentiments found on the corpora labeled manually

Corpus	Total count with RT	Only tweets	Anger	Happiness	Surprise	Disgust	Sadness	Fear	Ambiguous	Sentiments found
#TvAztecaMiente	5323	1133	37	11	102	1	41	1	5	193
#TodosSomosMexico	6340	2965	2	256	225	3	11	3	9	500
#Terremoto	8922	1078	15	182	148	1	639	71	107	1056
#TemblorMx	2648	752	24	26	49	0	242	22	25	363
#SomosMexico	8707	1292	0	92	16	5	11	0	6	124
#skyAlertMx	252	252	0	7	6	0	10	10	3	33
#SismoMX	9992	1440	13	175	178	26	181	3	29	576
#SismoMexico2017	10,000	2220	0	0	10	549	114	64	361	737
#Sismo	10,000	781	30	133	177	0	107	73	24	520
#RoboComoGraco	10,000	1470	251	75	54	36	69	27	38	512
#PueblaSigueDePie	3171	791	73	199	164	80	236	155	4	73
#PrayForMexico	10,000	1779	0	0	0	180	427	458	138	1065
#PartidosDenSuDinero	10,000	1695	0	0	114	45	107	0	20	266
#PartidosDenNuestroDinero	10,000	2699	123	52	102	0	33	4	9	314
#Millenials	4180	2838	33	166	149	7	15	8	52	378
#MexicoEstaDepie	10,000	1305	21	92	294	6	572	45	5	1030
#Jojutla	10,000	629	2	15	2	0	18	5	30	42
#FuerzaMexico	10,000	1458	3	72	6	0	23	3	40	107
#FuerteMexico	10,000	1207	32	202	213	82	123	95	473	747
#FinDelMundo	10,000	4247	40	150	125	108	38	28	62	489
#AyudaMexico	10,000	1871	38	151	157	216	191	71	223	824
#AyudaCdMx	10,000	1231	12	47	24	1	23	10	56	117
#AlertaSismica	10,000	1622	15	53	92	15	48	53	15	275
#SismoMX	9992	1440	13	175	178	26	181	3	29	676
Total	196,356	37,404	777	2132	2421	1307	3224	1057	1763	10918
Percentage			6.1%	16.8%	19.1%	10.3%	25.4%	8.3%	13.9%	

Table 8.3 Performances of classifiers on each corpus

Corpus	Classifier	Anger (%)	Happiness (%)	Surprise (%)	Disgust (%)	Sadness (%)	Fear (%)	Ambiguous (%)
#Todossomosmexico	NB	**99.22**	**53.91**	53.91	**98.44**	**96.88**	**100.00**	**99.22**
	DT	**99.22**	52.34	49.22	**98.44**	94.53	**100.00**	97.66
	SVM	**99.22**	48.44	**54.69**	**98.44**	**96.88**	**100.00**	**99.22**
#Terremoto	NB	**98.80**	**79.20**	**86.00**	**100.00**	**67.60**	**94.00**	**91.60**
	DT	**98.00**	76.00	82.00	**100.00**	66.00	91.60	86.80
	SVM	**98.80**	**79.20**	85.60	**100.00**	60.00	**94.40**	88.40
#Temblormx	NB	95.29	89.41	**83.53**	72.94	**96.47**	**92.94**	
	DT	91.76	**90.59**	**83.53**	**74.12**	92.94	82.35	
	SVM	95.29	89.41	81.18	70.59	95.29	**92.94**	
#Somosmexico	NB	**69.70**	**93.94**	**90.91**	**90.91**	**93.94**		
	DT	18.18	87.88	**90.91**	**90.91**	75.76		
	SVM	**69.70**	**93.94**	**90.91**	**90.91**	**93.94**		
#Skyalertmx	NB	66.67	66.67	77.78	66.67	88.89		
	DT	**77.78**	66.67	**88.89**	44.44	88.89		
	SVM	66.67	**100.00**	77.78	66.67	88.89		
#SismoMX	NB	**97.67**	**62.79**	**64.34**	**95.35**	**65.12**	**100.00**	**95.35**
	DT	94.57	60.47	63.57	90.70	59.69	99.22	89.92
	SVM	**97.67**	61.24	62.79	**95.35**	**65.12**	**100.00**	**95.35**
#Sismomexico2017	NB	**98.81**	**88.14**	**93.28**	**68.77**			
	DT	98.02	82.61	88.93	58.10			
	SVM	**98.81**	**88.14**	**93.28**	67.98			
#Sismo	NB	**96.95**	**70.23**	**68.70**	**78.63**	**88.55**	**93.89**	
	DT	94.66	67.18	59.54	75.57	83.97	90.08	
	SVM	**96.95**	67.18	**68.70**	**78.63**	**88.55**	**93.89**	

(continued)

Table 8.3 (continued)

Corpus	Classifier	Anger (%)	Happiness (%)	Surprise (%)	Disgust (%)	Sadness (%)	Fear (%)	Ambiguous (%)
#Robocomograco	NB	56.15	87.69	90.77	91.54	91.54	96.92	90.00
	DT	53.85	82.31	86.15	86.15	84.62	92.31	87.69
	SVM	54.62	87.69	90.77	91.54	91.54	96.92	90.00
#Pueblasiguedepie	NB	86.36	71.72	73.74	89.90	63.64	76.26	99.49
	DT	87.88	72.22	72.22	88.38	61.11	72.22	98.99
	SVM	90.40	79.29	76.26	91.41	70.20	79.80	99.49
#Prayformexico	NB	90.11	70.34	71.91	93.26			
	DT	82.70	64.27	62.25	88.99			
	SVM	90.34	71.01	71.91	93.26			
#Partidosdensudinero	NB	92.92	98.11	93.63	98.58			
	DT	87.74	96.70	90.09	97.41			
	SVM	92.92	98.11	93.63	98.58			
#PartidosDenNuestroDinero	NB	60.00	84.29	88.57	95.71	97.14		
	DT	65.71	77.14	80.00	95.71	95.71		
	SVM	48.57	84.29	88.57	95.71	97.14		
#Millenials	NB	89.52	80.95	69.52	100.00	95.24	99.05	84.76
	DT	94.29	75.24	79.05	100.00	93.33	98.10	85.71
	SVM	90.48	76.19	77.14	100.00	95.24	99.05	87.62
#Mexicoestadepie	NB	97.60	92.80	70.80	99.60	66.00	95.20	99.60
	DT	97.20	86.80	66.80	98.40	54.80	91.20	99.20
	SVM	97.60	93.20	71.20	99.60	60.00	95.20	99.60
#Jojutla	NB	94.44	77.78	100.00	77.78	94.44	50.00	
	DT	94.44	38.89	100.00	77.78	94.44	38.89	
	SVM	94.44	77.78	100.00	77.78	94.44	55.56	

(continued)

Table 8.3 (continued)

Corpus	Classifier	Anger (%)	Happiness (%)	Surprise (%)	Disgust (%)	Sadness (%)	Fear (%)	Ambiguous (%)
#Fuerzamexico1	NB	**97.30**	45.95	**94.59**	86.49	**97.30**	**72.97**	
	DT	**97.30**	**54.05**	**94.59**	83.78	**97.30**	67.57	
	SVM	**97.30**	51.35	**94.59**	86.49	**97.30**	**72.97**	
#Fuerzamexico2	NB	95.56	**95.56**	73.33	**100.00**	93.33		
	DT	**97.78**	88.89	**80.00**	**100.00**	93.33		
	SVM	95.56	**95.56**	75.56	**100.00**	93.33		
#Fuerzamexico3	NB	**98.40**	**92.00**	**86.00**	**100.00**	**97.60**	**98.80**	**99.20**
	DT	**98.00**	90.40	82.80	**100.00**	95.20	98.40	98.00
	SVM	**98.40**	**92.00**	85.60	**100.00**	**97.60**	**98.80**	**99.20**
#Fuerzamexico4	NB	92.00	**68.00**	**94.00**	**88.00**	92.00	**94.00**	**84.00**
	DT	**94.00**	62.00	86.00	84.00	92.00	92.00	76.00
	SVM	92.00	38.00	**94.00**	**88.00**	92.00	**94.00**	**84.00**
#Fuerzamexico5	NB	**97.50**	**57.50**	**77.50**	**97.50**	**82.50**	**90.00**	**90.00**
	DT	**97.50**	**57.50**	75.00	95.00	80.00	82.50	72.50
	SVM	**97.50**	**57.50**	**77.50**	**97.50**	**82.50**	**90.00**	**90.00**
#Fuertemexico	NB	**96.80**	**82.40**	**75.20**	**91.60**	**84.80**	**90.40**	**70.00**
	DT	95.60	70.80	67.60	90.80	82.40	86.80	64.00
	SVM	**96.80**	**82.40**	**75.20**	**91.60**	**84.80**	**90.40**	62.40
#Findelmundo	NB	92.86	**68.75**	**76.79**	**77.68**	**90.18**	**97.32**	83.93
	DT	92.86	60.71	66.07	71.43	87.50	95.54	**84.82**
	SVM	92.86	63.39	**76.79**	**77.68**	**90.18**	**97.32**	83.93
#AyudaMexico	NB	**96.15**	**86.15**	**86.92**	**80.77**	80.77	**93.08**	75.77
	DT	90.38	75.77	76.54	71.92	66.92	91.92	68.46
	SVM	**96.15**	**86.15**	**86.92**	**80.77**	**81.54**	**93.08**	**76.15**

(continued)

Table 8.3 (continued)

Corpus	Classifier	Anger (%)	Happiness (%)	Surprise (%)	Disgust (%)	Sadness (%)	Fear (%)	Ambiguous (%)
#Ayudacdmx	NB	**87.18**	**63.16**	**89.74**	**92.31**	**97.44**	**58.97**	
	DT	82.05	57.89	84.62	84.62	92.31	51.28	
	SVM	**87.18**	**63.16**	**89.74**	**92.31**	**97.44**	**58.97**	
#Alertasismica	NB	**97.26**	**80.82**	68.49	**93.15**	**80.82**	**80.82**	**97.26**
	DT	93.15	65.75	57.53	90.41	78.08	76.71	91.78
	SVM	**97.26**	79.45	**69.86**	**93.15**	**80.82**	79.45	**97.26**
#SismoMX	NB	**97.67**	62.79	**64.34**	**95.35**	65.12	**100.00**	**95.35**
	DT	95.35	**63.57**	61.24	90.70	60.47	99.22	89.92
	SVM	**97.67**	61.24	62.79	**95.35**	65.12	**100.00**	**95.35**
Average of the bests		91.44	78.81	81.29	89.76	76.84	66.06	50.32

Table 8.4 Predictions for each corpus

Corpus	Anger	Happiness	Surprise	Disgust	Sadness	Fear	Ambiguous
#Todossomosmexico	0	3748	0	0	0	256	0
#Terremoto	0	30	17	0	7564	0	236
#Temblormx	0	110	1	2221	162	2	0
#Somosmexico	9859	0	0	5	0	0	0
#Skyalertmx	21	0	10	0	0	0	0
#SismoMX	0	243	93	0	222	0	0
#Sismomexico2017	0	0	0	1608	0	0	0
#Sismo	0	264	594	42	21	0	0
#Robocomograco	4083	0	0	0	29	7	0
#Pueblasiguedepie	0	0	0	0	0	0	0
#Prayformexico	0	0	126	0	0	0	0
#Partidosdensudinero	0	0	130	0	0	0	0
#PartidosDenNuestroDinero	4344	1	0	0	0	0	0
#Millenials	342	100	1084	13	13	0	0
#Mexicoestadepie	0	0	0	0	6902	0	0
#Jojutla	0	0	0	6	0	0	0
#Fuerzamexico1	0	5842	0	1	0	0	0
#Fuerzamexico2	256	9821	1126	0	0	205	0
#Fuerzamexico3	0	9000	9	0	0	0	0
#Fuerzamexico4	283	1553	0	5	0	1	0
#Fuerzamexico5	0	9746	0	0	0	0	0
#Fuertemexico	0	3	11	0	11	9	4449

(continued)

Table 8.4 (continued)

Corpus	Anger	Happiness	Surprise	Disgust	Sadness	Fear	Ambiguous
#Findelmundo	8	501	384	15	0	0	1455
#AyudaMexico	0	6	5	43	0	0	0
#Ayudacdmx	0	5	0	1	1	17	0
#Alertasismica	0	259	0	0	0	0	0
#SismoMX	0	4150	93	0	222	125	0
Total	19196	45382	3683	3960	15147	622	6140

We use the best classifier for each sentiment on each data set to obtain the results shown in Table 8.4. It is interesting to mention that in most of these predictions, the classifiers with the best performances made a similar number of predictions about the presence of a sentiment, concerning the taggers. This allowed us to claim that the results presented in Table 8.4 agree with the data labeled manually (Table 8.2).

Based on Table 8.4 (row named "total"), it is possible to claim that almost half of the emotions in data correspond to happiness (48.2%). Anger represents 20.4% of all predicted emotions, and sadness represents 16.1%. This can be attributed to the fact that a large part of the tweets was related to solidarity, support, and help; another part with claims toward mass media, political parties, and governors; and others related to bad news about dead or disappeared persons.

There were very few emotions of surprise (3.9%), disgust (4.2%), and fear (0.7%) in all the analyzed data because much of the data were collected after the earthquake. Ambiguous tweets represented the 6.5% of data. Figure 8.4 shows a graphic summary of the sentiments found in the data analyzed through machine learning techniques.

The behavior of emotions is explained because the information was downloaded after the earthquake. Many of the Twitter users were happy because they were safe, and they found lost people or pets and also because of the actions carried out by the civil society. Unfortunately, there were many people who, due to the conditions of the event, could not interact on Twitter.

8.5 Conclusions and Future Research

Social media is an important source of data that nurture the big data every day. Most of the works on sentiment analysis only consider three possible cases for each opinion or post of users of social media. These are positive, negative, and neutral. The use of emotional frameworks, such as Ekman's model, allowed defining the emotions generated in social media more precisely, as they are related to human behavior.

It is useful to analyze identified emotions in order to study the reactions of social media users to certain kinds of events, people, services, or products, as well as their posture and the effects that they generate.

In this chapter, we explained a methodology to apply sentiment analysis on data from Twitter through machine learning techniques. As an example, we analyzed the emotions on tweets after an earthquake event in Mexico. Based on Ekman's model, we found that the most frequent predicted emotions were happiness, anger, and sadness, and 6.5% of predicted tweets were irrelevant. We built three classifiers to determine the emotions of tweets that belong to the same topic, and Naive Bayes and support vector machine were the classifiers with the best accuracy for predicting emotions.

Based on previous experience with sentiment analysis, we can suggest the following recommendations:

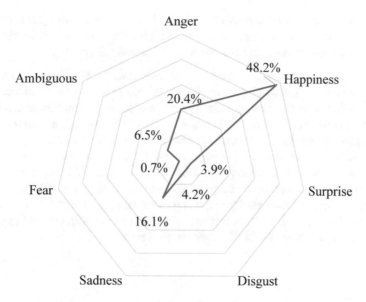

Fig. 8.4 Summary of sentiments in the analyzed corpora

- Preprocessing is a necessary phase for successfully applying machine learning methods for sentiment analysis. Most of the steps of preprocessing can be implemented by software.
- Labeling documents is another important phase before applying machine learning methods for sentiment analysis. The recommendation is to label enough data to produce accurate models. How much is enough data? Some authors suggest about 3000 labels [35]. Although they report that increasing this number damages the performance of classifiers, it is possible that this happens because if data of the same class have a different distribution, then the decision boundary is more complex. That is, the more data are added, the more concepts related to the same class must be discovered by the classification method. This causes the performance of the classifiers to be diminished.
- In general, the imbalance of classes affects the performance of classification methods dramatically. If certain sentiment or emotion predominates (majority class) and there are just a few of the other sentiments (minority class), the classifier will predict the majority class most of the time, or even always. In these cases, applying a balancing method, such as SMOTE [36], or labeling more data to balance the number of sentiments can be helpful.
- A larger number of sentiments or classes in data diminish the performance of classification methods. The greater the number of classes, the lower the performance of a classifier. The recommendation is to merge classes or get rid of the ones that are not relevant before using a supervised machine learning method. The approach used in this chapter is to create a classifier for each class, and then apply a binary classification method;

- It is recommended to build more than one classifier, each one of a different type, then measure their performances. If the classifiers are not achieving good results, extract more features from text and repeat the process.

The main purpose of this chapter is to provide insights into the use of sentiment analysis methods. A second contribution is the expansion of the emotions range, from three (negative, neutral, positive) to six in order to provide more elements to understand how users interact with social media platforms. A third contribution is the analysis of the data sets from Mexico's 2017 earthquake and expanding the understanding of Mexican emotions on social networks. Future research will include validation of the method with different data sets and emotions; the addition of new artificial intelligence techniques to improve accuracy; and also new research paths to clean data are considered along with other techniques for computing emotions. We hope this contribution will foster the use of methods and techniques to understand emotions in social media in the future.

References

1. Almarabeh T, Majdalawi YK, Mohammad H (2016) Cloud computing of e-government
2. Sasikala P (2012) Cloud computing and E-governance: advances, opportunities and challenges. Intl J Cloud Appl Comput (IJCAC) 2(4):32–52
3. Jadhav B, Patankar A (2018) Opportunities and challenges in integrating cloud computing and big data analytics to e-governance. Int J Comput Appl 180(15):6–11
4. Lohr S (2012) The age of big data. New York Times, 11 (2012)
5. Liebowitz J (2001) Knowledge management and its link to artificial intelligence. Expert Syst Appl 20(1):1–6
6. Fan W, Bifet A (2013) Mining big data: current status, and forecast to the future. ACM SIGKDD Explor Newsl 14(2):1–5
7. Wu X, Zhu X, Wu GQ, Ding W (2014) Data mining with big data. IEEE Trans Knowl Data Eng 26(1):97–107
8. Sandoval-Almazán R, Valle-Cruz D (2016) Understanding network links in Twitter: a Mexican case study. In: Proceedings of the 17th international digital government research conference on digital government research. ACM, pp 122–128
9. Shaikh S, Feldman LB, Barach E, Marzouki Y (2017) Tweet sentiment analysis with pronoun choice reveals online community dynamics in response to crisis events. In: Advances in cross-cultural decision making. Springer, Cham, pp 345–356
10. Vo BKH, Collier NIGEL (2013) Twitter emotion analysis in earthquake situations. Intl J Comput Linguist Appl 4(1):159–173
11. Sandoval-Almazan R (2019) Using twitter in political campaigns: The case of the PRI candidate in Mexico. In: Civic engagement and politics: concepts, methodologies, tools, and applications. IGI Global, pp 710–726
12. Sandoval-Almazán R, Valle-Cruz D (2018) Towards an understanding of Twitter networks: the case of the state of Mexico. First Monday, vol 23, no 4
13. Sandoval-Almazan R, Valle-Cruz D (2018) Facebook impact and sentiment analysis on political campaigns. In: Proceedings of the 19th annual international conference on digital government research: governance in the data age. ACM, p 56
14. Chen M, Zhang Y, Li Y, Mao S, Leung VC (2015) EMC: emotion-aware mobile cloud computing in 5G. IEEE Netw 29(2):32–38

15. Preethi G, Krishna PV, Obaidat MS, Saritha V, Yenduri S (2017) Application of deep learning to sentiment analysis for recommender system on cloud. In 2017 international conference on computer, information and telecommunication systems (CITS). IEEE, pp 93–97

16. Krishna PV, Misra S, Joshi D, Obaidat MS (2013) Learning automata based sentiment analysis for recommender system on cloud. In: 2013 international conference on computer, information and telecommunication systems (CITS). IEEE, pp 1–5

17. Assunção MD, Calheiros RN, Bianchi S, Netto MA, Buyya R (2015) Big data computing and clouds: trends and future directions. J Parallel Distrib Comput 79:3–15

18. Karyotis C, Doctor F, Iqbal R, James A, Chang V (2018) A fuzzy computational model of emotion for cloud based sentiment analysis. Inf Sci 433:448–463

19. Ali F, Kwak D, Khan P, Islam SR, Kim KH, Kwak KS (2017) Fuzzy ontology-based sentiment analysis of transportation and city feature reviews for safe traveling. Transp Res Part C: Emerg Technol 77:33–48

20. Sreeja PS, Mahalakshmi GS (2017) Emotion models: a review. Int J Control Theory Appl 10:651–657

21. Ekman PE, Davidson RJ (1994) The nature of emotion: fundamental questions. Oxford University Press, Oxford

22. Plutchik R (1965) What is an emotion? J Psychol 61(2):295–303

23. Munezero MD, Montero CS, Sutinen E, Pajunen J (2014) Are they different? affect, feeling, emotion, sentiment, and opinion detection in text. IEEE Trans Affect Comput 5(2):101–111

24. Nakamura (1993) Kanjo hyogen jiten [dictionary of emotive expressions]. Tokyodo, Teluk Intan

25. Yang Y, Jia J, Zhang S, Wu B, Chen Q, Li J et al (2014) How do your friends on social media disclose your emotions? In: Twenty-eighth AAAI conference on artificial intelligence

26. Wang Y, Pal A (2015) Detecting emotions in social media: a constrained optimization approach. In: Twenty-fourth international joint conference on artificial intelligence

27. Cambria E, Livingstone A, Hussain A (2012) The hourglass of emotions. In Cognitive behavioural systems. Springer, Berlin, pp 144–157

28. Zhang L, Hua K, Wang H, Qian G, Zhang L (2014) Sentiment analysis on reviews of mobile users. Procedia Comput Sci 34:458–465

29. Zhang L, Hua K, Wang H, Qian G, Zhang L (2014) Sentiment analysis on reviews of mobile users. Procedia Comput Sci 34:458–465

30. Prabowo R, Thelwall M (2009) Sentiment analysis: a combined approach. J Informetr 3(2):143–157

31. Kumar M, Bala A (2016) Analyzing Twitter sentiments through big data. In: 2016 3rd international conference on computing for sustainable global development (INDIACom). IEEE, pp 2628–2631

32. Subramaniyaswamy V, Vijayakumar V, Logesh R, Indragandhi V (2015) Unstructured data analysis on big data using map reduce. Procedia Comput Sci 50:456–465

33. Al-Kabi M, Al-Ayyoub M, Alsmadi I, Wahsheh H (2016) A prototype for a standard arabic sentiment analysis corpus. Int Arab J Inf Technol 13(1A):163–170

34. Kune R, Konugurthi PK, Agarwal A, Chillarige RR, Buyya R (2016) The anatomy of big data computing. Softw Pract Exp 46(1):79–105

35. Sidorov G et al (2013) Empirical study of machine learning based approach for opinion mining in tweets. In: Batyrshin I, González Mendoza M (eds) Advances in artificial intelligence. MICAI 2012. Lecture notes in computer science. Springer, Berlin, vol 7629

36. Chawla NV, Bowyer KW, Hall LO, Kegelmeyer WP (2002) SMOTE: synthetic minority over-sampling technique. J Artif Int Res AI Access Found 16:321–357

Chapter 9
Connection Handler: A Design Pattern for Recovery from Connection Crashes

Naghmeh Ivaki, Nuno Laranjeiro, Fernando Barros and Filipe Araújo

Abstract Creating dependable distributed applications for the cloud is a challenging task that heavily depends on the communication middleware. Such middleware will invariably depend on the Transmission Control Protocol (TCP), as TCP stands at the core of reliable communication on the Internet. Despite offering reliability against dropped and out-of-order packets, the ubiquitous TCP provides no recovery options when connections crash due to, for example, lost connectivity. Should this happen, developers must rollback the communication endpoints to some coherent state, using their own error-prone solutions. In fact, overcoming this limitation is a difficult and unsolved problem, and so far, no solution managed to gain wide acceptance, as they all impact TCP's simplicity, performance, or pervasiveness. In this chapter, we present the Connection Handler design pattern, a reusable design solution that allows the development of cloud communication middleware that is tolerant to connection crashes. Being a design pattern, it bears little or no dependence to the operating system, programming language, or external libraries, having minimal impact on any other cloud system layers. To demonstrate that the Connection Handler does not impair performance and involves a low programming complexity, we applied it to: (i) a stream-based TCP, (ii) an HTTP, and (iii) a message-oriented application. Our results show that our design pattern is efficient and of general use, thus being applicable to a wide range of cloud-based applications and services.

Keywords Reliable communication · Cloud computing · TCP · Connection crashes · Connection handler · Fault-tolerance · Design patterns

N. Ivaki (✉) · N. Laranjeiro · F. Barros · F. Araújo
Department of Informatics Engineering, CISUC, University of Coimbra, Coimbra, Portugal
e-mail: naghmeh@dei.uc.pt

N. Laranjeiro
e-mail: cnl@dei.uc.pt

F. Barros
e-mail: barros@dei.uc.pt

F. Araújo
e-mail: filipius@uc.pt

© Springer Nature Switzerland AG 2020
M. Ramachandran and Z. Mahmood (eds.), *Software Engineering in the Era of Cloud Computing*, Computer Communications and Networks,
https://doi.org/10.1007/978-3-030-33624-0_9

9.1 Introduction

While solutions akin to cloud computing have existed for many decades, the term "cloud computing" only became popular in the 2000s, as Amazon started to provide infrastructure services on demand. From there, cloud computing grew so much in importance that we can say without exaggeration that from basic grid services, such as electricity, water, and telecommunications, to business and leisure, it is increasingly difficult to come up with a human activity that does not rely on some form of cloud-provided service. Despite looking straightforward at surface, programming a correct and reliable cloud-based distributed application is a challenging task. Crashes in pretty much any component involved in the communication [1, 2] turn distributed programming into a complex and subtle task. Tolerating crashes that interrupt communication and recovering from them (which includes rolling back to a consistent state) is quite difficult, if possible at all, mainly due to incomplete information from peers [3, 4].

At first glance, TCP appears to be a simple and powerful solution against network unreliability, which is true to a certain degree, because TCP resends unacknowledged packets. However, despite emulating reliable communication, TCP connections can fail, even if both endpoints are still running. Technically, a TCP connection fails when the operating system aborts a connection, for one of the following reasons: (i) when data in the send buffer is not acknowledged after a given number of retransmissions; (ii) when the application waits for reading from the receive buffer for a period of time that exceeds a previously set timeout; (iii) when an underlying network failure is reported by the network layer; (iv) and when the IP address changes [5].

Connection crashes might be a problem for many applications, because rolling the peers back to a coherent state, is something entirely left for developers, which tend to create custom, ad hoc, and error-prone solutions. Moreover, TCP does not provide the application any information about the data it wrote or read to/from the peer, thus making the recovery process much more complicated.

The need to overcome this problem triggered a huge research effort, which we review in Sect. 9.2. This endeavor resulted in a vast body of solutions, including communication stacks, protocols, and middleware. Some of these try to use multiple alternative paths between client and server [6, 7]; others choose to replicate components [8, 9], checkpoint the state of the connections [10], or use a middle layer to intercept TCP system calls [5, 11, 12]. The problem with all existing solutions is that they either try to replace TCP, or they require special software or hardware that may not be readily available or mature for deployment in all platforms and languages. Thus, TCP remained as the only widely used protocol for implementing reliable communication, despite all its known limitations.

Nevertheless, it is quite common for cloud applications to need additional features TCP does not provide. In fact, ensuring reliable communication in a faulty and complex cloud environment is a very common problem occurring in many different contexts, such as data streaming or messaging over TCP (e.g., in communication between OpenStack components) or HTTP (e.g., in REST services that are widely used in cloud computing such as in the docker daemon API). Despite holding evident differences, these scenarios share the same basic reliable communication requirement. Thus, we argue that this problem should have a general repeatable solution. In this chapter, we propose a software design pattern [13], named Connection Handler, that leverages on TCP, and that can be used to ensure reliable communication, regardless of the operating system, programming language, middleware, or technology involved.

With the goal of building our design solution, in Sect. 9.3, we start by reviewing the general design of a connection-oriented application. This section presents the main contribution of this chapter: A minimalistic design that includes a client, a server, a transport handle, and a few more components that extend basic TCP to reliably send and receive data. This effort abstracts and subsumes our previous work [14–16], as we observed that different designs, tailored for different applications, actually share common components. For example, all our previous solutions share a buffer, to keep unacknowledged data, or a synchronizer, to enable race-free replacement of failed connections. Hence, the basic design must include these components. We now put together in a comprehensive solution all our previous work, by adding the notion of a session layer along with multi-thread support as in [14], and by solving the problem for HTTP [15] and for messaging [16].

In Sect. 9.4, we create the basic connection-oriented design, called Connection Handler. In Sect. 9.5, we specialize this design pattern and apply it to different types of applications, which are stream-based (e.g., media or file streaming), HTTP-based (e.g., Web applications), and message-based (e.g., chat). Each of these applications requires its own component adaptation, e.g., an extension or an interface implementation. The three kinds of applications use the same general design, with differences in the buffers, which must be more complex for HTTP, and in the extra components required by messaging.

We implemented the Connection Handler design pattern in Java, and in Sect. 9.6, we carried out an experimental evaluation where we applied it to the three distinct types of applications (stream-based, HTTP-based, and message-based), with the goal of obtaining new reliable versions. The results show not only its correctness and applicability to realistic systems, but also its overall good performance, low resource usage and complexity. The implementation of the design pattern is open-source and publicly available in Sourceforge [17].

9.2 Related Work

Several solutions have been proposed in the literature for supporting reliable communication in distributed systems. In this section, we discuss two kinds of solutions. We first review protocols or mechanisms implemented in the form of libraries that were proposed to tolerate connection crashes or to build reliable communication, we then review solutions at the design level (e.g., design patterns).

Concerning the *protocols and mechanisms*, we find different cases in the literature. Some of the solutions use redundancy via multi-homing in the transport layer to tolerate connection crashes. Multi-homing enables a peer to establish a session with another peer over multiple interfaces identified by different IP addresses. Multipath TCP [6] and SCTP [7] are examples of this technique.

We can also find session-based solutions which require changing the application programming interfaces (APIs) of the sockets or, to avoid this drastic change, use some software components that intercept current API calls. This is the case of Robust Socket (RSocket) [12], which changes Java core libraries, leaving the standard Java TCP interface untouched. Zandy and Miller proposed a similar approach, *rocks* [5], although based on a circular buffer, which we adopt in our own work. The work of Alvisi et al. in [11] is slightly different, as authors enable the server to stop and recover, but they also use the idea of wrapping TCP with additional layers, to tolerate server crashes transparently for the client and server implementation.

Some solutions imply modifying the server to enable replication. For example, ST-TCP [8] tolerates TCP server crashes using an active backup that keeps track of the TCP connection state, to take over whenever the primary fails. HydraNet-FT [9] replicates services across an internetwork, to provide a single view of a fault-tolerant server to the client. This requires a few modifications to TCP on the server side. In MI_TCP [10], servers in a cluster may create checkpoints of the connection state, to resume interaction after failures.

Despite the merit of the above solutions, none of them succeeded in gaining wide acceptance among developers, because they all somehow impact TCP's simplicity, performance, or ubiquity. Other solutions, such as JMS [18] and MSMQ [19], became quite popular instead. These solutions are not only widely used, but they are also quite rich in reliability features. Despite this, they are only suitable to be used in asynchronous message-based communication and do not fit the requirements of many applications. RPC [20] and RMI [21] are also quite well-known solutions, but they have poor reliability mechanisms. Moreover, their invocation model (i.e., blocking and non-pipelined invocations of remote objects) does not fit the requirements of many real applications.

With the increasingly important role of HTTP and also of specifications on top of HTTP, such as SOAP-based Web services, a few solutions were created to add reliability to HTTP-based communication. HTTPR [22], with the support of IBM, is one of the most well-known solutions. This specification, aimed at achieving reliable communication in the presence of failures in the communication channel or in endpoints. In short, the solution makes sure that each message is either delivered to the destination peer (exactly one time) or is marked as undelivered. Web Service Reliability [23] (WS-Reliability) is another solution that is used for exchanging SOAP [24] messages, typically over HTTP, with several reliability guarantees. The WS-Reliability standard specifies a set of reliability semantics that aim at guaranteeing message ordering, message delivery, elimination of duplicates, and finally message delivery and elimination of duplicates. WS-Reliable-Messaging [25] serves similar purposes and includes the same set of delivery guarantees. All of these HTTP-based solutions are message-oriented, which means that if a message does not reach the destination due to a connection crash, it has to be sent again. This is quite inefficient when a message is long (e.g., a file or the typically verbose SOAP message).

Rather than developing new libraries, we argue that a different approach to address reliability issues in distributed systems is needed. Research in *design-based solutions* (i.e., software design patterns) can definitely help to reduce the growing number of custom solutions that try to deliver reliable communication, by directing developers to design patterns [26]. The idea of using design patterns for software development started more than two decades ago. There is a huge number of design patterns that can be found in the literature [13, 27–30], some with applicability to dependability and security. However, research in design patterns that aim at addressing reliability issues in distributed communications has been largely disregarded.

In previous work, we proposed a session layer design solution for implementing reliable communication in distributed applications [14]. We resorted, in that work, to several design patterns that emerged for distributed programming in the last decades, namely the reactor [31] for dispatching the events to several handlers, acceptor–connector [32] for determining the basic interaction between the client and server, and the leader–followers [33], to provide support for highly concurrent applications. We then leveraged on our previous work and propose two stream-based fault-tolerant design patterns. The idea was to expose reliable communication as a session layer and to use this session layer in both blocking and non-blocking designs [14]. Then, we specialized our solution for HTTP-based applications, by addressing some key challenges in Web environment [15], and we further proposed design solutions to message-based applications [16].

Here, we use our previous experience [14–16] to abstract a design solution that allows recovering from connection crashes—the Connection Handler design pattern. Connection Handler is a solution that can be used in any distributed connection-based application (i.e., it uses a transport handle, like TCP Socket, for communication) to handle connection crashes. Thus, we show its applicability to three major types of distributed applications: plain stream-based applications, HTTP, and also messaging applications.

9.3   General Design of a Connection-Oriented Application

In this section, we review the general design of a connection-oriented application. The goal is to explain its components and basic behavior, so that we can first abstract its design and then in Sect. 9.4 specialize it in a reusable reliability solution. As we will see later in Sect. 9.5, and taking advantage of its reusability, this solution will be the object of further specialization, which will serve different types of applications (stream-based, HTTP-based HTTP, and Message-based). In a basic connection-oriented communication, two peers establish a connection before starting to exchange any data. One peer initiates the connection by sending a

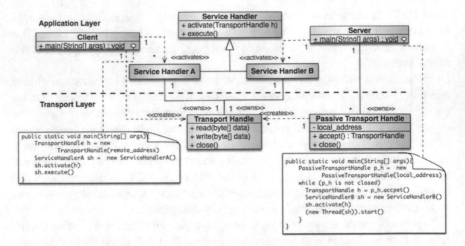

Fig. 9.1 General design of a connection-based application

connection request to the other peer. The initializer of a connection is named "client," and the peer that accepts the connection request is named "server." Figure 9.1, built based on the acceptor–connector design pattern [32], depicts a general pattern for connection-oriented communication. The remainder of this section describes the components of this design and their collaboration in detail.

As we can see in Fig. 9.1, in terms of design, client and the server are placed at the application layer. The other layer, identified as "Transport Layer," should be provided by the operating system or by some library in the form of a communication protocol, such as TCP. The remaining parts of the pattern are left to the programmer and determine the behavior of application, including the specification of message formats. Before any data exchange occurs between the application peers, the client must take the initiative of starting the connection, by sending a set up request to the server. Once the connection starts, the peers may engage in symmetric read and write operations with each other. In the end, any of the two might close the connection. To support these operations, connection-oriented implementations usually require the components visible in Fig. 9.1, where each of them fulfills the following specific purposes:

- The **Transport Handle** provides an interface to let applications establish a connection, write data, read data, and close the connection. This component belongs to the transport layer. A very well-known example of a transport handle is a TCP Socket. A fundamental problem of this transport handle model is the lack of tolerance to connection crashes. For instance, if peers are suddenly out of reach and, as a consequence, the operating system shuts down the transport handle, the application has no means to access information regarding the data already sent, received, and delivered to the application.

- The **Passive Transport Handle** is a passive mode transport handle that is bound to a network address (i.e., an IP address and a port number) below the application layer. It is used by the server to receive and accept connection requests from clients.
- The **Service Handler** implements application services and business logic, typically playing two different roles (e.g., sender or receiver) in the client and server sides. For this reason, we have two different components that extend the Service Handler, namely Service Handler A and Service Handler B. However, we may have applications whose peers play both roles (e.g., sender and receiver) in different circumstances during a session. In addition, each Service Handler owns a transport handle, such as a socket, to exchange data with its connected peer. It also implements an abstract operation (*execute()*) of the command pattern [26] that should be invoked to start the execution of an application service.
- The **Client** implements the actions to start the connection to the server, and then to initialize and activate a Service Handler.
- The **Server** keeps checking for the arrival of new connection requests and may own one or more Passive Transport Handles (a single passive handle can support many connections). Upon arrival of a new connection request, a new Transport Handle is created (one handle per connection). The server then initializes and activates the Service Handler, by passing the new transport handle.

Figure 9.2 shows the interactions between the components of the abovementioned basic model. As we can see, the Client starts establishing a connection to the server, by creating a Transport Handle (*h*) and passing the server's network address (*remote_address*). On the other side, the server, which must have already initialized a Passive Transport Handle, waits in a loop, for a connection request, by calling the method *accept()* of the passive handle. Upon reception and acceptance of the connection request on the server side, a Transport Handle is created (h_1).

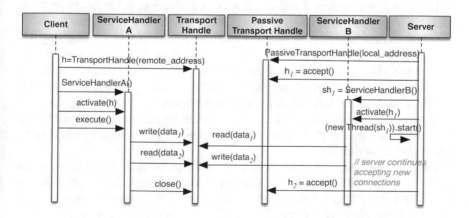

Fig. 9.2 Interaction between the components of the connection-based application

Afterwards, the Service Handler is initialized and activated, in both client and server (respectively, sh and sh_1), by passing the transport handle (respectively, h and h_1) through the method *activate()*. Then, once the connection is up, they both can start the execution of service and exchange of data ($data_1$ and $data_2$). This happens when the method *execute()* of the Service Handler is invoked by the client and server (or a new thread created by the server for this connection). Pseudocode presented in Fig. 9.1 is a simplified description and implementation of the process explained above for connection establishment and service initialization and execution.

9.4 Connection Handler Design Pattern

In this section, we build on the basic pattern described earlier and propose a design solution that targets the problem of connection crashes in connection-oriented communications. The final proposed solution was built based on our experience in designing solutions for different types of applications in previous work, in particular designing reliable stream-based TCP applications [14], reliable HTTP applications [15], and reliable messaging applications [16]. The solution presented in this section has been refined after being applied to these cases.

To recover from connection crashes, applications need to go beyond transport layer acknowledgments. They must buffer sent data until their peers deliver the data to the application layer. If the connection crashes, peers must then set up a new connection and resend whatever data did not reach the peer's endpoint application. According to the model of Fig. 9.1, and due to Network Address Translation (NAT) schemes and firewalls, the initiator of a connection is *always* the client. The server must then, somehow, distinguish a reconnection attempt from a new connection request, to replace the failed connection with a fresh one, if necessary. Once the successful reconnection phase finishes, peers have to retransmit lost data.

In Fig. 9.3, we present the Connection Handler design pattern, which allows performing the abovementioned actions to recover from connection crashes.

This pattern is independent from the application's logic, the programming language, or the platform where the application is running. Note that in our patterns, we use the blue color for components that are part of our reliability solutions (usually in the session layer) and the gray color to show components that exist already (usually in the application and transport layers). We also use a light blue, to distinguish the components that have already been explained from the new ones. The next paragraphs explain each of the components of the Connection Handler design pattern in detail.

The **Reliable Endpoint** owns a transport handle (*handle*), to exchange data with its remote peer, and keeps track of data sent (*data_written*) and received (*data_-read*). This component stores the sent data into a buffer, to enable retransmission, should a connection crash occur. To reestablish a connection and retransmit lost data, the reliable endpoint needs to implement some extra actions that are defined

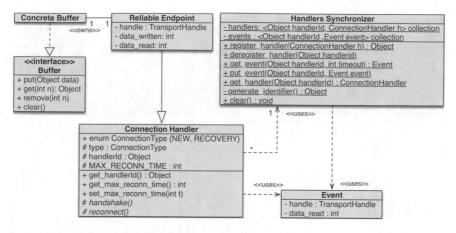

Fig. 9.3 Connection handler design pattern

and encapsulated in a new component, named Connection Handler. According to the design of the connection-based application (Fig. 9.1), the reliable endpoint can be implemented inside the Service Handler (i.e., in application layer) or separated as a different component underneath the Service Handler (i.e., in session layer). Each reliable endpoint owns a **Concrete Buffer**, to keep the data sent, because all data in transit (e.g., data in the sender's send buffer and receiver's receive buffer) may be lost due to a connection crash. The Concrete Buffer implements the interface of **buffer**, which allows saving, retrieving, and removing the acknowledged data, respectively, through the methods *put()*, *get()*, and *remove()*. The method *clear()* is used to remove all data from the buffer when it is not needed anymore (e.g., when the connection is closed by the application). The Concrete Buffer must be implemented properly, depending on the type of data (e.g., bytes or message) that the peers use for communication.

The **Connection Handler** implements all actions required to establish a connection and reestablish a failed one. Each instantiated Connection Handler has a unique identifier that is generated by the server, to distinguish a brand-new connection from a connection that was established for recovery purposes. The unique identifier is exchanged between peers during the handshake process, in the method *handshake()*, once the TCP connection starts.

The **handshake process** used to exchange the identifier works as follows. Upon establishment of a new TCP connection, the client sends 0 (or another symbol, which is defined by developer for this case) as its identifier, which allows the server to identify the connection as new. Then, the connection *type* is set to *NEW*, and a unique identifier is generated and sent back to the client. When the client establishes a connection for recovery purposes, it sends this identifier (*handlerId*), to let the server replace the old connection. In this case, the connection type is set to *RECOVERY*. The actions to reconnect and resend the lost data must be implemented in the method *reconnect()*. Moreover, the Connection Handler allows the

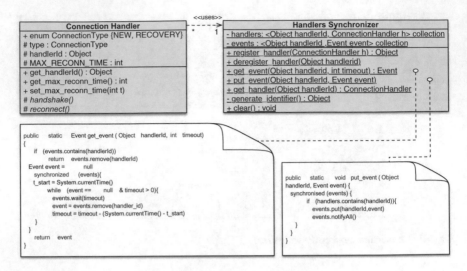

Fig. 9.4 Handler synchronizer in connection handler design pattern

application to define the maximum time (*MAX_RECONN_TIME*) permitted for the recovery process through the method *set_max_reconn_time()*.

To enable the recovery of broken connections, the server keeps the list of all open Connection Handlers together with their identifier (in handlers), in a component named **Handlers Synchronizer** which is depicted in Fig. 9.4 along with a partial implementation in pseudocode. The Component Handlers Synchronizer ensures that only one of the connections to the client will actually work at a time. The Handlers Synchronizer provides an interface, allowing Connection Handlers to: (1) register and deregister themselves into/from the list of the handlers; (2) put an event for another handler (i.e., used by the Connection Handler of the recovered connection); and (3) wait for an event coming from a new handler (i.e., used by the Connection Handler with a failed connection). When a Connection Handler is being registered, the Handlers Synchronizer generates a unique identifier and returns it back to the Connection Handler.

The registered handlers can wait for an event, by calling the method *get_event()* and passing their identifier and a timeout, which represents the maximum time allowed for a new event. The Connection Handlers can also leave events for other registered handlers, by calling the method *put_event()* and by passing the information of the new connection and the identifier of the destination handler. These two operations are synchronized over a collection that includes the events and their destination handler (*events*). Once the method *get_event()* is invoked, the calling handler blocks until a new event, including a new connection, is inserted into the collection *events* for it, or the timer goes off. Once an event is inserted by the new Connection Handler, created for recovery, the blocked handler is notified to get the event from the *events*. The Handlers Synchronizer only keeps the last event for each handler, which means that if a client's Connection Handler attempts several reconnections, only the last one will succeed. In summary, the main objective of the Handlers Synchronizer in the Connection Handler design pattern is to synchronize Connection Handlers upon connection crashes and reconnection processes.

The events exchanged between the Connection Handlers through the Handlers Synchronizer in the server are of the type **Event**. The Event contains a Transport Handle and information about the data read in the remote peer.

When a connection is established for recovery purposes, a new Transport Handle is generated and a new Connection Handler is initialized. Upon initialization of the Connection Handler, a handshake request, including the identifier of the connection and the information about the data read, is received. At this point, the Connection Handler builds an Event, out of the new Transport Handle and the information about the data read, and asks the Handlers Synchronizer to give this event to the right Connection Handler, by providing its identifier. The handshake will be completed (i.e., a handshake response is sent back) by the old Connection Handler (whose connection failed) after receiving the event and replacing the failed connection.

9.5 Design of Reliable Applications Using the Connection Handler Design Pattern

In this section, we show how the design proposed in Sect. 9.4 can be reused and further specialized to support three different types of applications having reliable communication requirements. The different types of applications considered are as follows:

- Stream-based applications;
- HTTP-based applications;
- Message-based applications.

The design specializations are inspired in our own previous work [14–16], which aimed to present preliminary design solutions for each of the three cases. As with the Connection Handler design pattern, also the solutions for these three cases were changed to reach a homogeneous final design.

File and multimedia systems are examples of **stream-based applications** that transmit data without an envelope (i.e., without message boundaries), thus there is no concept of discrete messages, there is a flow of bytes instead. Our goal, in this case, is to ensure that this flow of bytes is not interrupted, even if the underlying connection crashes. In the case of **HTTP-based applications**, and although HTTP is a messaging protocol, we proposed a stream-based solution for HTTP applications, for the sake of efficiency. Thus, the HTTP-based solution relies on our reliable stream-based solution and addresses some extra challenges that are specific to the Web environment. Finally, we consider **message-based applications**. In these applications, discrete messages are placed into an envelope when sent. Thus, this type of applications includes middleware on top of TCP, with which our solution must interact.

In all of these cases, the Connection Handler design pattern is used to recover from connection crashes and retransmit the lost data. To use this pattern, developers need to implement the buffer and reliable endpoint classes of Fig. 9.3, based on the type of application being targeted. Thus, in the next sections we will describe different implementations of these two components, which will be using different names, depending on the application type being targeted: (i) buffer and reliable transporter; (ii) HTTP Buffer and Reliable HTTP Transporter; and (iii) Message Buffer and Reliable Messenger.

9.5.1 Reliable Stream-Based Applications

Here, we design a reliable distributed stream-based application using the Connection Handler design pattern in the session layer. As previously mentioned, we need to go through the implementation of the buffer and the reliable endpoint, which are explained in the following subsections.

Stream Buffer When a TCP socket fails, the connection state, including the sequence number and the number of bytes sent or received, is lost, because operating systems usually lack standard means to provide the contents or the number of bytes available in internal TCP buffers. Therefore, to obtain this information, we need to implement our own layer of buffering over TCP. To avoid explicit acknowledgments, we take advantages of TCP's reliability mechanism and resort to a sort of circular buffer (only with start index) [34], which is based on Zandy and Miller's idea [5]. We name this buffer as Stream Buffer.

To explain how this works, we depict three buffers in Fig. 9.5. These being: a sender application's buffer, a sender's TCP send buffer, and a receiver's TCP receive buffer. As shown in Fig. 9.5, we assume that the receiver got m bytes so far, whereas the sender has a total of n bytes in the buffer, and the connection fails right

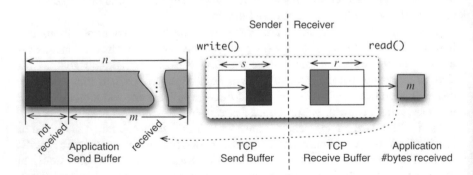

Fig. 9.5 Sender and receiver's buffers

at this point. Since the contents of both TCP buffers disappear due to crashes, the receiver needs to send the value m to the sender after reconnection (the size of the green part in the figure), in order to let the sender, determine the number of bytes that were successfully received. The sender must then resend the last $n - m$ bytes in the buffer (the blue and red parts in the figure).

If the application knew the number of bytes read by the receiver, it could shrink the size of its own buffer. This would be quite convenient, and, fortunately, TCP can help us, in the following manner. Let us begin by assuming s bytes and r bytes, respectively, as the size of the TCP send buffer of the sender and the size of the TCP receive buffer of the receiver. Let us also assume $b = s + r$. When the sender writes $w > b$ bytes to the socket, we know that the other peer received *at least w b* bytes, which means that the sender only needs to store the last $b = s + r$ sent bytes in a circular buffer, and may overwrite data that is older than b bytes. Using this mechanism, we can avoid explicit acknowledgments for the received bytes. Note that we can avoid any modulus operation, by using two's complement arithmetic over standard 32 or 64-bit counters that keep the sent and received bytes on each side, for buffer sizes strictly smaller than 2^{32} and 2^{64}, respectively. Note that apart from these limits, the buffers can have arbitrary sizes (equal to or greater than b), according to the sender plus receiver TCP buffer sizes.

To implement this idea in practice, peers have to exchange the size of their receive buffer, through a handshake procedure, right after establishing the connection and before exchanging any data. It lets the peers initialize their Stream Buffer in compliance with the minimum size limit (b). During the communication, data can be stored into the Stream Buffer only after being successfully written to the socket. Since there is no explicit acknowledgment, the old data is not removed from the buffer (i.e., move the end index of the circular buffer forward) but is overwritten with the data recently sent, therefore, there is no need to keep the end index of the Stream Buffer. **Reliable Transporter** Since we aim to implement our solution in the session layer, the implementation of reliable endpoint, the central component of the Connection Handler design pattern, must not only support the aforementioned buffering mechanism and data tracking, but also provide an interface for the application layer, with crash-oblivious read and write operations. Given the basic model of a connection-oriented application, presented earlier in Fig. 9.1, the design of a reliable stream-based application using the Connection Handler design pattern can be made as depicted in Fig. 9.6. As shown in this design, only the session layer is aware of operations that are necessary to recover from connection crashes.

The application layer includes a Service Handler and interacts with its connected peer through the session layer, which includes a Reliable Transporter, the extended implementation of the reliable endpoint. In this session-based design, we also need a Passive Reliable Transporter, which is used to transparently initialize a Reliable Transporter on the server side, upon establishment of a new connection. Each Reliable Transporter owns one Stream Buffer and extends the functionalities of the Connection Handler, to enable recovery from connection crashes. It implements the actions necessary to establish a connection for the first time and also after a crash, including the handshake, reconnection, and retransmission of the lost bytes.

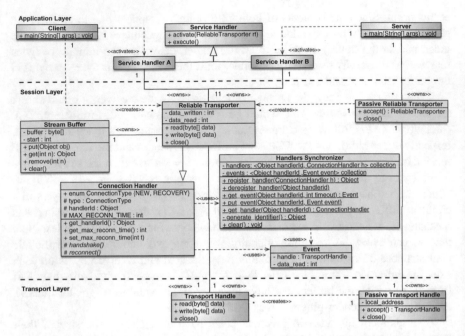

Fig. 9.6 Reliable transporter: design of a reliable stream-based connection-oriented application

To transparently accomplish the recovery process, without intervention from the application layer, the Reliable Transporter is inserted between the transport layer (Transport Handle) and the application layer (Service Handler). Thus, besides owning a Transport Handle, to exchange application data, the Reliable Transporter provides read() and write() operations that perform the following actions: (1) store the data sent by the application into the Stream Buffer; (2) count the number of bytes written and read; (3) intercept the read and write operations for detecting a connection crash; and finally (4) reconnect and retransmit the lost data, when a connection crash is detected. In addition to these actions, the handshake process is also performed transparently from the application layer, once a Reliable Transporter is initialized.

In this stream-based solution, the handshake is used, not only to exchange the identifier of the connection, but also the size of the TCP receive buffer, which is necessary for calculating the size of the Stream Buffer. The handshake messages follow a predefined configurable format, which is shown in Fig. 9.7. It includes a header line that can be configured differently on the client and server sides, depending on the application layer protocol. Then, we can have several lines that carry the necessary information for the handshake. Each line is separated from the other lines using a separator (e.g., \r\n).

Considering the client handshake message when a new connection is established, the field FT Identifier carries the identifier of the connection, which is used on the server side to identify whether the connection is new or recovered. For

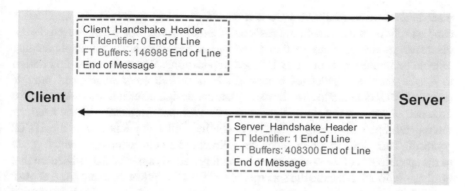

Fig. 9.7 Handshake message format

setting up a new connection, the client sets the identifier to 0 (or another symbol, which is defined by developer for this case), and the server generates a new immutable identifier for the connection in the response (1 in Fig. 9.7). The field FT Buffers carries the size of the TCP receive buffer, which is used in the server, to calculate the size of its Stream Buffer. The server replies with a similar message.

Once a reconnection occurs, the client sends its connection identifier with the number of bytes it received up to the connection crash in the field FT Recovery. The server identifies that it is a recovery connection, thus, it replaces the failed connection with the new one and sends a similar message back to the client. Finally, both client and server send the buffered data that the other peer did not receive due to the connection crashes.

9.5.2   Reliable HTTP-Based Applications

To the best of our knowledge, all current solutions for building reliable HTTP-based applications (e.g., HTTPR [22], WS-Reliability [23]) are message-oriented, requiring, for instance, buffering (or logging) and retransmission of messages to ensure reliable delivery. These solutions involve resending whole messages, which is quite inefficient when messages are large. Also, message-based solutions cannot easily offer reliability to long-standing connections. As an example, in AJAX [35] environments, the server often needs to keep the connection open for a long time, to push updates to the browser. If this connection fails, orchestrating a workable solution can be very difficult, as the HTTP client must be able to repeat requests to obtain the missing parts of responses, whereas the HTTP server must be able to identify and handle repeated requests.

Our perspective is that a stream-based solution, that buffers and resends only unconfirmed data, is a much cleaner one, as it can be implemented without requiring changes to the application's semantics. From the application perspective, there is no need to explicitly store and resend complete messages; the application can just rely on the channel (and associated middleware). The stream-based design discussed in the previous section provides us the foundation for implementing reliable communication in all TCP-based applications. We now extend that design to handle the new challenges brought in by the Web environment and provide reliable HTTP communication. In particular, the design takes into consideration the presence of proxies (frequent elements in the Web environment) and the need for interoperability with legacy software, including full compliance with the HTTP protocol. Thus, this design includes the following key characteristics, which relate to the specificities of the environment being targeted: (i) a handshake procedure that has been tailored to handle the specificities of HTTP applications; and (ii) a control channel per client (shared by all the connections to the server) which is used in communication scenarios that involve proxies. Moreover, all data exchanged comply with the HTTP protocol and we ensure that reliable and non-reliable HTTP peers can still interoperate.

HTTP Buffer

Despite being simple and efficient, the buffering scheme presented for stream-based applications cannot withstand proxies. These intermediate nodes can actually keep an arbitrary amount of data outside their own buffers, thus causing the data in transit to exceed the $b = s + r$ bytes available in the Stream Buffer. This means that data can be lost if the connections that have the proxy as endpoint crash (or if the proxy itself crashes).

Figure 9.8 shows a simple sender–receiver scenario, which involves a proxy, and depicts the internal data buffers involved. As we can see, there is extra buffering of data at the proxy. While our main idea stands on having a Stream Buffer as large as the TCP send and receive buffers combined, now we have a total of five points that can serve as buffers: the sender TCP send buffer, the proxy TCP receive buffer,

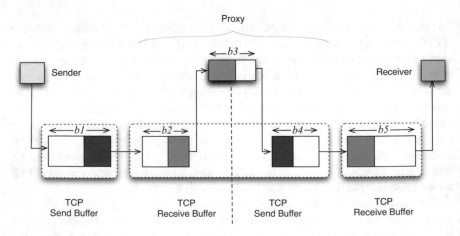

Fig. 9.8 Buffers in a client-server model with proxies

the proxy internal state, the proxy TCP send buffer, and the receiver TCP receive buffer. The size of the buffers is now $b1 + b2 + b3 + b4 + b5$, much more than the $b1 + b5$ that the Stream Buffer was prepared to take. Unfortunately, we cannot easily determine $b2$, $b3$, and $b4$, and thus cannot know how much data should be kept in the Stream Buffer.

To solve the extra-buffering problem, we use a combination of explicit and implicit acknowledgments. When no proxy exists, client and server can rely on the previously explained implicit acknowledgment mechanism. In contrast, when a proxy exists, the buffering and acknowledgment scheme must become explicit, because the sender must never allow the amount of data in transit to exceed the size of its buffer. Thus, to store HTTP messages, we use HTTP Buffer that implements the interface of the buffer and uses a standard circular buffer (with start and end indexes) [34]. In fact, to enable explicit acknowledgments, we need to mark the end of buffer, to identify whether it is full or if it has space for new data. In scenarios with a proxy, whenever an HTTP Buffer is becoming full, the peer should acknowledge the reception of data, to allow the sender to release some space in its buffer and thus keep sending data without interruption. To enable early acknowledgments, once a peer receives a number of bytes equal or greater than half the size of the peer's HTTP Buffer, an acknowledgment should be sent. This allows the sender to clean its buffer, thus allowing it to proceed. To exchange acknowledgment messages when there is a proxy, we use a control channel.

Figure 9.9 presents the design details of the HTTP Buffer. Each HTTP Buffer owns an array of bytes (buffer), pointers to the start and end of the buffer, and a boolean attribute, named *write_constraints*, which indicates if the buffer needs to keep the pointer to the end of the buffer, when proxies are present. The *put()* and *get ()* methods are used to save and retrieve data, respectively. Methods *has_space()* and *release_space()* are used in scenarios with proxies, respectively, to check

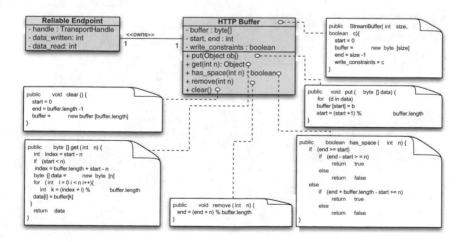

Fig. 9.9 HTTP Buffer

whether the buffer has enough space for new data and to delete acknowledged data from the buffer.

Reliable HTTP Transporter

The Reliable Transporter must also be refined, so that explicit acknowledgments are supported, in our case, through a control connection. We call this "Reliable HTTP Transporter." In the resulting design pattern of Fig. 9.10, each Reliable HTTP Transporter owns one HTTP Buffer and extends the functionalities of the Connection Handler, to enable recovery from connection crashes. Moreover, each Reliable HTTP Transporter owns one control connection, when the communication involves proxies. The control connection is shared by all connections created from the same client. In the scenarios with proxies, the Reliable HTTP Transporter also needs to keep the size of the remote HTTP Buffer (*remoteBufferSize*) and the number of bytes read so far, after the last acknowledgment (*numOfBytesReadAfterLastAck*). The information needed for calculating the local and remote buffer sizes are exchanged through the handshake process.

The handshake process should also address two other important issues. First, we must not expect endpoints to adhere to a specific reliable communication mechanism—any solution for reliable communication should ensure interoperation with legacy software, given the number of legacy endpoints that exist. Second, for

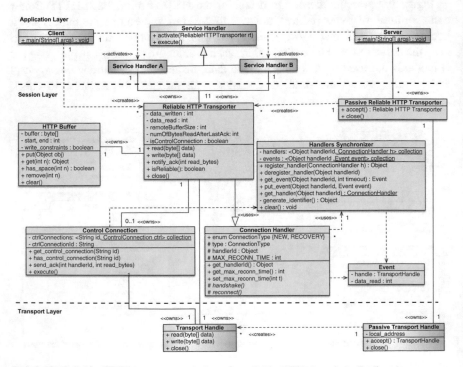

Fig. 9.10 Reliable HTTP transporter: design of a reliable HTTP-based application

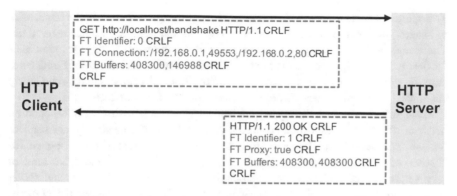

Fig. 9.11 Handshake messages adapted for HTTP protocol

security reasons, proxies may filter non-HTTP messages, something that would make our critical handshake step fail. Thus, the handshake aims to identify: (1) if the peer is legacy software that does not support our reliability mechanism; (2) if the connection is brand new, or created for recovery purposes; (3) if there is any proxy in the middle of the connection between the endpoints; (4) the size of the local buffer, when both peers implement the reliability solution and the connection is new; and (5) the size of the remote buffer, when there is a proxy.

Figure 9.11 illustrates a handshake message configured for the HTTP protocol. Peers start by sending the identifier of the connection. The *FT Connection* header carries the network address of the client and server. This enables the server to check if the peer connection comes from the client or from a proxy. The *FT Buffers* carries the size of the TCP send buffer and TCP receive buffer, which is used to calculate, if necessary, the size of peer's HTTP Buffer. The *FT Proxy* header is used by the server to inform the client whether a proxy was detected or not. The server detects the presence of a proxy if the source address sent in the *FT Connection* header is different from the destination address of the TCP connection it owns. In this case, the client creates a new control connection to the server, for exchanging acknowledgment messages. As mentioned, each client uses a single control connection to the same server. A control connection is identified by the server's address in the client and by the client's address in the server (*CtrlConnectionId*). A handshake message, including the *FT Control* header, is sent by the client, allowing the server to distinguish a data connection from a control connection (*isControlConnection* is set when a connection is created for exchange of acknowledgments).

The control connection keeps the references to the existing control connections in *ctrlConnections* and provides an interface to let the Reliable HTTP Transporter check the existence of a control connection to a specific peer (*has_control_connection()*) and access it (*get_control_connection*). It also provides an interface for sending acknowledgment messages (*send_ack()*). A control connection also checks for the arrival of acknowledgment messages and delivers them to the appropriate handler through the method *notify_ack()*, provided by the Reliable HTTP

Transporter. The acknowledgment messages have the same format as the handshake messages, but include an *FT ACK* header, carrying the number of bytes read so far. The Reliable HTTP Transporters need to count the number of the bytes read after the last acknowledgment message sent (*numOfBytesReadAfterLastAck*) and compare it with the size of the remote buffer (*remoteBufferSize*), to send early acknowledgment messages before the remote buffer becomes full.

It is worth noting that the above handshaking mechanism ensures interoperation between reliable and non-reliable nodes. A legacy client will simply not send or receive any handshake message from the server. In contrast, the response (or absence of it) from a legacy server will always tell the client about the kind of server it is talking to. Hence, all combinations of legacy/reliable client and server work. Moreover, the application layer can explicitly check whether the communication is reliable or not, by calling the method *isReliable()*.

Interaction Between Components in Reliable HTTP-Based Applications

To better understand the details of the handshake and reconnection procedures, we now review the interaction of the components in reliable HTTP-based applications. Figure 9.12 presents a failure-free scenario.

To initialize connections, the server creates one (or more, depending on the number of ports defined and assigned to the application server) Passive Reliable HTTP Transporter and binds it to the local network address (IP address and port number). Then the server waits for a new connection, by invoking the method *accept()* of this passive handle. The client initializes a Reliable HTTP Transporter, by giving the network address of the server, to establish a new connection. This will internally create a Transport Handle.

Upon reception and acceptance of a connection request in the server, a Reliable HTTP Transporter is generated. Then, the client starts the handshake procedure, to complete the initialization of the Connection Handler. The handshake request includes the identifier of the connection (zero in this scenario), the local and remote address of the connection, and the size of the client's TCP send and receive buffers. The server's Reliable HTTP Transporter identifies that the connection is new (because the identifier is zero), and registers itself into the Handlers Synchronizer, through the method *register_handler()*, which returns a unique identifier. A handshake reply is sent back to the client, including the unique identifier of the handler, the size of the buffers on the server side, and information about the existence of a proxy. At this point, both, client and server, can initialize their HTTP Buffer with the appropriate configuration, depending on the information exchanged between them.

When no proxy exists, peers initialize and activate Service Handlers, by passing the previously created Reliable HTTP Transporter (*rt* in the client and rt_1 in the server). This means that the client and server's Service Handler can start writing and reading data. After a successful write operation, the Reliable HTTP Transporters put the data into the HTTP Buffer and update the value of *written_-data*. After a successful read operation, they update the value of data_read (please refer to part (a) of Fig. 9.12).

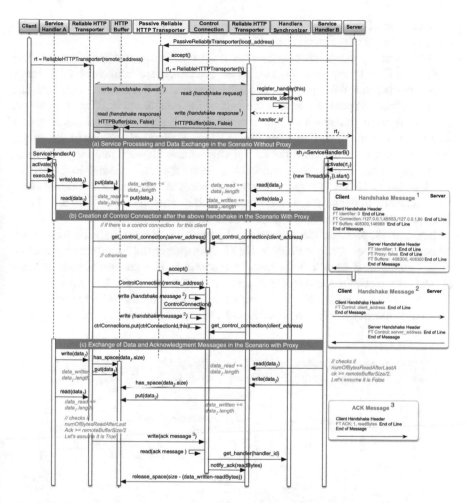

Fig. 9.12 Component interactions on in a failure-free scenario

In contrast, when there is some proxy, the Reliable HTTP Transporters require a control connection to exchange acknowledgment messages in both sides (please refer to part (b) of Fig. 9.12). Since the control connection is shared between several connections created by the same client, peers check the existence of a control connection, by specifying an identifier that is equal to their peer's address. If a connection already exists, they simply get it from the list and use it, otherwise the client must create a new one. When a control connection is successfully created, the client sends a handshake request including the *FT Control* header with the local address of the client, which will be used by the server as the identifier of the control connection. The server sends a handshake reply back to the client including the *FT Control* header, with the IP address used by the server, which will be used as the identifier of the control connection on the client side. Both client and server store the reference of the control connection in a list (*ctrlConnections*), to be used with other Reliable HTTP Transporters, if necessary.

When a proxy exists, client and server must change the way they interact. The Reliable HTTP Transporter checks if there is enough space in the HTTP Buffer before writing the data and checks if the number of bytes read, after the last acknowledgment message, exceeds the half of the remote buffer. If so, it sends an acknowledgment through the control connection. Figure 9.12, part (c), shows a scenario where an acknowledgment is sent from the client. As shown in the figure, this message carries the identifier of the Connection Handler and the number of bytes read so far. The control connection delivers the read message to the appropriate Reliable HTTP Transporter, which is accessed by means of the Handlers Synchronizer, through the method *notify_ack()*. This lets the Reliable HTTP Transporter release some space from the HTTP Buffer.

Figure 9.13 presents the component interactions present in a scenario with connection crashes. Once a Reliable HTTP Transporter fails completing a read or write operation, it transparently tries to reconnect. The reconnection is accomplished differently in the client and server. As shown in the figure, neither the client's Service Handler, nor the server's are involved on the recovery procedure. When a connection crashes, both sides will eventually start the reconnection phase,

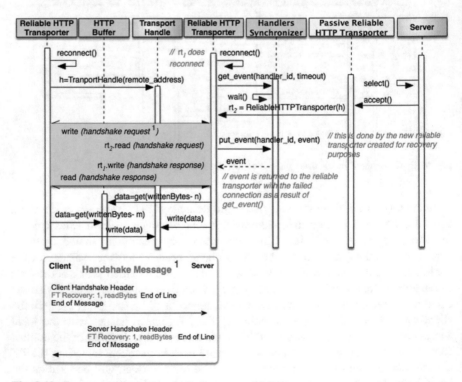

Fig. 9.13 Component interactions in the presence of failures

by calling the method *reconnect()*. Upon invoking this method, the client's Connection Handler tries to create a new connection to the server during a pre-defined period of time. On the other side, the server's Connection Handler waits for a new connection, by giving the connection identifier and a waiting time to the Handlers Synchronizer, through the *get_event()* method. After the new connection is established, the Connection Handlers start the handshake process. The handshake message sent by the client includes an FT Recovery header, which carries the identifier of the handler whose connection crashed, and the number of bytes received up to the crash. This lets the server distinguish fresh connections from reconnections. The Connection Handler created on the server is responsible for notifying the waiting handler and delivering an event, including the new transport handle and the number of bytes read so far, through the method *put_event()* of the Handlers Synchronizer. Then, the server's Connection Handler completes the handshake procedure, by sending a handshake message back to the client, including the same header with corresponding information. At this point, both sides can start retransmitting data that had been lost due to connection crashes.

9.5.3   Reliable Message-Based Applications

In order to present the design of a reliable message-based application using the Connection Handler design pattern, we start with the basic case of a synchronous message-based communication, and then add reliability to this design.

General Design of a Connection-Oriented Message-Based Application

Figure 9.14 presents the general design of a message-based application. Since TCP is stream-oriented, for a TCP-based message-oriented application, we need an additional layer, comparing to Fig. 9.1, for message formatting and encapsulation. This layer includes the message, which is a serializable data structure encapsulating application data and any associated metadata. A message can be interpreted as data, as the description of a command to be invoked or as the description of an event that occurred (e.g., a mouse click). Each message includes two parts, a header to carry meta-data and a body to carry data. The header of a message contains metadata about the message (e.g., identifier, size) and any information required for communication, many times depending on the protocol used between the application peers. This information is stored into a structure comprised of various fields and their corresponding values. While the header can be used by the application and session layer, the body contains the application's data and is ignored by the session layer.

The Messenger is dedicated to take the necessary actions for sending and receiving the application's messages through the Transport Handle. Thus, the Messenger is responsible for sending a message as an array of bytes through the stream-based Transport Handle, and also for delivering an array of bytes, read from the Transport Handle, to the Service Handlers as a message.

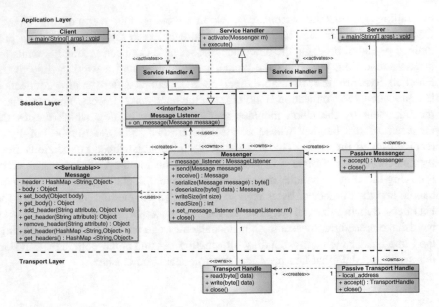

Fig. 9.14 Messenger: design of a basic message-based application

When a message is given to the Messenger through the method *send()*, this component converts (or serializes) the message to an array of bytes, writes its size to the stream, and then sends the serialized bytes. There are other mechanisms to determine the end of each message (e.g., defining a unique marker in the beginning or end of each message), but we use the size of message for simplicity. On the receiving end, when the method *receive()* is invoked by the application, the receiver reads the size of the incoming message, receives, and deserializes it from an array of bytes to a message, before delivering it to the Service Handler.

Similarly to current technologies, such as Java Message Service, messages can also be delivered to the application using a callback method passed to *set_message_listener()*. In this case, the Service Handler must implement the method *on_message()* of the Message Listener. When this happens, the Messenger internally dedicates a new thread for reading the messages and delivering them to the Service Handler through the method *on_message()*.

Design of a Reliable Message-Based Application

In this section, we advance the Messenger's design to tolerate connection crashes. For this, we resort to the Connection Handler design pattern. The resulting design is presented in Fig. 9.15. As shown, the Reliable Messenger extends the functionalities of the Messenger and the Connection Handler. In the following paragraphs, we explain how the Connection Handler design pattern is incorporated and integrated with the Messenger to ensure recovery from connection crashes.

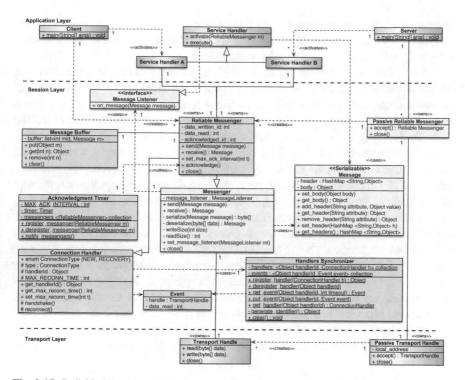

Fig. 9.15 Reliable Messenger: design of a reliable message-based application

To be able to recover from connection crashes, we need a reliable endpoint that inherits the properties of the Connection Handler and implements its handshake and reconnection processes. This reliable endpoint, which is called Reliable Messenger, also extends the functionalities of the Messenger to enable exchange of messages.

We also need a simple mechanism of buffering and retransmission of messages, to keep both peers in a consistent state after recovery. Thus, each Reliable Messenger owns one Message Buffer that implements the interface of the buffer component of the Connection Handler design pattern. Furthermore, the Reliable Messenger must modify the *send()* and *receive()* operations of the Messenger, to implement the actions that are necessary for buffering the messages (before sending them), removing the acknowledged messages from the buffer (after receiving an acknowledgment) and intercepting a connection crash (while writing or reading into/from the channel).

Upon creation of a connection and after initialization of both client and server, the Service Handlers start exchanging messages. They send their messages by invoking the method *send()* of the Reliable Messenger, which in turn assigns a unique identifier to each message. The message identifiers are sequential integers starting from one (the value of the last identifier is kept in *data_written*). The Reliable Messenger, in addition to the unique identifier, piggybacks the acknowledgment and updates the value of *acknowledged_id*.

To receive messages, the method *receive()* of the Reliable Messenger is invoked. This method, after reading the message, updates the identifier of the last message received (*data_read*), removes (sent and) acknowledged messages, if any information is piggybacked, and delivers the message to the Service Handler. The Reliable Messenger also sends acknowledgments periodically, if some messages remain unacknowledged. The Reliable Messenger uses a central timer, named acknowledgment timer, to efficiently perform periodical asynchronous acknowledgment. The acknowledgment intervals can be defined by the application through the method *set_max_ack_interval()*, possibly depending on its messaging rate (i.e., the number of messages exchanged per unit of time). The acknowledgment timer, which implements the observer design pattern [36], is used to periodically trigger all Reliable Messengers (belonging to different concurrent connections that may exist, especially on the server side) for sending an acknowledgment if there are any unacknowledged messages. We dedicated just one central timer for all Messengers to reduce the memory utilization especially in the server. We must emphasize that the acknowledgments can also be piggybacked in the header of the application's messages to reduce overhead on the network caused by extra messages. We omit the description of connection crash recoveries, as it is very similar to the stream and HTTP cases.

9.6 Experimental Evaluation

In this section, we describe the experimental evaluation, composed of a set of experiments, defined to illustrate several key aspects regarding the deployment and general characteristics of our Connection Handler design solution. We apply and evaluate the solution under the form of the three specializations described in this chapter, i.e., the Reliable Transporter, the Reliable HTTP Transporter, and the Reliable Messenger. We first show the **applicability** of the proposed solution, with the experiments focusing on four key aspects: correctness, performance, resource usage, and complexity. To evaluate **correctness**, we check whether our solution tolerate connection crashes; to evaluate **performance**, we measure latency (round-trip-time of a request-response interaction) and throughput (number of operations per time unit); to evaluate **resource usage**, we measure the resource utilization, in terms of memory and CPU; and finally, to evaluate **complexity**, we use three metrics: lines of code (LOC), cyclomatic complexity, and nested block depth [37].

9.6.1 Experimental Setup

We implemented the Connection Handler design pattern in Java, in the following three middleware solutions that match the three major types of applications previously described:

- **FSocket (plain version)**: (Fault-Tolerant Socket) implements the *Reliable Transporter* and offers support to build reliable stream-based applications;
- **FSocket (HTTP version)**: implements the Reliable HTTP Transporter and offers support to build reliable stream-based HTTP applications;
- **FTSL**: (Fault-Tolerant Session Layer) implements the Reliable Messenger design pattern and supports the creation of reliable message-based applications.

We used this middleware in the corresponding three types of applications: (i) a stream-based application; (ii) an HTTP-based application; and (iii) a message-based application. We kept two versions of each of these three applications: a reliable one, using our (appropriate) reliability solution, and an unreliable one, without any reliability mechanism. Thus, this adds up to a total of six application types involved in the experiments (three types two versions).

The applications involved in the experiments support three main operations, which we can consider to be a service that the client can trigger by sending the appropriate request. The operations are named Invoke1, Invoke2, and Invoke3 and all receive a 10-byte string and return another 10-byte string. The major difference between the operations is that Invoke1 replies immediately, Invoke2 sleeps 1 ms (ms) before replying, whereas Invoke3 sleeps 2 ms. The reason why we put the server threads to sleep is to minimize interference with our results. One should be aware that putting a thread to sleep and waking it up takes around 0.08 ms on the machine where we ran the server (and 0.15 on the client machine). To determine this number, we ran a single-threaded program that slept for 1 ms 1000 times.

Regarding the infrastructure for the experiments, we used two computers sharing the same Local Area Network (LAN) in order to run two endpoints of the applications in the role of client and server. Table 9.1 presents the characteristics of the computers used in the experiments.

In each experiment, we increase the number of clients (usually from 1 to 1000), to evaluate the effect of concurrent connections on the performance and resource utilization. For this, we ran all the clients on a single process, using different threads. The results obtained and discussed in the following sections are the results

Table 9.1 Systems used in the experiments

Endpoint	OS	CPU	Memory
Client	Mac OS X version 10.10.5	2.4 GHz Intel Core 2 Duo	4 GiB RAM, 3 MiB cache
Server	Linux version 2.6.34.8	2.8 GHz Intel(R) 4 Cores (TM) i7	12 GiB RAM, 8 MiB cache

of 100 executions, except when otherwise noted (e.g., the performance experiments involve 1000 executions). To minimize environmental effects on the experiments and possible warm-up periods, before the 100 executions, we ran each test 30 times and discarded these warm-up results.

9.6.2   Applicability Evaluation

With the goal of showing the applicability of our solution, we performed the following three deployments. We deployed our FSocket plain in an open source FTP server named ANOMIC [38]). We named the reliable version of this application ftANOMIC and made its source code available online [39]. We deployed our FSocket HTTP in the Apache Tomcat 7.0.13 HTTP connector [40], included in JBoss AS 7.1.1 [41]. HTTP server. We also integrated FTSL in a custom messaging application.

Table 9.2 presents the key functions that developers need to use when creating distributed applications. For each function, we show the standard call in Java and also the call when using the FSocket API.

As we can see in the table, the modifications necessary to deploy our solution are trivial. Essentially, we need to replace every Socket object by an FSocket object. As explained before, each server in a connection-based communication owns one passive handle to accept new connections. In Java TCP, this passive handle is called ServerSocket. We have an equivalent passive handle in our implementation, named ServerFSocket, and a developer needs to replace the ServerSocket with this object. Upon accepting a new connection, the ServerFSocket returns an FSocket instead of a Socket. Moreover, all the read and write operations done on the TCP socket's InputStream and OutputStream must be replaced with the read and write operations on the FSocket objects.

Table 9.2 Comparison between the Java Sockets API and the FSocket API

Function	Code example
Connection creation	Socket socket = new Socket (server,port)
	FSocket fsocket = new **FSocket** (server,port)
ServerSocket creation	ServerSocket serverSocket = new ServerSocket(port)
	ServerFSocket serverFSocket = new **ServerFSocket**(port)
Connection acceptance	Socket socket = serverSocket.accept()
	FSocket fsocket = **ServerFSocket**.accept()
Read	int read = inputStream.read(data)
	int read = **fsocket**.read(data)
Write	outputStream.write(data)
	fsocket.write(data)

In addition to the above changes, we needed to do one more modification to the FTP server, as the server may listen on more than one port, to accept control and data connections. To make this need clear, we briefly explain the active and passive modes of FTP servers. In the active mode, the client connects from a random port N to the FTP server port (usually 21). Then, the client starts listening on port $N + 1$ and sends a control message to the server, with the number $N + 1$. The server will then connect back to the client's specified data port. In contrast, in the passive mode, the client initiates both connections to the server. After opening an FTP connection, the client sends the *PASV* command. The server then opens a random port (above 1023) and sends the number back to the client. The client responds by initiating a new data connection to the server on that port [42].

The modification required applies to the FTP server for the passive mode. In this mode, although the server continuously checks on the command port (e.g., 21) for new control connections, it checks the port dedicated to the data connection only once. This would cause a problem should the connection crash, because the client attempts to reconnect would fail. To solve this problem, we force the server to listen on the data port, until the data connection is closed.

9.6.3 Evaluation of Correctness

A first aspect to verify is the correctness of the implementation of the Connection Handler design pattern. This essentially means that the applications should be able to reconnect when in presence of a connection crash and that they should be able to communication without losing messages. For this purpose, we let the client and the server (in all three kinds of applications) continuously exchange data for 5 min (as mentioned, each test was repeated 100 times). We then used tcpkill to cause connection crashes at random instants during each test (three crashes per test). We then verified if all data arrived correctly at the destination, which was the case for all stream-, HTTP- and message-based applications. The results showed that our middleware was able to reconnect and revealed no failure in the delivery of data. For the 100 repetitions of the test, we also observed that, while the first connection establishment to the server took 15 ms in average, reconnection plus sending lost data (bytes or messages) took an average of 26 ms.

The HTTP version has, as discussed in Sect. 9.5, a few other correctness aspects that should be evaluated (e.g., dealing with legacy software and proxies). To evaluate its correctness in the presence of legacy software and proxies, we considered different HTTP client–server communication scenarios. In each scenario, we refer to reliable and non-reliable peers (i.e., client or server), respectively, as using or not using our reliable communication solution. The scenarios are as follows: (1) a reliable HTTP client communicating with a non-reliable (legacy) JBoss AS; (2) a non-reliable HTTP client communicating with a reliable JBoss AS; (3) a reliable HTTP client communicating with a reliable JBoss AS, without any proxy in the middle; (4) a reliable HTTP client communicating with a reliable JBoss AS via a proxy. Scenarios (1) and (2) are used to show that our solution is compatible with legacy and unreliable software, and scenarios (3) and (4) are used to show that our design pattern is able to tolerate connection crashes with and without proxies.

We first used a browser to generate HTTP requests for a set of typical Web resources deployed in the non-reliable JBoss AS. We used these requests within our custom HTTP client and also used the responses as an oracle for comparison with the responses obtained from the reliable JBoss AS during the tests. For each of the four scenarios, we let client and server exchange messages during 5 min (each test was repeated 10 times). We observed that reliable and non-reliable peers were able to communicate perfectly in scenarios (1) and (2). To evaluate the ability to recover from crashes (scenarios 3 and 4) without and with proxy, we emulated connection crashes and observed that all interactions worked correctly as all expected data was correctly received, even in the presence of crashes.

9.6.4 *Evaluation of Performance*

With the goal of evaluating performance, we selected two very typical attributes: latency and throughput, which are used as performance indicators in many other contexts [43]. In these experiments, the computed performance results are the average of 1000 trials. Latency refers to the round-trip-time of the request-response interaction. To examine the latency of our reliability solution, we send a request from a reliable client to a reliable server and calculate the time taken from sending the request to receiving the reply from the server. We also measure the latency for the unreliable version of each application tested, to demonstrate the performance degradation of the reliable version. The latency degradation is calculated by the following equation:

$$(\text{Latency}_{\text{reliable}} - \text{Latency}_{\text{unreliable}})/\text{Latency}_{\text{reliable}}$$

The throughput is defined as the number of requests processed per unit of time. To examine throughput, we send a large number of requests (1000 from each client), to the server without waiting for any response (a different thread is responsible for waiting for the responses), and calculate the time taken from receiving the first request to sending the last reply. As with latency, we calculate the throughput degradation as follows:

$$(\text{Throughput}_{\text{unreliable}} - \text{Throughput}_{\text{reliable}})/\text{Throughput}_{\text{unreliable}}$$

Figure 9.16 shows the latency (a) and throughput (b) for different numbers of clients (from 1 to 1000) for both unreliable and reliable **stream-based applications**. For both latency and throughput, results show that the reliable application is almost on par with the unreliable application. As shown in the plots, the maximum degradation observed for latency is less than 1 percent (0.52%). Throughput, for the

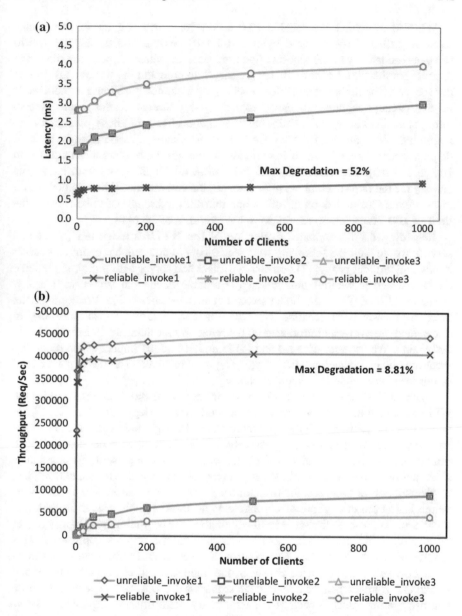

Fig. 9.16 Latency, throughput and performance degradation with reliable stream-based solution (plain FSocket)

slower invocations Invoke2 and Invoke3, is pretty much the same in both applications. Invoke1 shows a higher degradation, but still below 10%. This latter invocation is the worst case for measuring the performance degradation, because Invoke1 does not take any time at the server, thus exposing all the communication overhead.

We also evaluated our reliable FTP server for a growing number of clients requesting files of two sizes: 6 bytes and 1 GiB. We use the former file size to compute the latency of the requests (the time since the client requests the file to the time it gets the file), while the latter file serves to compute the throughput (in bits per second). The higher complexity of setting up a connection should be noticeable in the latency, whereas memory copies to the Stream Buffer could impact throughput. However, our results show that the effects of these operations are negligible. We downloaded files from 1 to 50 clients, observing only a small degradation of latency, which is common to the non-fault-tolerant version (from 100 to 111 ms), whereas throughput held on at 89 Mbps, again, with no visible penalty for the fault-tolerant version. A small, but relevant detail here is that we do not write the files to disk on the client and this allows throughput to be close to the limit of 100 Mbps. These results were the averages of 10 trials.

Regarding the performance evaluation of the **HTTP application** (client and server), we defined the following four scenarios: (1) non-reliable client and server interacting without proxy; (2) non-reliable client and server with proxy; (3) reliable client and server without proxy; and (4) reliable client and server with proxy. Scenarios (1) and (2) (non-reliable scenarios) are used as baseline. We compare the behavior measured in scenario (1) with the one observed in scenario (3), to understand the overhead introduced by the reliability mechanisms in a direct client–server link. We use scenarios (2) and (4) to understand the impact of the reliability mechanisms, in a situation where there is a proxy involved. The proxy server used in our tests was Squid 3.1 (squid-cache.org).

Figure 9.17 shows the results obtained for latency (a) and throughput (b) for the HTTP application. The plots also show the performance degradation of the reliable version, in comparison with the unreliable one on the right-side vertical axis. As we can see, latency increases progressively in all cases; the same happens with throughput. As expected, latency is higher when a proxy is present. Throughput in all scenarios increases rapidly in the beginning and then stays at the same level, i.e., no degradation is observed for the number of clients we tested. The main observation is that the throughput of unreliable applications in both scenarios, with and without proxy, reaches the same level, although in the beginning it is slightly higher when there is no proxy. This does not happen for the reliable application. This difference is caused by the extra control connection and extra actions taken in FSocket, when a proxy exists. However, the important aspect for both latency and throughput is that, when we compare the scenarios that use reliable peers with those that use the non-reliable peers, even with proxy, performance degradation shows low values (about 3 percent). In fact, although we have all necessary mechanisms for reliable communication in place and in operation, performance degradation is quite small.

Finally, we evaluated the performance of the message-based applications. Figure 9.18a shows the observed latency of the unreliable and reliable versions of the message-based application (i.e., the versions, respectively, using Messenger and Reliable Messenger). Figure 9.18b shows the results obtained for throughput of these applications. The results obtained for the application using Messenger are

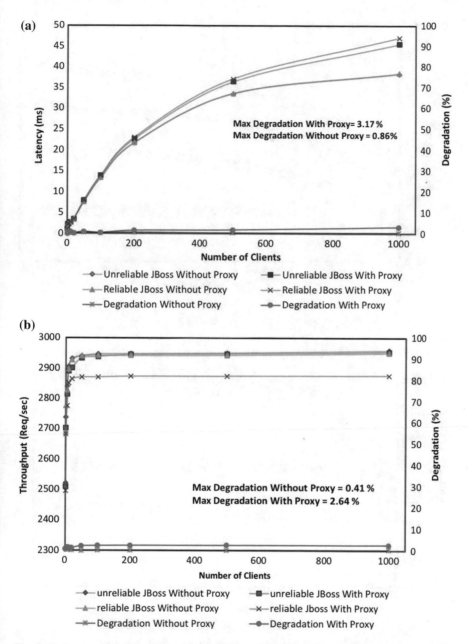

Fig. 9.17 Latency and throughput with unreliable and reliable (HTTP-supported FSocket) HTTP servers

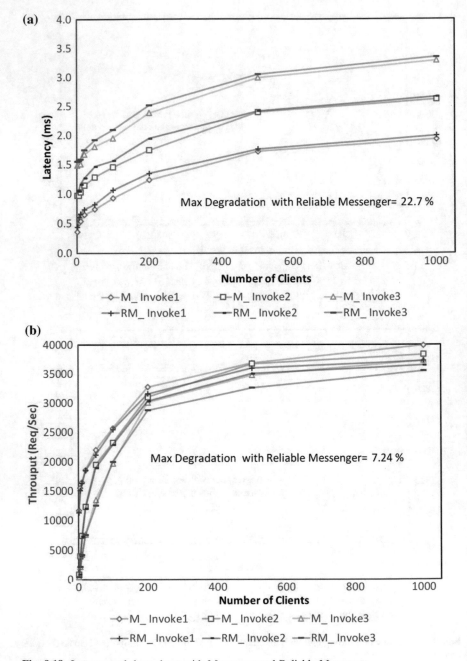

Fig. 9.18 Latency and throughput with Messenger and Reliable Messenger

used as baseline to measure the overhead and performance degradation of the Reliable Messenger.

In all scenarios and with all invocations, latency increases smoothly with the number of clients. We see three different levels of latency for different invocations, due to the different processing times of the invocations. We calculated the latency degradation of the Reliable Messenger in comparison with the Messenger. The overhead of Reliable Messenger is mainly associated to the extra operations it has for buffering the messages, piggybacking the acknowledgment information, extracting the acknowledgment information, periodically acknowledging, and removing the acknowledged messages from the buffer. It is worth mentioning that the worst case for performance degradation (22.7%) occurs for Invoke1 (no delays in the server), when the number of the clients is very low. In fact, for all invocations, the degradation decreases as the number of clients increases. For 1000 clients, for example, the maximum degradation in all scenarios is 3.09%. This observation shows that in highly concurrent applications, the latency imposed by Reliable Messenger will be very low.

Unlike latency that increases smoothly, the throughput increases rapidly in the beginning, by increasing the number of clients, and then pretty much levels out. Although our resources did not allow us to increase the number of clients to more than 1000, the figures show that the slope of the plots starts to decrease rapidly, when the number of clients increases to more than 200. Throughput for Invoke1 starts at a much higher value of more than 11000 requests per second, Invoke2 at about 900 requests per second, and Invoke3 at about 450 requests per second). The plots show that this difference remains until the end. An interesting observation is that the performance degradation is very low in all scenarios, with a maximum observed of 7.24%.

9.6.5   Resource Usage

Regarding resource usage, we again ran experiments where the number of clients increase and used each client to send 100 requests per second during 5 min, which we experimentally observed to be enough to show the usage of resources. We used the *ps* command to periodically read memory and CPU usage at the server. Since the function of HTTP-based FSocket in scenarios without proxy is pretty much similar to the plain FSocket, we simply used our reliable (with HTTP-based FSocket) and unreliable HTTP server in two scenarios, without and with proxies, to measure resource usage for plain stream-based and HTTP-based applications. Figure 9.19 shows that the overhead in terms of (a) memory and (b) CPU is kept under acceptable limits. The memory used by our reliable server is, as expected, higher than the non-reliable one, with a maximum overhead of 60%, due to the extra buffering placed on top of TCP. The CPU overhead is again quite low (maximum of 15%), which is an excellent indication, as this resource can be many times of critical importance. Moreover, we can see that both CPU and Memory

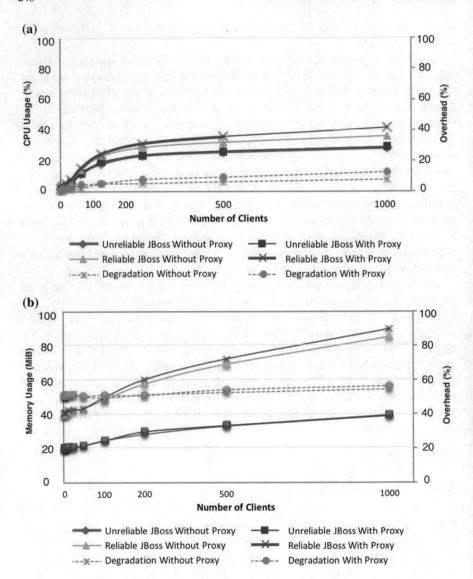

Fig. 9.19 CPU and memory usage in unreliable and reliable HTTP servers in the scenarios with and without proxy

Usages are higher in the scenarios with proxy. This overhead is caused by the extra control channel and extra messaging (e.g., acknowledgment messages) of FSocket, when proxies exist.

Regarding the message-based solution, we again measured CPU and memory usage in applications with the Messenger and with the Reliable Messenger. Figure 9.20a presents the results obtained for CPU usage. As shown, the extra

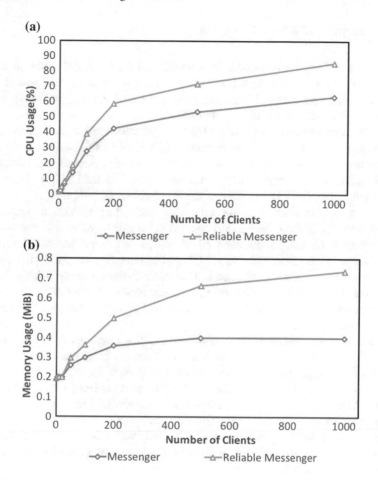

Fig. 9.20 CPU and memory usage with the Messengers

complexity of Reliable Messenger, in comparison with Messenger, imposes an expectable extra price in the CPU utilization (a maximum observed of 37%). The results for memory usage (in Fig. 9.20b), show that the Reliable Messenger has a much higher overhead (a maximum observed of 68%) than Messenger, due to the extra buffering of messages. Also, the results show that the memory overhead in the Reliable Messenger increases with the growing number of clients. Adjusting the time difference between periodical acknowledgments, which allows the peers to release some space in the buffer, can help to improve memory usage, but may impact on other parameters like CPU usage.

9.6.6   Implementation Complexity

We used three complexity metrics, to evaluate the implementation complexity of our designs: lines of code (LOC), cyclomatic complexity (CC), and nested block depth (NBD). To perform these measurements, we used the Eclipse Metrics plugin [44]. Table 9.3 summarizes the results obtained.

The measurements for the stream-based applications show that we used 520 extra lines of code in the reliable application (with plain FSocket) in comparison with the unreliable application. In addition, the average cyclomatic complexity per method in both cases is around 1.87, while the depth of nested blocks of the reliable application is 1.4, close to the 1.25 of the unreliable application.

The results also show that we used 801 extra lines to turn a non-reliable HTTP-based application into a reliable HTTP application. If we consider the average cyclomatic complexity per method, we can see that it increases by a small amount from 1.8 to 1.9 for reliable HTTP applications. Finally, the depth of nested blocks of the non-reliable application is 1.26, close to the 1.4 of the reliable version. These results show that providing reliable communication for HTTP applications is quite inexpensive, especially when considering the huge gains that our solution brings for developers.

In the case of message-based applications, the measurements presented in Table 9.3 show that we used 750 lines of code to implement the unreliable message-based application, and 485 additional lines of code, to implement the Reliable Messenger. The cyclomatic complexity increases from 1.33 to 1.98 in the Reliable Messenger, which is a very good indication of low complexity in our designs and implementations [37]. The difference for nested block depth, which increases from 1.38 in Messenger to 1.58 in Reliable Messenger, is also small. In general, these results show the simplicity of our design solutions, with respect to the functionalities provided.

Table 9.3 Implementation complexity of the unreliable and reliable stream-based applications

Applications	Lines of code (LOC)	Cyclomatic complexity (CC)	Nested block depth (NBD)
Unreliable stream-based application	537	1.871	1.25
Reliable stream-based application (with plain version of FSocket)	1057	1.875	1.40
Unreliable HTTP-based application	572	1.87	1.26
Reliable HTTP-based application (with HTTP-supported version of FSocket)	1373	1.95	1.40
Unreliable message-based application	750	1.872	1.38
Reliable message-based application (with FTSL)	1235	1.98	1.58

9.7 Conclusion

This chapter presented a software design pattern [13], named Connection Handler, which can be used to ensure application-transparent reliable communication, regardless of the operating system, programming language, middleware, or technology involved. We specialized them to provide reliable communication to three major types of applications in cloud computing: stream-based, HTTP, and message-based applications.

Our experimental evaluation showed the negligible performance overhead and relatively small resource usage cost of our solutions. Most of all, we showed that the Connection Handler design pattern can be easily implemented and correctly used by stream-based, HTTP, and messaging applications. This kind of solution, which is not merely a new mechanism or library, can help reducing the number of custom, ad hoc solutions created by developers, to assure reliable communication. In future work, besides evolving and further simplifying the Connection Handler design pattern, we intend to research techniques to automatically verify the correctness of a given implementation against the pattern.

References

1. Birman KP (1997) Building secure and reliable network applications. Springer, Berlin
2. Zhao W, Melliar-Smith PM, Moser LE (2010) Fault tolerance middleware for cloud computing. In: IEEE 3rd international conference on cloud computing, pp 67–74
3. Gray JN (1979) A discussion of distributed systems. IBM Thomas J. Watson Research Division, Cambridge
4. Halpern JY (1987) Using reasoning about knowledge to analyze distributed systems. Annual Rev Comput Sci 2(1):37–68. http://www.annualreviews.org/doi/pdf/10.1146/annurev.cs.02. 060187.000345
5. Zandy VC, Miller BP (2002) Reliable network connections. In: Proceedings of the 8th annual international conference on Mobile computing and networking, ACM, New York, NY, USA, MobiCom '02, pp 95–106. https://doi.org/10.1145/570645.570657
6. Barre S, Paasch C, Bonaventure O (2011) MultiPath TCP: from theory to practice. In: Domingo-Pascual J, Manzoni P, Palazzo S, Pont A, Scoglio C (eds) NETWORKING 2011, no. 6640 in Lecture notes in computer science. Springer, Berlin, pp 444–457
7. Stewart R (2001) SCTP: new transport protocol for TCP/IP. IEEE Internet Comput 5(6): 64–69
8. Marwah M, Mishra S (2003) TCP server fault tolerance using connection migration to a backup server. In: International conference on dependable systems and networks (DSN) pp 373–382
9. Shenoy G, Satapati SK (2000) HYDRANET-FT: network support for dependable services. In: International conference on distributed computing systems
10. Jin H, Xu J, Cheng B, Shao Z, Yue J (2003) A fault-tolerant TCP scheme based on multi-images. In: IEEE Pacific Rim conference on communications computers and signal processing (PACRIM), Victoria, Canada, pp 968–971. https://doi.org/10.1109/pacrim.2003. 1235945
11. Alvisi L, Bressoud TC, El-Khashab A (2001) Wrapping server-side TCP to mask connection failures. In: IEEE international conference on computer communications (INFOCOM)

12. Ekwall R, Urbán P, Schiper A (2002) Robust TCP connections for fault tolerant computing. In: The 9th international conference on parallel and distributed systems (ICPADS), pp 501–508
13. Gamma E, Helm R, Johnson R, Vlissides J (1993) Design patterns: abstraction and reuse of object-oriented design. Springer, Berlin, p 707
14. Ivaki N, Araujo F, Barros F (2014) Session-based fault-tolerant design patterns. In: 20th IEEE international conference on parallel and distributed systems (ICPADS 2014), Hsinchu, Taiwan
15. Ivaki N, Laranjeiro N, Araujo F (2016) A design pattern for recovering from TCP connection crashes in HTTP applications. Intl J Serv Comput 4(1):39–54
16. Ivaki N, Laranjeiro N, Araujo F (2017) Design patterns for reliable one-way messaging. In: 2017 IEEE international conference on services computing (SCC), pp 257–264. https://doi.org/10.1109/scc.2017.40
17. SourceForge (2019) Fault-tolerant socket: an implementation of connection handler design pattern. https://sourceforge.net/projects/fsocket/
18. Richards M, Monson-Haefel R, Chappell DA (2009) Java message service. O'Reilly Media, Newton
19. Horrell S (1999) Microsoft message queue. Enterprise Middleware
20. Birrell AD, Nelson BJ (1984) Implementing remote procedure calls. ACM Trans Comput Syst (TOCS) 2(1):39–59. https://doi.org/10.1145/2080.357392
21. Downing TB (1998) Java RMI: remote method invocation, 1st edn. IDG Books Worldwide Inc, Foster City
22. Banks A, Challenger J, Clarke P, Davis D, King RP, Witting K, Donoho A, Holloway T, Ibbotson J, Todd S (2002) HTTPR specification. IBM Software Group 10
23. Evans C, Chappell D, Bunting D, Tharakan G, Shimamura H, Durand J, Mischkinsky J, Nihei K, Iwasa K, Chapman M et al (2003) Web services reliability (WS-Reliability), ver. 1.0. Joint specification by Fujitsu, NEC, Oracle, Sonic Software, and Sun Microsystems
24. Cerami E (2002) Web services essentials: distributed applications with XML-RPC, SOAP, UDDI & WSDL. O'Reilly Media, Inc
25. Davis D, et al (2006) Web services reliable messaging (WS-ReliableMessaging). Technical report, OASIS. http://docs.oasis-open.org/ws-rx/wsrm/200702/wsrm-1.2-spec-os.pdf
26. Gamma E, Helm R, Johnson R, Vlissides J (1994) Design patterns: elements of reusable object-oriented software. Addison-Wesley Professional, Boston
27. Daigneau R (2011) Service design patterns: fundamental design solutions for SOAP/WSDL and RESTful Web Services, 1st edn. Addison-Wesley Professional, Boston
28. Gawand H, Mundada R, Swaminathan P (2011) Design patterns to implement safety and fault tolerance. Intl J Comput Appl 18(2):6–13
29. Yoshioka N, Washizaki H, Maruyama K (2008) A survey on security patterns. Progr Inf 5:35–47
30. Laverdiere MA, Mourad A, Hanna A, Debbabi M (2006) Security design patterns: survey and evaluation. In: CCECE'06. Canadian conference on electrical and computer engineering, 2006. IEEE, pp 1605–1608
31. Schmidt DC (1995) Reactor: an object behavioral pattern for concurrent event demultiplexing and dispatching
32. Schmidt D (1996) Acceptor-connector: an object creational pattern for connecting and initializing communication services. Pattern Languag Progr Des 3:191–229
33. Schmidt D, Ryan C, Kircher M, Pyarali I, Buschmann F (1998) Leader-followers. In: Pattern languages of programs conference (PLoP)
34. Nievergelt J, Hinrichs K (2011) Algorithms and data structures with applications to graphics and geometry. Lulu.com. 15 Sept 2014
35. Garrett JJ et al (2005) Ajax: a new approach to web applications
36. Hohpe G, Woolf B (2003) Enterprise integration patterns—designing, building, and deploying messaging solutions. Addison-Wesley Professional, Boston

37. Jorgensen PC (2008) Software testing: a craftsman's approach, 3rd edn. Auerbach Publications, Boston
38. Github (2019) Anomic FTPD: a freeware ftp server in java. https://github.com/Orbiter/anomic_ftp_server
39. SourceForge (2019) Fault-tolerant AnomicFTPD: a freeware ftp server in java (2019). https://sourceforge.net/projects/ftanomic/
40. Goodwill J (2002) Apache jakarta tomcat, vol 1. Springer, Berlin
41. Fleury M, Reverbel F (2003) The Jboss extensible server. In: Proceedings of the ACM/IFIP/USENIX 2003 international conference on Middleware. Springer, New York, pp 344–373
42. Tools Ietf (2019) RFC 959: file transfer protocol (FTP). Internet engineering task force (2019). http://tools.ietf.org/html/rfc959
43. Zhang J, Sivasubramaniam A, Wang Q, Riska A, Riedel E (2006) Storage performance virtualization via throughput and latency control. ACM Trans Storage (TOS) 2(3):283–308
44. Sauer F (2013) Metrics 1.3.6. http://metrics.sourceforge.net

Part III
Cloud Testing and Software Process Improvement as a Service

Chapter 10
A Modern Perspective on Cloud Testing Ecosystems

V. Vijayaraghavan, Akanksha Rajendra Singh and Swati Sucharita

Abstract The Cloud testing market share is expected to be over 10 billion USD by 2022. Cloud migration for applications has become an attractive phenomenon, and end users have, as a result, achieved various benefits such as autonomy, scalability and agility and improved return on investment by migrating to Cloud. Cloud environment is inherently elastic with respect to applications, infrastructure and platform resources which consequentially translate to the benefits mentioned. The rapid consumer adoption of Cloud paradigm mandates software testing in order to ensure that services over the Cloud are working as expected. In addition to the need for testing Cloud services and applications, the emergence of Cloud computing has opened new avenues for providing testing services over the Cloud in the form of testing as a service (TaaS). Many quality assurance (QA) processes which have a direct impact on testing cycles, like test environment management and test data management, can be provisioned via the Cloud, resulting in immense additional benefits. This chapter sets the context of Cloud computing and its growing significance for the software industry before focusing on Cloud TaaS. Additionally, different types of Cloud deployments in testing ecosystem are discussed in this chapter including: testing on the Cloud, testing models, test processes relating to the Cloud, tools and frameworks.

Keywords Cloud testing · TaaS · Functional testing · Testing as a service · QA processes · Security

V. Vijayaraghavan (✉) · A. R. Singh
Infosys Limited, Bangalore, India
e-mail: Vijayaraghavan_V01@infosys.com

A. R. Singh
e-mail: Akanksha_R@infosys.com

S. Sucharita
Infosys Limited, Bhubaneshwar, India
e-mail: Swati_Sucharita@infosys.com

© Springer Nature Switzerland AG 2020
M. Ramachandran and Z. Mahmood (eds.), *Software Engineering in the Era of Cloud Computing*, Computer Communications and Networks,
https://doi.org/10.1007/978-3-030-33624-0_10

10.1   Introduction

Although Cloud computing has been in existence for the last two decades, its prominence has increased over the past few years as most industries are moving their information technology (IT) infrastructures to Cloud environments. The Cloud computing market is expected to cross 200 billion dollars in 2019 [1]. Cloud first has also become the expected approach to IT for quick business growth, business agility, cost optimization, service resilience and scalability. AI, machine learning (ML), Internet of things (IoT) and many other next-generation trending technologies are now readily available by the leading Cloud providers as *services*. There are relevant salient aspects which separate the Cloud paradigm from traditional forms of computing which cedes the upper hand to the Cloud. The next subsections explain the distinct features of the Cloud vision.

10.1.1   *Cloud Computing Versus Traditional Computing*

Cloud computing has enabled enterprises to focus on their key business areas by taking care of the required IT infrastructures by leveraging the Internet. Until recently, in a conventional setting also known as on-premise, a business would take care of their computing resources in terms of software license, procurement, electricity, maintenance and servers. But with the advent of Cloud, an organization can take care of its IT needs by using the Cloud mechanisms and harness all the computing resources required over the Internet without worrying about procurement, utilization, demands and maintenance.

The benefits of Cloud computing become clearer when cost and effectiveness are taken into account. The total cost of ownership is very high for on-premise infrastructure as it incurs maintenance and procurement costs to the end user. But in case of the Cloud, the total cost of ownership is drastically reduced and businesses can be free from demand–supply concerns regarding infrastructure. Furthermore, when it comes to resource utilization, Cloud emerges as the clear winner. With on-premise servers, forecasting models are required to predict the IT needs of an organization and procure the resources. A sudden spike in demand would take a considerable amount of time to fulfill the infra needs, and if the requirement is not as much as the servers commissioned, then the servers would be underutilized. Pay as you go (PAYG) and on-demand model with the Cloud ensure that resources at disposal are utilized properly. Additionally, if need arises, more resources can be availed easily via the Cloud.

While the advantages of Cloud easily trump the disadvantages, special attention must be paid with respect to security and compliance mandates, when creating or migrating an application on Cloud. Sensitive data needs to be protected during Cloud migration and also stored on Cloud. On-premise applications have their own servers and infrastructure, so the data and applications are comparatively safe.

Businesses might want to keep their sensitive data and critical applications with them (i.e., within the organization) rather than on the Cloud. In the Cloud environment, the data and security checkpoints should be strongly enforced; else, there might be a strong risk of security breaches, as the Cloud might be public, private or hybrid. With increasingly stringent data privacy and compliance regulations across the globe, special care must be taken to ensure data regulations are adhered on Cloud. With Cloud, the servers and infra could be located in any corner of the globe, which makes it essential that the respective data policies for an application are abided by. The caveats attached to Cloud computing in terms of data and security render Cloud testing essential in any Cloud-based deployment.

10.1.2   Importance of Cloud Testing and Applications

As per a press release [2], *the Cloud testing market is expected to grow from USD 5.55 billion in 2017 to USD 10.24 billion by 2022, at a compound annual growth rate (CAGR) of 13.01% during the forecast period. The increasing adoption of the Cloud technology across businesses, reduced cost of ownership, scalability and flexibility offered by Cloud-based testing platforms are some of the major driving factors for the growth of Cloud testing market.* It is evident that there is a huge demand for Cloud testing and with the right Cloud testing tools and frameworks, the demands of this market opportunity can be fulfilled.

Data security is a major threat to Cloud. A huge amount of customer and organizational data are sent/received or stored on the Cloud which is outside the control of the organization. With multi-tenant model of Cloud provision, the risk of data loss and data theft is high, due to lack of data visibility of the storage. It can be negated by proper data testing and security testing of Cloud infrastructure. Apart from security testing, validation of performance, availability testing and integration testing are the key elements of Cloud testing.

The chapter is organized as follows: Sect. 10.2 provides a detailed introduction to Cloud testing and salient features differentiating Cloud testing from traditional testing. In Sect. 10.3, various types of Cloud testing models with specific examples of QA processes on Cloud are explained. Section 10.4 discusses market tools for Cloud testing and proposes a framework to test software on the Cloud; Sect. 10.5 provides the conclusion.

10.2   Cloud Testing

"Cloud testing is an area of software testing in which readiness of Cloud-based apps and environments is tested, including Web applications which use Cloud computing environments to simulate real-world user traffic" [3]. The prevailing digital wave has brought in new technologies such as artificial intelligence, automation, machine learning, IoT and blockchain; but most importantly, it has rendered Social platforms, Mobile apps, Analytics and Cloud (SMAC) a basic necessity for fast-growing businesses. SMAC has created a major change in testing focus in QA teams, that is, from application testing toward appraising the end user desires. Today, organizations are looking for new approaches to testing which should be dynamic and flexible enough to accommodate the demands for speed of digitization. There is no better option than leveraging Cloud for testing to meet the demands of the digital age. Cloud testing platforms which offer testing as a service (TaaS) are also expected to be driven by demand for more Cloud services across domains, resulting in reduced expenditure for the end user, no maintenance for the businesses owning them, ease of scaling and agility [4]. An effective QA strategy with a good repository of tools and frameworks can aid in overcoming the challenges around Cloud testing. This would also be fruitful in tackling the massive demands for Cloud testing business that would arise in the near future [5].

Figure 10.1 shows an understanding of how Cloud has become instrumental for testing and as a result the organizational growth.

Two decades ago, testing was confined to remain in the same organization or maybe to the same team who developed the software. It was more like debugging or unit testing with set of market tools available from product-based vendors. With increase in complexity and size of software, validation became critical as well as challenging. IT arms of organizations reaped benefit by outsourcing validation needs to independent vendors with skill-based testers and set of tools.

Over the past few years, the rapid development of new products, mobile apps, customer-centric sites and utilities has brought attention to usability and customer sentiment. End user experience has become a key factor for success or failure factor of an application. Big customer-oriented organizations like Google use crowd-sourced testers over Cloud to validate their apps by simulating end user environment. This has given opportunity for easy end user testing and crowd testing over Cloud, where a tester can try out products and get rewarded based on defects found. With virtualization, both outsourcing and crowd source testing have become effortless because of on-demand resources from Cloud.

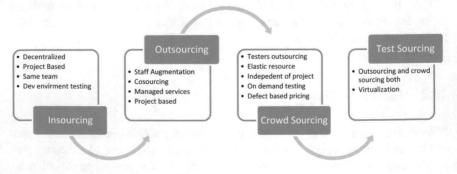

Fig. 10.1 Evolution of software testing

Moving to a Cloud testing model requires a testing team to follow a specific testing life cycle tailored for Cloud environment which is different from the normal testing life cycle. Difference between the conventional and Cloud testing life cycle is explained in the following subsections.

10.2.1 Traditional Software Testing Life Cycle (STLC)

A software testing life cycle, in traditional sense, begins with requirement analysis, followed by test strategy and test planning phase which then leads to the test case design phase. In parallel, an environment management team would be setting up the test environment where the test scenarios would be executed. After the afore-mentioned preparatory stages, comes the test execution phase which can last for many weeks depending on the testing model (agile or waterfall), release type (independent or integrated) and types of testing involved (ad hoc/functional/regression). Each stage of the STLC has an entry and exit criteria. There is then a test closure phase where the testing is signed off by the programming teams. A traditional STLC does not use Cloud to enhance its test operations because of which the test phases sometimes are protracted, testing is below expectations due to inadequate test coverage, and handling ad hoc requirements becomes a hassle.

10.2.2 Cloud Testing Life Cycle

A Cloud testing life cycle is similar to conventional software testing life cycle (STLC) in initial stages, but later the testing life cycle is modified to leverage the Cloud resources optimally. In case of the Cloud testing, test laboratories can be provisioned for all users at any time on demand, anywhere in the world with the same availability and network speeds. Figure 10.2 depicts a typical Cloud testing life cycle which is anytime, anywhere model.

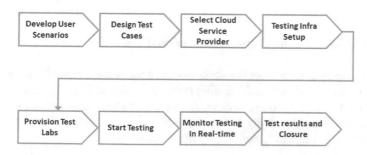

Fig. 10.2 Cloud testing life cycle

The similarities between Cloud testing life cycle and STLC cease at the stage of designing test cases and bifurcate to utilizing the Cloud for testing. Instead of sourcing the infrastructure, the testing team needs to select only the Cloud Service Provider (CSP) who provides the test laboratories as per their needs. Once test laboratories are set up in the Cloud, the testing teams can use the Cloud servers and begin testing. Sometimes, to save the cost and resources, the testing need not be on the applications itself but can be executed on the virtual forms of an application. Cloud testing life cycle helps to utilize shared resources via virtualization.

10.2.3 World of Virtualization

Virtualization technology is the answer to typical testing challenges due to environment, time and cost. Below are typical problems which can be addressed with Cloud virtualization:

- Development environment and test environment mismatch which lead to a lot of time wastage in reproducing bugs—On Cloud, the same configuration images can be used to easily create virtual machines (VMs) for development and test environments.
- Time lapse due to setting up the environment—Pre-built configurations are available which can be instantly procured for deployment in the Cloud.
- Location-based application testing challenges and tester availability—Cloud is spread across multiple geographic regions. VMs can be configured and maintained in those regions by Cloud providers. Testers from different locations can do localization testing by accessing applications on Internet.
- Cost for setup and maintaining a test laboratory with required testing tools— Cloud gives flexibility due to pay as you go approach. Instead of procuring high-end servers on Cloud, testing organizations can register for pre-configured hardware and testing tools for only the test duration.
- Test box remains unutilized after testing unless it is a continuous testing process —Pay only for what you use and how long you use is the key benefit of Cloud source environment.

10.2.3.1 Cloud Testing Versus Conventional Testing

Cloud testing inculcates both the functional and the non-functional aspects of traditional testing along with niche testing areas applicable only to the Cloud: latency testing, multi-tenancy testing, backup, restore and disaster recovery testing, secure access testing, interoperability testing. In Table 10.1, the main differences between conventional testing and Cloud testing are enumerated.

Table 10.1 Cloud testing versus traditional testing

Parameter	Conventional testing	Cloud testing
Primary testing objectives	The functionality and performance of the application based on the given specifications should be working well: Check usability, compatibility, interoperability	Assure the quality of functions and performance of the application, on Cloud and applications by using a Cloud environment. Assure the quality of Cloud elasticity and scalability based on the SLA
Testing environment	A pre-configured test environment	An open crowd-sourced test environment with computing resources. Scalable private test environment in a test laboratory
Testing costs	Server costs, hardware costs and software (license) costs	Based on pre-defined SLAs—pay as you test (Cloud testing cost). Engineering cost will be applicable based on SaaS/Cloud/application vendors
Test simulation	Online user access, online traffic and online data deluge need to be simulated	Virtual/online user access simulation and virtual/online traffic data simulation
Functional testing	Validating functions (unit and system) as well as features	Cloud testing functions, functional and feature validations
Integration testing	Function-based—component-based—architecture-based	SaaS-based integration in Cloud, and SaaS integration testing in public, private and hybrid Clouds. End-to-end integration testing on Cloud
Security testing	Security features to be tested: user privacy, client-/server-based security, process-based security	Cloud security features to be tested: user privacy and access, Cloud API testing, connectivity testing, security testing with virtual/real time in vendor's Cloud
Scalability and performance testing	Fixed test environment with simulated user access. Test data with online monitor and regular evaluation needed	Scalable test environment based on SLA. Can use both online and virtual data

10.2.4   Challenges of Cloud Testing

Cloud testing is a solution to many of the issues in traditional testing, but it also has its own set of challenges. All of these challenges can be overcome with proper planning, smart service-level agreement (SLA) and training.

- **Data Concerns**: Meticulous planning of assuaging security concerns and securing data should be done before starting to test on the Cloud. Data migration when moving from one Cloud provider to another can be a major challenge depending on the databases and warehouses used.

- **Application Updates**: Making sure the application UI/functionalities are not lost during Cloud updates or changes is a very important part of Cloud testing. Customers might continue to use the legacy version of an application without being aware of the new version or not receive timely update patches for their app. This is a serious issue involving security and customer satisfaction.
- **Integration Issues**: Integration testing at enterprise level is another huge challenge in Cloud. Integration testing of a hybrid application where it is shared between a Cloud and an on-premise server is a different challenge altogether.

10.3 Cloud Testing and Deployment Models

Many business applications are moving to the Cloud whether it is public, private or hybrid. This creates an enormous opportunity for testing application readiness for Cloud. Furthermore, with the emergence of Cloud vision as a core technology for organizations, this can be leveraged to provide testing as a service (TaaS). In this respect, there are three main models in Cloud testing:

- Testing in the Cloud
- Testing as a service (TaaS) in the Cloud
- Test support as a service.

10.3.1 Testing in the Cloud

Testing in the Cloud refers to the testing of an application readiness on the Cloud environment. There are three key reasons why testing an application before deploying it on the Cloud is essential.

- **Customers Using Online Applications**: Hosting an application on Cloud suggests that it will be accessible online from anywhere and anytime. Today's applications are majorly customer facing with high availability. The target system or application to be tested might be a software developed and deployed on Cloud or software exposed as a service by Cloud vendor, but it must be tested to ensure the end customer experience is not affected. Testing scope would consist of functional, non-functional and focused Cloud testing.
- **Diversifying the Deployment Model**: Cloud deployment model could be public, private and hybrid. So, infrastructure and platforms are hosted across different deployment models of the Cloud that must be a major consideration while formulating test strategy.
- **Testing of the Cloud Itself**: Before application testing, it is important to validate if the infrastructure services used to host the application on Cloud can support the required performance, availability, security and scalability.

Different types of testing should be executed on Cloud at various phases of deployment/update of application depending on the type of Cloud (private, public or hybrid) as represented in Fig. 10.3. At service level, routing and network testing, service integration testing and API testing should be performed. At the authentication stage, identity and access testing, security testing, multi-tenancy testing and compliance testing should be conducted. A hybrid model should be tested for API functionalities, high availability, data security and service integration. If the application is deployed on a public Cloud, network testing, load testing, scalability testing, data security testing and performance testing should be done. Elasticity on Cloud should be tested for elastic cache, high availability, chaos and resilience, compliance and data security. Finally, post an application deployment, live testing and resilience testing must be performed at regular intervals, especially during application updates or patch deployments.

Testing inside the Cloud is essentially testing the Cloud-based applications that are hosted and deliberated in a Cloud environment and consists of three different categories of testing:

- Functional testing
- Non-functional testing
- Focused Cloud testing.

Functional Testing

Functional testing is basic testing to ensure services on Cloud are running smoothly and as per user's requirement. This can be validated manually or by using a Cloud testing tool.

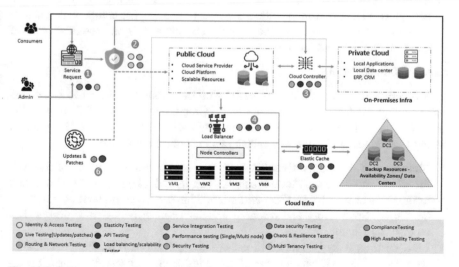

Fig. 10.3 Different types of testing inside the Cloud

Non-functional Testing

Typical non-functional testing, e.g., performance testing, data security testing, load testing, stress testing and compatibility testing, is applicable to applications that are hosted on Cloud or already are Cloud native. Brief description of each type is given below:

- **Performance Testing**: Cloud provision means one identify how systems act when a specific workload occurs. Performance testing is executed specifically to check the response speed and stability of the system to withstand peak traffic as well as to validate some quality attributes of the system like simplicity, scalability and availability [6].
- **Load Testing**: This is performed to generate traffic from multiple users and then calculating the system response under this traffic. For Cloud application, stress testing is required to check how the application behaves when there is a break in one of the services used by the application. Stress testing and load testing both should be done before hosting an application as it will test the Cloud application availability, robustness and completeness when extreme conditions occur.
- **Compatibility Testing**: Compatibility testing is performed to check the capability of the system or application to work on cross-browsers and multiple operating systems. Furthermore, compatibility testing can provide a yardstick on the ease of application migration from one vendor to another, and can be used to fix compatibility problems that are important for the system. Compatibility tests cover compatibility between a different hardware, various operating systems' compatibility, networks, computers and application environments.
- **Data Security Testing**: This is highly critical in case of deployment in Cloud. Security testing discovers the weakness of the software on Cloud. Testing unauthorized access from unauthenticated user to a specific component, and also regulatory testing with respect to data compliance is an essential part of non-functional Cloud testing.

Focused Cloud Testing

Both functional and non-functional testings on the Cloud are similar to traditional testing. But when an application is hosted on Cloud, there is a set of Cloud-specific niche testing that needs to be conducted. This section highlights a few focused testing techniques applicable to applications hosted on the Cloud.

- **Identity and Access Testing**: This refers to role-based access control (RBAC) for users and admins. Access can be configured at user/identity level and resource level with read/write permission. This can be configured as policy on the Cloud. So, identity and resource policies need to be tested to ensure data security on Cloud.
- **Live Testing**: A key driver of Cloud-based applications is the high availability provided. If there is any failure of any kind such as network outage, break-through due to load and system failure, it should be back online with minimum

adverse effect on business. So, it is important to measure how fast the failure is indicated and if any data loss occurs during this period.

- **Compatibility Testing**: A Cloud application must be capable to work and be executed across multiple environments and various Cloud platforms. Accessibility and multi-browser/platform/device testing are some of the testings that should be carried out for Cloud applications.
- **API Testing**: Cloud providers also expose different APIs for its services to build applications. In such cases, testing of those API integrations is critical for success/failure of the application. Connectivity and invocation testing, API load testing, security testing, etc., are the API testings that should be mandatorily performed on Cloud.
- **Network and Routing Testing**: Performance of any Internet-facing application is very critical. Measuring the latency between the action and the corresponding response for any application after deploying it on Cloud comes under network testing. The tests are executed, using the agents, from multiple locations around the world.
- **High Availability Testing**: Availability testing must be done to assure uninterrupted service or accessibility of the application without any abrupt downtime for business continuity.
- **Elasticity Testing**: Elasticity of Cloud is degree to which a system is able to adapt changes by provisioning or de-provisioning resources based on load. Cost is a key consideration on Cloud. Effective management of the resources according to business peak period is vital. Elasticity in the Cloud environment should be ensured by testing the following:

  - **Resource Acquisition/Release Time**: Test ramp up and ramp down time of dynamically allocated resources.
  - **Provisioning on the Go**: Test ability to provision resources on need basis.
  - **Load Testing for Elastic Load Balancing**: Elastic load balancing (ELB) is a load balancing solution that automatically scales its request-handling capacity in response to incoming application traffic. Some elastic load testing scenarios that should be tested during this phase are unpredictable bursts, predictable bursts, periodic usage and hyped usage.
  - **Multi-tenancy Testing**: Multiple clients and organization use on-demand services activated at a given time. Cloud services should be customizable for each client keeping security at data and service-/resource-level compliance to avoid any access-related issue or data leak.
  - **Chaos and Resilience Testing**: Cloud verification must be done to ensure the service is back online with minimum adverse effect on business. This includes testing robustness of the platform against the disaster, measuring the recovery time in case of disaster and self-healing ability.

10.3.2 Testing as a Service (TaaS)

High-speed engineering and faster delivery to market are key success factors of digital business growth. Testing as a service (TaaS) is an on-demand testing delivery model through Cloud-based environment. It is an ecosystem of methods, tools and people synchronized to deliver service. TaaS can be evaluated for testing demands that are fractional in nature, e.g., performance, security testing and usability testing, testing that needs complex infrastructure, e.g., device-specific testing and SOA testing. It can also be considered for effective test asset management and capacity utilization through shared infrastructure.

10.3.2.1 Approaching TaaS

Moving all the quality assurance (QA) activities at once to the Cloud is not the objective. Moving an organization's testing to Cloud has to be decided based on current and future road map and has to be rolled out in phases. To gain confidence, it is suggested to first carry out pilots with defined objectives and analyzing ROI before making the leap. Feasibility study must be carried out before moving testing to the Cloud. Comparison of the in-house provisioning cost and cost of using Cloud needs to be done as part of strategic planning. It is important that testing organizations should be clearly aware of the benefits of adopting Cloud testing model to their business.

There are three different forms of testing as a service (TaaS) in a Cloud environment [7]:

- **TaaS for Web-based Software on the Cloud**: Web-based applications deployed on Cloud must be tested with large-scale test simulations and elastic computing of resources by TaaS provider.
- **TaaS on the Cloud**: In this form, Cloud-based applications integrated with SaaS systems are tested to check scalability and multi-tenancy of SaaS systems.
- **TaaS on the Cloud**: SaaS applications crossing hybrid Clouds are deployed and validated based on different Clouds. In a hybrid Cloud, infrastructure and diverse on-demand test services are provided and delivered by TaaS vendor.

In the next section, specific examples of TaaS that can be leveraged in testing to realize immense benefits are discussed.

10.3.2.2 TaaS Examples: QA Processes for the Cloud

In the most recent World Quality Report, slowing of QA processes due to lack of proper test data or test environment has been cited as a huge pain point in moving to an agile model of delivery [8]. With the help of Cloud testing, all these issues can be resolved to a great extent, thereby continuing to support a QA processes' transition to agile and DevOps modes of operation.

Another aspect of testing as a service is to provide end-to-end QA processes on the Cloud. Cloud testing has helped QA organizations to become agile while at the same time helping to reduce costs. QA processes like Test Environment Management (TEM) and Test Data Management (TDM) can be executed efficiently from Cloud. Advantages of provisioning TDM and TEM on the Cloud are as follows:

- **Accessibility Anytime and Anywhere**: Cloud provision means one can access the data/test laboratory from anywhere and anytime over the Internet without any concerns about the network speed. Testing teams are going global. Cloud-based TDM/TEM will provide access to test laboratories or device laboratories on Cloud anywhere/anytime around the globe.
- **Low Cost and Ease of Setup**: Test environment provisioning on Cloud costs much less on Cloud than on-premise. TEM and TDM on Cloud are comparatively easier to set up than on-premises.

Test Environment Management (TEM) in the Cloud

As shown in Fig. 10.4, a typical Test Environment Management provisioning moves from analyzing the requirements of a particular release/sprint, creating parallel environments for testing, provisioning the test environment, test environment booking/allocation, code build, environment shakedown and environment termination.

Test environment setup includes many steps right from estimating demands, to analyzing requirements and taking service requests [9]. A conventional test environment would come with prefixed servers with all configurations and setup done in advance. Setting up a testing environment locally on machines and testing if the configurations are set up properly slow down the whole QA process. There is

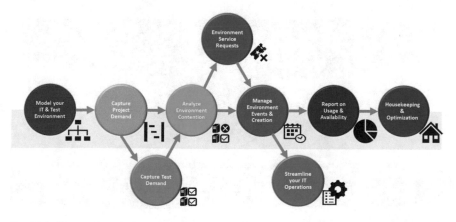

Fig. 10.4 Test environment management

always the risk of instability and late delivery of environment. During a testing life cycle, the most common complaint of a tester is that the test environment is not ready while the testing cycle has already started or there are frequent crashes of the test environment. Furthermore, the cost of provisioning a test environment, the conventional way, is expensive. With an on-premise traditional TEM center, it would be difficult to commission a new environment based on demand, whereas on Cloud there can be a public or private test environment with access to unlimited resources. The question of resource allocation or demand–supply need when it comes to test environment provisioning flies out of the window when done via Cloud.

Test Environment as a Service

A test environment in the current DevOps world should be convenient to set up, easy to scale on demand and faster to access and could be terminated as soon as the need for it ends. A test environment hosted on Cloud accomplishes all of these while at the same time being cost effective. As depicted in Fig. 10.5, Cloud-based TEM offers users many advantages like ease of use and setup. A Cloud test management service would provide various test laboratories to be used by testers as and when required.

10.3.2.3 Test Data Management (TDM) in the Cloud

Proper Test Data Management (TDM), an important function of test environment provisioning, is a major activity in the software testing life cycle. A complete TDM cycle consists of many phases like TDM planning which begins with defining data requirements and data provisioning plans for a testing life cycle followed by an analysis where the TDM team takes stock of the current data existing in all environments and databases while also working on defining the new data profiles

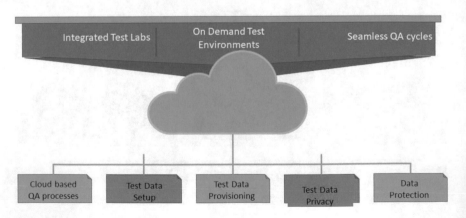

Fig. 10.5 Test environment management on Cloud

required for the next phase of testing. In the analysis phase, data security, backup and storage will also be planned. Next comes the design phase where the data distribution and data sources and tools for TDM will be designed with initial builds. In the next stage, ideally, a self-service TDM portal should be set up, where the ad hoc data requirements of the testing team with defined SLAs should be raised to the TDM team. Post the testing cycle, the TDM team should continue to take in change requests, maintain the existing data, create synthetic copies of data and delete the data which is redundant or no longer in scope.

Challenges of Test Data Management include additional time to set up data manually and the admin effort required in TDM. There might be expenses in terms of hardware and storage of data. If the TDM portal is hosted on the Cloud and care is taken of the data sensitivity and privacy, Cloud-based TDM will become more robust than the traditional TDMs. TDM coupled with TEM on Cloud can help testing teams with the provisioning issues freeing up the testing team's time. This ensures that the focus of testing team is on finding bugs and not on setting up data and laboratories.

10.3.2.4   Achieving TaaS Maturity

A TaaS setup requires four key stakeholders: (1) Cloud vendor who can provide computing and storage resources, (2) test solution vendor who can provide diverse testing tools, (3) test process managers who have necessary solutions and (4) test-ware vendor who can provide testers to build test scripts. Testing service vendor provides, manages and controls TaaS infrastructure and platforms and delivers testing services based on provided resources, test tools and test-wares from other vendors. The following can help to achieve TaaS maturity:

- Scalable Test Environment: Provisioning TaaS users to select and configure desired test environment with free and licensed test tools and solutions.
- Digital Test Management Services: Enabling TaaS clients with essential test management services for clients, e.g., Test Project Management, Test Process Management and Test-ware Management solutions and services.
- Large-Scale Test Simulation: TaaS vendor must offer large-scale test simulations for SaaS performance validation and scalability validation.
- On-demand Test Automation Service: Offering of on-demand automatic test execution, test control and bug reporting to support test automation needs.
- Multi-tenant Test Service Analysis: Delivering test adequacy analysis for each SaaS tenant. It will enable to monitor and evaluate the quality of its underlying test services for each test project.

10.3.2.5 TaaS Benefits

Testing as a service (TaaS) in the Cloud can deliver many key benefits, as briefly mentioned below:

- Reduction in test laboratory setup cost: Typical huge investment in hardware procurement, management and maintenance can be avoided by shared infrastructure on the Cloud. Also, software license and support cost are another major investment for the organization which can be reduced with pay as you use or pay as long you use models of Cloud: This reduction in initial investment can help organizations to realize faster ROI on new digital business.
- Significant reduction in maintenance of testing environment.
- Faster time-to-market: With easy setup of infrastructure on Cloud and setup of test laboratories with adequate testing tools, faster time-to-market is possible.
- Reduction in business risks: Availability of most of the leading testing tools and quality management solutions helps to reduce business risk.
- Better collaboration and team efficiency: Since everything is accessed via Internet, it increases team efficiency and collaboration even if teams are not colocated.
- Anywhere, anytime accessible infrastructure: This ability of the Cloud ensures high accessibility with 24×7 support systems.

10.3.2.6 Market View of TaaS

Besides predicting compound annual growth with digital journey of business in TaaS model, the market expects to reap a massive ROI via TaaS within the following areas:

- TaaS can complement the existing in-house QA team with scalable Cloud-based solutions for different types of testing, from functional testing, performance testing, load testing, all powered by real-time monitoring of the application and environment.
- Reduction in license cost of heavy testing tools, infrastructure cost for setup and support-related costs.
- Ease of setting up test laboratories and ready to use environment with testing tools and quality management capabilities are expected to contribute to budget rationalization of QA spending.
- Since the testing environment is available anywhere and anytime, the concept of crowd sourcing for testing continues gaining traction. It helps organizations leverage their talents from outside as well as internally from any corner of the globe.

10.3.3   Test Support as a Service (TSaaS)

Test support as a service (TSaaS) leverages the open test APIs in far-flung Cloud platforms for testing. If Cloud Platform A has a testing service T1 over Cloud and has APIs which expose the test functions, then any other Cloud providers P1, P2, P3 can leverage T1 via open APIs to test their own application under test. TSaaS reduces the burden of building automated frameworks and tools to test Cloud applications. Any QA team can leverage test support as a service over Cloud using TSaaS APIs to fulfill their testing needs [10]. With the help of TSaaS, multiple Cloud partners can avail the Cloud testing services across platforms, across Clouds and even within the organizations. auto-monitoring and reporting of test processes over Cloud become intuitive and simplified with help of TSaaS. TSaaS can serve as a good testing bed for deployment of self-healing, zero-touch artificial intelligence-based QA services over Cloud.

It should be noted that all the processes and deployment models provided by the Cloud can benefit only if there are proper tools and frameworks in place which is explored in the next section.

10.4   Tools and Frameworks for Cloud Testing

Good testing on the Cloud can be achieved through selecting the right testing tool that suits the needs of the QA organization and also by having a well-defined framework in place for Cloud testing.

10.4.1   Market Tools for Cloud Testing

There are many Cloud testing tools in the market, but there are just a few that are both efficient and powerful, e.g., SOASTA Cloud Test, LoadStorm and BlazeMeter which are widely used for Cloud testing. Watir is a powerful open-source Cloud testing tool. It is strongly advised that when selecting a Cloud testing tool, the following factors are carefully evaluated: licensing costs, tool vendor's road map, QA team's needs and application life cycle management aspects of the testing tool [11]. Three of the available Cloud testing tools are briefly discussed below.

- **SOASTA Cloud Test**: SOASTA Cloud Test is the leader in Cloud testing covering all end-to-end testing needs over the Cloud. It is available in various editions like Cloud test, Cloud test lite and Cloud test on demand. Cloud test lite is the community edition of SOASTA where users can test up to 100 virtual units (VU) from their own server [12].

- **LoadStorm and BlazeMeter**: These are commercial performance testing tools used to simulate user traffic over Cloud and execute various types of performance testing over Cloud [13].
- **Watir**: Watir stands for Web Application Testing in Ruby. It is one of the most powerful open-source Cloud testing tools used to automate Web browser testing. Watir can be integrated with SauceLabs for mobile testing and Applitools for visual testing [14].

10.4.2 A Framework for Cloud Testing

Although there are many tools in the market by major industry players like SOASTA Cloud Test, Xamarin, BlazeMeter, LoadStorm and Nessus, a framework for Cloud testing is mandatory to ensure optimal testing. One of the methods to ensure optimal testing is test case prioritization which could be based on code or test diversity or risk [15]. An end-to-end framework covering all aspects and features of Cloud testing eases the test coverage of Cloud-based testing. As depicted in Fig. 10.6, Cloud testing when done end to end requires testing of various components like functional and non-functional components which are specific only to Cloud.

A Cloud testing framework in line with the ideal testing framework specified by ISTQB [16] has been illustrated in Fig. 10.7.

The Cloud testing framework depicted in Fig. 10.7 is explained in some detail below:

1. The first step is to develop the test scenarios which would be dependent on the type of testing to be performed. If the goal is to test an application or platform readiness on Cloud, then all layers of Cloud with respect to the application or

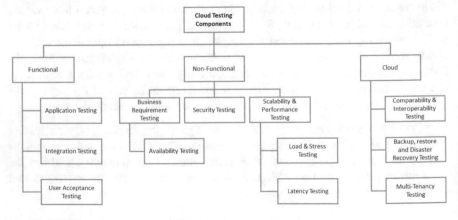

Fig. 10.6 Different components of Cloud testing

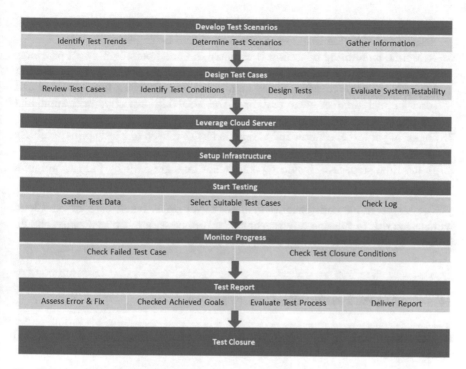

Fig. 10.7 Cloud testing framework

platform, namely network, infrastructure and platform, need to be tested depending on the Cloud play involved. If the type of methodology to be followed is TaaS, basic due diligence testing pertaining to Cloud ensuring test coverage should fulfill the Cloud testing goals.

2. Next comes creating the test case documents and the design documents for testing which is built upon the test scenarios developed.

3. Leveraging Cloud resources optimally is important to ensure success of a Cloud testing venture. Many a time, Cloud testing services are not properly utilized by the team due to lack of understanding of Cloud and its billing per usage, which may lead to more cost. Test planning team should plan and optimize utilization of resources to realize cost benefits.

4. The Cloud infrastructure and test laboratories should be set up in line with requirements of the testing life cycle. At this stage, ground zero testing should be executed to ensure the test laboratories are fully functional and ready to be deployed for the testing cycle.

5. Testing begins at this step where the new functionality is tested first followed by regression scenarios. Documenting the test progress and raising defects with a good test management tool are very important in any test life cycle whether Cloud or conventional. In Cloud testing, it is even more pertinent because many

users in different parts of the globe will be accessing the same Cloud. So, a good defect and execution tracking mechanism should be in place.

6. Real-time monitoring of testing is essential in Cloud paradigm. Most providers provision real-time insights for the hardware and services registered. During testing, it is recommended to use monitoring tools in order to understand and react in real time or near real time for resolving capacity, utilization and performance-related issues.

7. Reporting of test execution and various statistics related to a test sprint or release should be consolidated and shared with the concerned stakeholders.

8. In the final step, test closure reports would be signed off by the test program managers.

The above framework ensures that test coverage is taken care of on Cloud without the risk of missing any scenarios. Nevertheless, a QA organization should follow best practices for Cloud along with the framework.

10.4.3   Cloud Testing Best Practices

It is advisable that the following best practices must be adhered to during Cloud testing processes:

Understand Business Needs

Defining the objective of moving a testing project should be the first step of the testing strategy. It is essential to understand your business needs as well as advantage and restrictions of the Cloud. Skilled developers and testers are required for conducting unit, functional and integration testing. Moving testing to Cloud does not mean eliminating tester requirement.

Develop Cloud Inclusive Test Strategy

Test strategy should clearly call out the intent to be achieved including cost, reduction in time, high availability and accessibility. It should include what has to be tested, why testing has to be conducted on Cloud, the risk associated with this type of testing. An ideal Cloud testing framework journey begins with analyzing the requirements and design documents to come up with the test strategy and test scenarios. The test strategy would consist a high-level overview of testing in scope (functional, non-functional and niche). The test strategy and design document would be Cloud inclusive, viz: cover the focused Cloud testing aspects.

Choose Right Cloud Service Provider

Selecting a considerably experienced provider is wiser in initial phase. They typically help the QA team with quick setup, while providing a wide range of end-to-end service. Once the QA team is comfortable with the model, cheaper options in the market should be explored. While selecting provider, it is important to scrutinize their platform for security, reliability and performance.

Acquire Optimal Cloud Infrastructure

After the test design phase, the QA team should take stock of the infrastructure needs required for the release or sprint in question. The Cloud servers will be procured from the Cloud service provider based on the perceived need. Note that this could be increased or decreased as per the real-time spike or fall in demands.

Carefully Plan Test Environment

Major benefit of moving to Cloud is cost-effective infrastructure. Test strategy should define infrastructure requirement for setting up test laboratory on Cloud. Organization should carefully plan their test environment including hardware, software and testing tools and determine when and how long they need it to optimize their investment.

10.5 Conclusion

To summarize, good tools, proper framework and best practices are the keys to a successful Cloud testing project. Cloud has enabled organizations to truly focus on their business without worrying about infra and IT. With applications and resources increasingly deployed on Cloud, testing an application's readiness for Cloud in various areas is essential for meeting the end customer expectations. Using Cloud for QA can help the teams across IT organizations to obtain licenses, test laboratories and tools at a low cost without the intense effort in setting up the environments as well as not having to be concerned about infrastructure utilization or sudden demands. Cloud testing helps fix the problems associated with traditional testing.

Security on the Cloud is a concern that should be addressed effectively in the future leading to better adaption and faster business operations [17]. With the development of smart cryptography, Cloud security will help ensure safety. While still under the process of construction, autonomic testing systems on the Cloud, powered by artificial intelligence, will exponentially improve Cloud capabilities by providing self-healing, self-adjusting and completely automated Cloud testing services. Advancements in AI and automation in Cloud testing will finally pave way for QA which will be streamlined, seamless and optimized for maximum performance and efficiency.

References

1. John E (2018) Cloud computing 2019: the cloud comes of age. https://www.informationweek.com/cloud/cloud-computing-2019-the-cloud-comes-of-age/d/d-id/1333442
2. MarketsandMarkets (2019) Press release, cloud testing market worth 10.24 Billion USD by 2022. www.marketsandmarkets.com/PressReleases/cloud-testing.asp
3. Wikipedia (2019) Cloud testing. https://en.wikipedia.org/wiki/Cloud_testing
4. Ps-Market-Research (2019) Cloud testing market by component (tools, services, applications, geography, market size, share, development, growth, and demand forecast, 2014–2024. https://www.psmarketresearch.com/market-analysis/cloud-testing-market
5. Challenge Curve (2016) Cloud QA quality framework white paper. https://www.cloudindustryforum.org/content/how-qa-and-testing-framework-could-be-key-avoiding-cloud-thunderstorm
6. Mylavarapu VK (2013) Taking testing to the cloud. www.cognizant.com/InsightsWhitepapers/Taking-Testing-to-the-Cloud.pdf
7. Gao J, Bai X, Wei-Tek Tsai; Tadahiro Uehara (2013) Testing as a Service (TaaS) on clouds. In: IEEE seventh international symposium on service-oriented system engineering
8. Capgemini and Microfocus (2018) World quality report. https://www.capgemini.com/service/world-quality-report-2018-19/
9. Enov (2016) Test environment management. http://enov8.over-blog.com/2016/12/test-environment-management.html
10. King TM, Ganti AS (2012) Migrating autonomic self testing to the cloud. In: Third international conference on software testing, verification, and validation workshops
11. Sharma M, Keswani B, Pathak V (2017) Cloud testing: enhanced software testing framework. Intl J Eng Sci Math 6
12. Akamai Documentation (2019) Load test creation tutorial. https://learn.akamai.com/en-us/webhelp/cloudtest/load-test-creation-tutorial/GUID-96813E58-C506-4B62-86D2-056A2EDB469A.html
13. Huma Warsi (2018) https://geekyduck.com/7-important-cloud-based-tools-for-load-testing/, https://geekyduck.com/7-important-cloud-based-tools-for-load-testing
14. Smartbear Cross Browser Testing (2019) https://help.crossbrowsertesting.com/selenium-testing/frameworks/watir
15. Hossain M, Abufardeh S, Kumar S (2018) Frameworks for performing on cloud automated software testing using swarm intelligence algorithm: brief survey. Adv Sci Technol Eng Syst J 3(2):252–256
16. Hosseini S, Nasiri R, Shabgahi GL (2015) A new framework for cloud based application testing. Intl J Sci Eng Appl Sci (IJSEAS) 1(3)
17. Zenker E, Shahpasand M (2018) A review of testing cloud security. Intl J Internet Technol Secur Trans (IJITST) 8(3)

Chapter 11
Towards Green Software Testing in Agile and DevOps Using Cloud Virtualization for Environmental Protection

D. Jeya Mala and A. Pradeep Reynold

Abstract Among the software engineering activities, software testing is a crucial one which consumes more than 50% of total cost and time needed in the development process. As quality is the most important criterion for successful delivery of the software, complete testing is the only way to achieve it. The various surveys conducted during the past few years reported not only the problems of exhaustive testing but also the problems associated with energy consumption and the overall impact on the environment due to dedicated hardware and other infrastructure resources utilized for testing. In traditional test environment, the quality management and testing activities are performed using the dedicated environmental set-up. This in turn alarmingly increases the amount of carbon emission in the environment. Hence, this chapter provides a key solution to make this higher energy-consuming task into a less energy-consuming one. The objective of this chapter is twofold: firstly, to provide a green software testing framework using cloud-based virtualization, and secondly, to apply cloud-based testing in Agile and DevOps-based software development environments. This chapter focuses on an important paradigm shift from traditional testing with dedicated resources to a cloud-based testing solution to achieve environmental protection. Hence, the testing activities which include the test case generation and execution to deliver quality software can now be achieved by means of cloud-based virtualization and by means of service on the cloud termed as TaaS (testing as a service).

Keywords Software testing · TaaS · Testing as a service · Green software testing · Cloud virtualization · Agile · DevOps

D. Jeya Mala (✉)
Fatima College, Madurai, Tamil Nadu, India
e-mail: djeyamala@gmail.com

A. Pradeep Reynold
Hubert Enviro Care Solutions Pvt. Ltd., Chennai, India

© Springer Nature Switzerland AG 2020
M. Ramachandran and Z. Mahmood (eds.), *Software Engineering in the Era of Cloud Computing*, Computer Communications and Networks,
https://doi.org/10.1007/978-3-030-33624-0_11

11.1 Introduction

The importance of environmental impact assessment and protecting the environment from industrial set-ups has gained much attention in recent past as it highly affects the human life. This is applicable not only to the manufacturing industries but also to the information technology (IT) and other industries. In the IT scenario, dedicated infrastructure resources such as servers, client machines and other needed hardware in a software development environment are the most significant contributors to environmental pollution.

This has made the IT industries to think about a paradigm shift from traditional resource set-up to a green computing environment to reduce the carbon emission in the environment. This leads them to moving towards "green computing" or "green IT" environment. These new buzzwords helps us to think about reducing energy consumption and their greenhouse gas emissions. Besides saving energy and aiming for efficiency, green computing is a complex trade-off between efficiently using any required resources and keeping the environmental impact lower.

"Greenness" in the software development process is an emerging quality attribute that takes into account each and every phase of the development life cycle of the software to be developed. This requires methods and techniques that support the identification, realization and measurement of software solutions that helps make the infrastructures to have virtual processes, dematerialization, smart grids and clouds [1, 2]. Typical examples are applications that help reduce energy consumption in production and testing in software engineering process. The analysis of all factors that have an environmental impact and the search for the optimal trade-off therefore has to be included in software development methods.

Although there are several ways to achieve this green computing environment, this chapter focuses on applying cloud computing-based solutions to achieve the objective of providing a green software testing framework. As per a report by Accenture [3], performing business applications on the cloud will significantly reduce the carbon footprint of an organization. This report also states that carbon emission is dramatically reduced up to 90% in small businesses when cloud-based virtualization is used. Similarly, for large and mid-size businesses this emission has been reduced from 30 to 90% [2].

According to Gartner [4], within the time of next 5 years, almost all the software development organizations will move to cloud computing-based solutions for one or the other development activities. As the industries are currently applying Agile and DevOps-based software development process models for their application development, this move will tremendously reduce the carbon emission in the environment [5]. As these development models require continuous testing to achieve continuous integration and continuous deployment (CI-CD), the cloud virtualization and cloud-based services will help the environment to be safe and better in every way.

Even though the software engineering activities performed for software development are time, cost and resource consuming, software testing is a crucial one which consumes more than 50% of total cost, time and resources needed in the development process. As quality is the most important criterion for successful software development, complete testing is the only way to achieve it. But, as Phil McMinn briefed in his paper [6], exhaustive enumeration of a program's input is infeasible for any reasonably sized program, yet random methods are unreliable and unlikely to exercise deeper features of software that are not exercised by mere chance. This survey not only indicates the problem of exhaustive testing but also reveals the problems associated with energy consumption needed to perform this process. The overall impact on the environment due to hardware resource utilization to perform various quality management and testing activities is enormous in traditional development environment.

In these circumstances, testing as a service (TaaS) in the cloud environment is an important paradigm shift today to achieve green software testing. Hence, the testing activities which need the test case generation and execution based on the data from the input domain can now be performed by means of cloud-based virtualization. Some of the testing activities that can be done using cloud and which can be provided as a service on the cloud are: unit testing, load testing, regression testing, system testing, etc. Some of these are later discussed in this chapter.

Also, it will be easier to motivate the IT industries, to move to green software testing environment because of their lower capital costs. Especially, where tests that involve databases, operating systems, software or memory configurations not currently owned by or available in the company can be run by tapping into cloud infrastructure, cloud platform and cloud-based software services [2]. Here, the organizations benefit by only paying for the cloud services used. This means that if the tests show that the applied configurations are not suitable, only limited time and expense will be lost compared to lending, renting or purchasing, setting up, utilizing, breaking down and returning the same in a dedicated resource set-up. Cloud providers are normally geared towards quick access to hardware, platform and software services. Once the company has established relationships with the cloud providers, subsequent similar requests can be serviced even faster. This reduces the overall impact over the environment as the idle resources which are available in the form of cloud are used to do this more importantly, time and cost consuming activity.

This chapter has been divided into the following sections: Sect. 11.2 is about the characteristics of cloud computing and types of cloud services; Sect. 11.3 describes the features of green software engineering; Sect. 11.4 demonstrates green software testing on the cloud with types of testing that can be done on the cloud; Sect. 11.5 provides details about how the cloud vendors are providing testing as a service; Sect. 11.6 focuses on how cloud testing is beneficial in modern software development models such as Agile and DevOps; finally, Sect. 11.7 concludes the chapter on the importance of cloud virtualization to protect the environment from huge carbon emission due to heavy process loads in the dedicated hardware platforms.

11.2 Cloud Computing and Services on the Cloud

Cloud computing can be viewed from two different perspectives: cloud application and cloud infrastructure. The goals of cloud computing are to have high scalability and high availability of resources based on the need and as required.

A cloud is a combination of infrastructure, software and services that are distributed over a network and is not local to a user [1]. In this scenario, the infrastructure and services can be accessed through a web interface or a mobile interface. The cloud exhibits a number of highly desired characteristics, namely location independence, scalability, extensibility, reduced costs, less maintenance and on-demand access. Some of the most crucial benefits of cloud computing include [2, 3]: (i) easy to use and efficiency, (ii) provision of software as a service and (iii) on-demand, pay-as-you-go provision of computing-related resources. The cloud model helps to create a flexible and cost-effective means to access computer resources. A cloud-based IT service management is shown in Fig. 11.1.

At the top-level layer, the user interface (UI) components and UI process components are provided. Using them, the users can get the services from the cloud. The next (middle) layer contains service-related components such as service bus, service registry, orchestration, common services and service agent to get the appropriate services from the cloud.

At the bottommost layer, the databases/middleware and virtual execution engine are housed. Here, the hypervisor is present to provide cloud virtualization for any operating system-related services. In this layer, service-level agreement (SLA), configuration management, infrastructure management and identity and access management components are present to provide the services according to the clients needs.

Garg and Buyya [7] have identified that the green policies to use cloud environment for software development activities reduce the carbon emission by almost 20% in comparison with profit-based policies. Their observation states that, based on the users' needs, the cloud resources can be used which helps to achieve on-demand-based service which is the backbone of green computing. So, the organizations should devise green policies so that major computational activities that consume more energy can be shifted to the cloud.

Traditionally, there are three different service/delivery models (Saas, PaaS and IaaS) which help organization in adopting the right one as per their business/testing requirement. Now, as testing has been identified as a service on the cloud, TaaS has also been added as a cloud-provided service for organizations. These delivery models are briefly mentioned below.

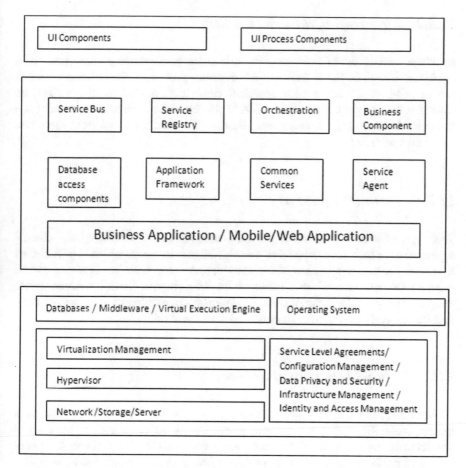

Fig. 11.1 Cloud-based IT service management

11.2.1 SaaS (Software as a Service)

In this type, a multitenant architecture is provided in which software resides as a service on the cloud. Here, more than one customer can access the same software without compromising the privacy and security of their data. By using this architectural model, the users can use their Web browsers to access the needed software available on the cloud. There is no need to invest in customer-owned costlier software products or in licensing fees; there is no need to invest in huge infrastructures such as server-side software. This helps in reducing the need for dedicated infrastructure to have the required software for independent utilization.

11.2.2 IaaS (Infrastructure as a Service)

In this provision, the IT infrastructure such as hardware devices with pre-defined configuration can be leased from the cloud service providers. The companies which are in need of hardware with specified configuration can get the virtualized resources on pay per use and time basis. In this case, the companies need not have the dedicated hardware infrastructure; instead, they can use them from the cloud based on their need.

11.2.3 PaaS (Platform as a Service)

This type of service is used to provide virtualization on executing the client's applications in their target operating system (OS) environment which is provided as a service on the cloud. Also, the clients can develop their applications on the required OS platform without having it in their internal working environment. Hence, PaaS provides a distributed development environment for an organization that has several development branches all over the world.

11.2.4 TaaS (Testing as a Service)

Cloud testing is termed as TaaS (Testing as a Service). This includes functional testing, scalability testing and non-functional testing activities such as the ones to test security, stress, load, performance and interoperability of numerous applications and products. Cloud testing not only helps in the redundant testing activities such as regression testing but also helps in stress and load testing cases too in which the product's load level can be easily calculated using a number of user requests received by means of cloud service requests.

11.3 Green Computing

This section elaborates the green computing characteristics and software testing in the green computing paradigm.

11.3.1   Characteristics, Promise and Benefits

Green computing is a buzzword which is used nowadays to indicate the need for reducing carbon footprint and its impact on the environment. It is used in devising efficient ways of increasing the practices of using computing resources efficiently with the objective of reducing their impact on the environment.

The enormous use of servers, desktop machines and other computing resources, to perform from a simple computational activity to large computationally high complex activities, is on the increase [1, 7]. In these circumstances, instead of having independent computational resources for every computational activity, if the resources are available in the cloud environment and getting these resources based on the need, will tremendously reduce the total carbon emission on the environment.

This cloud-based solution is becoming highly attractive to organizations to achieve distributed resource sharing, on-demand services, effective utilization of resources and reduction of idle computational resources. However, the major impact of this cloud-based virtualization lies in green computing and its positive effects on the environment. Several industries are offering their resources using a cloud-based environment either as a private, public or a hybrid cloud. This helps in increasing the companies to concentrate more on business severity areas in computation which in turn helps them to grow faster in a different direction without wasting their resources idle and unnecessarily emitting carbon to the environment. The efficient use of cloud services drastically improves the distribution of workload when more jobs are requesting the resources at the same time. In any of these cases, the computational resources are effectively utilized which thus paves the way for green computing.

11.3.2   Green Software Engineering

Green computing requires countermeasures to reduce or remove the threats in the environment. Using this assertion and the IEEE definition of software engineering, one can suggest that *green software engineering is software engineering in the green computing paradigm*. In green software engineering, the following best practices need to be adopted:

- Developing code in such a way that they consume less CPU processing cycles
- Reducing the use of hardware resources and reduction of energy consumption by computing centres
- Reducing the hard copy printouts and multiple posts to various devices
- Adapting cloud computing environment and cloud-related tools.

The energy-aware computing and the software-intensive technology plays a crucial role in reducing the environmental impact of computing. The green software engineering best practices help to reduce the environmental impact of the software product development processes [8].

As per a survey given in a white paper, the ICT industry emits 2% of global CO_2 emissions, most resulting from the power consumption of PCs, servers and cooling systems [9]. It has also been observed that green software development activities will greatly improve energy efficiency, lower greenhouse gas emissions, encouraging efficiency, usage of less harmful materials and encouraging reuse and recycling [7].

11.3.3 Environmental Impact of Software Testing as a Process

Software testing is an activity employed to reveal any errors in the software product before it is delivered to the customer side [6]. The industries are spending more than 50% of their cost and time in this most important activity as otherwise a single fault will make the entire software to fail. To achieve the near-zero-defect quality software, the testing team and the developers will generate enormous amount of test cases to exercise the code and the final product to reveal the errors in it. Each execution of such test cases requires infrastructure, computing resources and other related software and hardware components. This in turn emits carbon footprint in the environment for each of the test case executions.

In case of regression testing which aims to re-execute the same set of test cases, whenever a change or a defect fix has been made in the released build, it requires running of large amount of test cases from the test repository and thus consumes considerable amount of computational resources.

As the testing activities are recurrent in nature, they need to be done repeatedly using different sets of test cases for the same software. The traditional testing methods generally consume high resource utilization even though the activity is same but being done from different geographical locations. If the number of testers or developers using this software is dynamically increasing, the amount of computational resource consumption will also be proportionally increasing. This has the adverse effect on the environment as each of these users will have their own dedicated set of resources which will include hardware and software resources to do the testing activity.

11.4 Green Software Testing on the Cloud

Green software testing can be achieved by taking advantage of the cloud computing paradigm. It helps in concurrent execution of test cases on the cloud. The usage of cloud virtualization in terms of hardware, software and other infrastructures and services makes the green testing process to migrate the legacy testing assets to the cloud. In software testing, the software development industries generally dedicate infrastructures to do effective testing of software in order to release quality software product to the customer side. This not only increases their capital expenditure but also increases the carbon emission in the environment.

Under these circumstances, cloud computing offers a way to reduce this impact by providing the required resources and services to be accessed from the cloud on need basis [1]. So, in the organizations even though the offices are geographically distributed, the testing operations can be done on the cloud without any problem in synchronized execution. As the resources are used only on need basis, the administrative and maintenance costs are also considerably reduced.

Performing such a complex and redundant testing activity on the cloud results in green testing. It provides the following benefits:

- Cost effective and efficient testing on-demand
- Standardized test processes based on the type of testing
- No need for individual set-up of test tools and test environments
- Reduction of dedicated infrastructure
- On-demand based test case execution reduces carbon emission.

As testing can now be migrated to the cloud, the on-premise need of all the testing resources is considerably reduced. This helps in achieving the on-demand-based access to the cloud depending on the current testing need. This has been shown in Fig. 11.2 which depicts a complete use case diagram to perform software testing on the cloud.

As illustrated in Fig. 11.2, the developers can load the software under test (SUT) to be tested on the cloud. The developers and testers can generate test cases using either manual testing or automated testing. Then to execute the test cases, the resource identification process will be done.

At this point, the decision making on choosing cloud to perform the redundant test activities with on-demand based resource utilization is made. Once, the cloud service has been identified, the task is assigned to the cloud and then the cloud server will execute the test cases based on the need. The test results can be stored in the test repository for further use. Also, test reports are generated and viewed by the stakeholders as shown in Fig. 11.2.

This helps not only to greatly reduce the cost but also to reduce the environmental impact of having dedicated resources for these kinds of redundant activities. In the traditional testing environment, the organizations need to establish the required infrastructure to perform testing in an efficient manner. Also, because of the cloud set-up, the organizations are free from having a dedicated infrastructure and the required resources to perform the testing process in their premise. Rather, by means of service-level agreement (SLA), they can move the most important and crucial life cycle phase in the software development process such as testing to the cloud.

Nowadays, as the cloud vendors are providing a high level of security in providing the service, the organizations can redeem the testing service without any

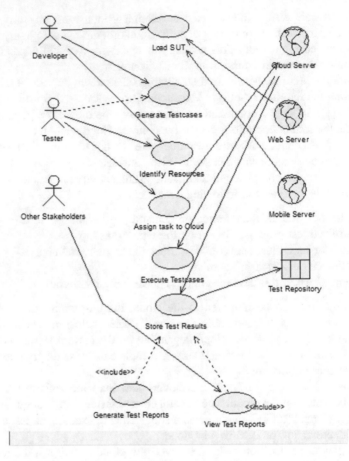

Fig. 11.2 Green software testing on the cloud

worry about data breaching. This helps the organizations to get the service as fast as possible as the cloud is already equipped with all the processes such as initialization, processing and execution of testing operations.

There are two types of cloud testing services provided by the organizations, namely: (i) on-premise—in which case, cloud testing can be used for validating and verifying different products owned by individuals or organizations using private cloud, and (ii) on-demand—which is getting increasingly popular nowadays, and it is used to test on-demand software. The possible solutions given by cloud-based green software testing are: (i) storage area to store the testing tools, (ii) expected target configuration to test the software and (iii) distributed testing environment for geographically distributed teams.

Thus, cloud virtualization reduces the execution time of testing of large applications and leads to cost-effective solutions.

11.4.1   Types of Testing Performed on the Cloud

This section provides an overview on how different types of testing can be done on the cloud.

Load Testing

This is used to evaluate the performance of an application when a huge amount of user requests is entering the server at the same time. For instance, on a university website, the number of candidates who try to download the applications during the time of admission may be huge. So, the server may go down if it cannot handle that many numbers of requests.

To avoid these situations, the load testing helps in measuring the maximum limit of user requests that can be serviced by the server without any problem in the response speed and the response itself. Cloud-based testing helps in achieving this in real time as the service requests will be received at any time and so it is easy to measure the server's capacity to handle the heavy load instead of artificially creating network traffic for evaluation.

Performance Testing

The performance level of an application and its limitations should be identified to provide a hassle-free operation for the clients. This can be tested using performance testing in which the performance of an application is tested using different workloads.

The cloud testing environment helps to create such a test bed by varying the clients' requests based on their need. This helps to evaluate the performance level of the application without dedicating a personalized server for it and evaluate the performance by varying the number of client requests. This improves the test efficiency in terms of cost and also decreases the carbon emission on the environment.

Functional Testing

This form of testing helps to evaluate the system functionalities against the specification document. The software requirements specification (SRS) document helps to derive the system requirements and user requirements, and these are used to evaluate the functionality of the developed product.

This process is generally termed as black box testing or validation testing. The cloud testing helps to achieve this functional testing, by executing the test cases generated from the requirements against the developed software. In traditional working environment, the execution of test cases on the same software under the same set-up is a time and resource consuming process, whereas in the case of cloud-based environment, this is overcame by providing this execution process as a service.

Stress Testing

This testing is performed to evaluate the stability of an application when an excessive number of requests have been sent to it. To evaluate it, the organizations are using simulators to artificially increase the loads and find out the efficiency level of the application.

This testing is used to validate whether the application works under heavy loads and the sustainability of the application. As the artificial creation of such loads in a dedicated in-premise environment will be costly and also incurs more cost, the industries are going for a cloud-based solution in which such a huge number of requests can be easily generated in the cloud as real-time requests. This reduces the carbon emission for such testing of varying loads on the dedicated servers.

Compatibility Testing

To ensure that the software works equally well in different operating systems, the cloud based solution is recommended as a cloud service. In this case, the cloud provides different operating system virtualizations and so the software compatibility can be easily tested.

Browser Performance Testing

The website providers can validate their website's performance in various web browsers using cloud. The cloud service supports the validation of browsers when a particular website is opened in them.

Latency Testing

Latency refers to the delay in time taken between the request and response. This time can be easily reduced by means of cloud testing as the applications' executable file can be sent to the cloud service, the request and response sequences will be evaluated and the latency can be determined to assess the response time of the application.

System Integration Testing (SIT)

This type of testing is used to validate the working of all the developed software components in the current infrastructure and under the stated environments when integrated together to produce a complete build.

In SIT, a cloud service helps to provide the expected environment and provides evidences that the current integration in a particular working environment does not affect the same integration to work in some other working environment.

User Acceptance Testing (UAT)

Both alpha and beta testing can be done on the cloud as the development team needs the satisfaction level of the customer/customer representative or the end-users. This helps to assess whether the business needs have been provided by the software.

Security Testing

Cloud testing also helps to ensure whether the data and functionalities provided over the cloud are secure. This helps to protect the crucial information processed in the system. This testing also helps to verify the privacy and secured access to the application data.

11.5   Cloud Vendors' Provision of TaaS

As shown in Fig. 11.3, user scenarios are collected and the test cases are designed by the organizations that need testing as a service. Once this is completed, the test service providers leverage cloud servers provided by cloud platform vendors to do the appropriate testing. For instance, to do load testing, in order to find the website's performance on heavy load times, the test service providers need to generate the web traffic artificially and test the load level of the software. In the case of cloud testing environment, the clients' requests to use the web service are used as test data which is usually the actual traffic data and not the artificial one. Hence, it is highly efficient to test the application for its performance level at high load times using cloud testing.

Cloud testing is often used for performance or load tests; however, as discussed earlier, it also covers many other types of testing. To perform the test execution, the testing tool will offer a test service that the client can utilize, as testing tool services provided by vendors can generate the test reports for further analysis.

If any of the desktop or mobile application needs to be tested, then the corresponding testing tool set-up will be done as a cloud service. The service provider can provide access to the client or the software organization to test their stand-alone applications in this target environment without any issues.

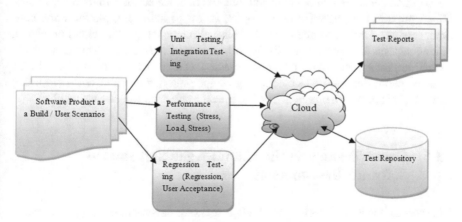

Fig. 11.3 Testing as a service on the cloud

11.5.1 Benefits and Issues of Cloud Testing

Some of the benefits of cloud testing are improved reliability, scalability and quality assurance, as well as reduction in cost, time and capital expenditure. Cloud testing helps the development team to work together with the testing team as they can access the resources on the cloud at any time. The bugs which are reported by the testing team can be easily assessed, and defect fixing can be done by the developers in the infrastructure provided by the cloud without any overlapping. Hence, testing efforts are improved and the status can be easily tracked by both the teams.

Some of the issues in cloud testing are: (i) high initial set-up cost; (ii) legacy systems and services that need to be modified in order to be tested on cloud; (iii) security issues during data transfer; (iv) lack of completeness and correctness of test results; (v) some cloud virtualization set-up may be specific to some application types and cannot be generalized. Test data management is also critical due to variations in regulations across different geographical regions. Companies might be resistant due to budgets. Cloud-based test service providers should provide transparent pricing models so that customers are equipped with sufficient information for budgeting and cost estimation. The issues in cloud testing may be crucial in some aspects, but in several of the cases, the cloud testing helps the organizations to have the testing environments at lower costs even though the initial set-up cost will be high.

For tests that involve testing of applications on different database servers, operating systems, memory and processor configurations, the cloud testing appears to be a boon. As it does not require the organizations to establish all the resources on their premise, the total cost will be very low. Also, this in turn reduces the total carbon emission on the environment due to individual deployment of such servers which needs to be active always.

Hence, the organizations gain benefits by using the cloud services based on their need. The tests are going to be done whenever a change has been incorporated into the product. Also, the redundant executions can be done on the cloud on need basis. Cloud providers are geared towards quick access to hardware, platform and software services. Once a company has established relationships with cloud providers, subsequent similar requests are likely to be serviced even faster. This reduces the overall impact on the environment as the idle resources which are available in the form of cloud are used to do this more importantly, time- and cost-consuming activity. This in turn helps in achieving greener effect on the environment.

11.6 Green Testing on the Cloud: Agile and DevOps Software Development

This section focuses on Agile and DevOps software development and green testing in the cloud computing environment.

11.6.1 Agile Software Development with Green Software Testing

Agile software development is the most promising software development process model employed by industries to achieve greater customer satisfaction. Agility refers to embracing change. This indicates that the changes given by the customers and by the development and testing teams will be accommodated in the software development process to achieve quality software product.

The traditional software development processes are too heavyweight or cumbersome. Too many things are done that are not directly related to software product being produced. The current software development is too rigid to adopt the changes. There is a great difficulty with incomplete or changing requirements to accommodate in traditional development models. The current need is to have short development cycles with more active customer involvement. Agile software development imposes self-organizing teams. The product development generally progresses in a series of month-long "sprints". The requirements are captured as items in a list of "product backlog". As there are no specific engineering practices prescribed, the approach uses generative rules to create an agile environment to deliver projects.

Scrum is an agile process that allows to focus on delivering the highest business value in the shortest time. It allows to rapidly and repeatedly inspect actual working software (every two weeks to one month). The business sets the priorities. The teams self-manage to determine the best way to deliver the highest priority features. Frequently, anyone can see real working software and decide to release it as is or continue to enhance for next iteration.

When the industries focus on changing to aggressive software development such as via Agile, naturally that poses a question of how to do the testing process as testing does not occur at the end as in waterfall model or even at the end of each phase. Rather, the testing of the software, in this approach, is done whenever some change is incorporated into the software.

In this case, the same set of testing activities need to be performed every time a change request or a new feature addition has been done during the development. If this testing process is done as in traditional industrial set-up having dedicated infrastructure with required hardware and software resources, the amount of carbon emission in establishing such permanent set-up in the premises becomes alarmingly high. Also, whenever the organizations are involved in the development of their own products, the testing process is the same for different types of testing.

In this case, instead of having dedicated infrastructure in each of the organizations to perform same testing operations every now and then based on the need, it is a better initiative to migrate to cloud-based testing. Now, IaaS and TaaS come into the picture and the organizations are free to use the resources and performing the testing activities according to their need. This pay-per-use approach will highly reduce not only the cost but also the carbon emission on the environment because of the dedicated infrastructural aspects. Hence, Agile software development can also provide a complete support to the environment if the paradigm shift is to the cloud-based resource sharing and on-demand-based testing.

11.6.2 DevOps Development—Overview

As discussed earlier, Agile approach is focusing on the delivery of a quality software product by following the test cycle model. The term DevOps is a current industrial buzzword that couples two important principles: continuous integration and continuous delivery (CI-CD) [10, 11].

According to an industry whitepaper [12], DevOps is defined as a software development and delivery process which considers the production of software from end to end, from concept to customer satisfaction. DevOps is a new challenging industrial strength software development model, in which all the stakeholders are involved in the software development. Here, the software development will be continuous until the customer satisfaction is achieved. As per the definition, the term DevOps can be considered as:

$$DevOps = Development(Dev) + Operations(Ops)$$

In DevOps, the software development (Dev) [13] is a process for creating software product using the various life cycle phases such as requirement analysis, design, coding, testing, implementation and user acceptance testing and software operations (Ops) [13] is a process of making the developed product to be in use in an operational environment and supporting the users in using it through the processes such as installation, upgrade, migration, operational control and monitoring, configuration management and support. The key characteristics of DevOps include [11, 4]:

- Automation of development, testing and deployment
- Monitoring the progress by means of metrics and measures
- Performance analysis by means of analytics using performance data and customer behaviour analysis
- Collaborative development environment.

As shown in Fig. 11.4, DevOps requires CI and CD, which are essential to develop and test the software frequently as and when it is needed to add a new feature or a new defect fix or a change has been done in the software. At this juncture, the industries are looking in for cloud-based solutions for automation in development and testing [12].

As indicated in Fig. 11.4, the CI includes various stages including the following:

- Requirement elicitation
- Module development

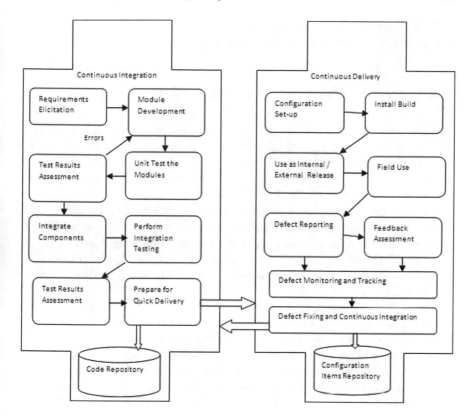

Fig. 11.4 CI and CD in DevOps development

- Unit test modules
- Assessment of test reports
- Debugging in module development
- Component integration
- Integration testing
- Test report assessment
- Quick delivery of the developed product and move to CD phase
- Store code in the repository for maintenance and support.

In CD, the following phases are present as shown in Fig. 11.4:

- Configuration set-up
- Initial build preparation
- Internal/external release
- Field use of the released product
- Defect reporting
- Feedback management
- Defect monitoring and tracking

- Defect removal and continuous integration
- Configuration management using repository.

11.6.3 DevOps with Green Software Testing

Nowadays, the industries are employing cloud computing as a driving force of DevOps, as it is used to establish the above characteristics faster with higher efficiency level [10]. According to Schneider [12], the connectivity between the development and the back-end infrastructure can be achieved by means of cloud services such as IaaS, PaaS, SaaS and TaaS.

For cloud applications, some of the areas that can benefit from the cloud vision include: (1) software deployment testing, namely system testing under different configurations, (2) software design and implementation using software as a service on the cloud, (3) software defect reporting and fixing on the cloud, (4) monitoring the data transfer for security concerns and (5) usage of necessary tools available on the cloud [14].

Many of the software development companies are using DevOps in cloud-based application development. Many of them have adopted the use of cloud in performing redundant activities done in the DevOps process [4].

To achieve greener effects in the software development process in DevOps, the cloud-based solution helps in monitoring the changes incorporated in the software using a version control system provided in the cloud of the organization. This, in turn, leads to the reduction of dedicated infrastructure at the organizations' offices which in turn reduces the environmental impact of such resources.

The development and testing parts are done in parallel in DevOps in which the construction of the build and integration of various components based on the need are done at the development site itself, whereas testing can be migrated to the cloud which helps in doing the automated evaluation of the software using test scripts and testing tools which are available as services on the cloud.

The next stage in DevOps prior to actual production and release of software product is performing system testing and user acceptance testing activities. As discussed earlier, system testing involves various non-functional quality attributes-based testing. Some of these typically include security testing, performance testing and compatibility testing. These can be done more effectively on the cloud which greatly reduces the impact on the environmental aspects.

11.6.4 Automation in DevOps Principles

In DevOps, the processes included are: continuous test, integration, delivery and deployment. The teams in DevOps working environment are divided as development, operations, testing and quality assurance. As per the SEI standard, the DevOps principles include collaboration, infrastructure as code, automation and monitoring. All these principles are achieved by means of the tools. For instance, collaboration which aims at efficient communication is achieved using Azure DevOps, JIRA, Slack, etc.

The infrastructure as code focuses on version control for development, testing, production, etc., and normally includes tools such as OpenStack, AWS, Docker and Chef. The automation principle aims at providing guidelines for automating all internal processes in order to increase the frequency of testing, continuous integration, delivery and deployment. It is achieved using Jenkins, AWS DevOps, Docker, etc. Finally, the monitoring process provides data needed for development and operational decisions and is performed using tools such as CloudWatch, CelloMeter, OpenStack, etc.

The processes such as continuous integration, continuous delivery or continuous deployment are required to be done continuously and hence need to be automated. If these are done manually, it will consume more time and also error prone. The sub-processes that can be automated in integration are automated submission of individual tested units for integration, visibility of reports to the developers, testers and operation professionals for security and management. In continuous delivery, changes need to be pushed to the customer for their evaluation and feedback or a mock product has to be provided in the target environmental set-up for experimentation and evaluation. In continuous deployment, the changes are pushed into production continuously.

All these processes need automation; the DevOps working team uses tools to perform the processes quickly and efficiently. To ensure that automation is successful, cloud-based virtualization helps in providing an efficient solution. Tools such as virtual machines (VMs), Containers and Dockers help with infrastructure-level virtualization.

For containerization, OS-based virtualization in terms of sandboxed instances helps in achieving shared OS kernel which removes the need for guest OS or hypervisor. The container instances are built from distinct OS images and run natively in host OS where VM runs virtually as guest. Since the images contain only the application and library dependencies, the runtime loading of all the relevant files is avoided which thus improves execution time and resource utilization. The execution environment is now lightweight and portable.

Docker is an open platform used for development, shipping and running of the application using container technology. Generally, the Docker engine has server, REST API, command-line interface (CLI) and registry, and the automation in the cloud helps reduce the energy consumption.

Orchestration tools and micro-services such as Docker engine, Docker Compose and Docker Swarm help in providing virtual hosts, defining YAML (YAML Ain't Markup Language) files and cluster management as service model.

Cloud orchestration employs OpenStack which adopts template-based service execution. The infrastructure resources that can be described include: servers and floating point IPs. The tools used to achieve it are Chef and Puppet. They are very much compatible with AWS cloud formation. They also support containers using OpenStack.

Hence, DevOps is greatly utilizing cloud computing in all its processes which thus reduces the overall impact on the environment as the utilization of the resources is only based on demand.

11.7   Conclusion

Recent studies have indicated that, compared to the traditional data centres that are placed in premise of an organization, the cloud computing solutions emit less carbon footprints. Due to cloud virtualization, resource sharing and distributed workload, the cloud data centres are providing energy efficient solutions, thus reducing the environmental impact.

The traditional data centres need to handle all the Web requests in their dedicated Web servers which need to be available 24×7. During lower workloads, this results in low resource utilization and energy wastage, whereas cloud data centre can reduce the energy consumption by means of sharing of workloads in cloud-based virtual environments and not in dedicated physical servers for each such request.

Many of the software development companies are using DevOps and Agile approaches in cloud-based application development. Many of them have adopted the use of cloud in performing redundant activities done in the DevOps and Agile processes.

To achieve greener effects in the software development process in DevOps and Agile, the cloud-based solution helps in monitoring the changes incorporated in the software using a version control system provided in the cloud of the organization. This leads to the reduction of dedicated infrastructure at the organization offices which in turn reduces the environmental impact of such resources.

As discussed earlier, when cloud computing is used for testing, it further reduces the carbon emission in the environment as the testing activity is a redundant activity which requires more infrastructure, software and platforms to evaluate the software. By using this cloud migration for testing activity, the testing can be done more efficiently with low energy consumption compared to the traditional dedicated in-premise set-up. For a better cloud testing experience, test team should adopt a robust strategy that better caters for their business needs.

References

1. Beloglazov A, Buyya R, Lee, YC, Zomaya A (2011) A taxonomy and survey of energy-efficient data centers and cloud computing systems, advances in computers. In: Zelkowitz M (ed). Elsevier, Amsterdam, ISBN 13: 978-0-12-012141-0
2. Buyya R, Yeo CS, Venugopal S (2008) Market-oriented cloud computing: vision, hype, and reality for delivering it services as computing utilities. In: Proceedings of the 10th IEEE international conference on high performance computing and communications, Los Alamitos, CA, USA
3. Accenture Microsoft Report (2010) Cloud computing and sustainability: the environmental benefits of moving to the cloud. http://www.wspenvironmental.com/media/docs/newsroom/Cloud_computing_and_Sustainability_Whitepaper_Nov_2010.pdf
4. Callanan M, Spillane A (2016) DevOps: making it easy to do the right thing. IEEE Softw 33 (3):53–59
5. Anderson E, Lim SY, Joglekar N (2017) Are more frequent releases always better? dynamics of pivoting, scaling, and the minimum viable product. In: Proceedings 50th Hawaii international conference on system sciences, pp 5849–5858
6. McMinn P (2004) Search-based software test data generation: a survey. Res Artic Softw Testing Verif Reliab 14(2):105–156
7. Garg SK, Buyya R (2012), Green cloud computing and environmental sustainability. Available at: http://www.cloudbus.org/papers/Cloud-EnvSustainability2011.pdf
8. Kashfi H (2017) Software engineering challenges in cloud environment: software development lifecycle perspective. Intl J Sci Res Comput Sci Eng Inf Technol 2(3):251–256
9. Gartner (2019) Gartner forecast. Available at: https://www.gartner.com/en/newsroom/press-releases/2018-09-12-gartner-forecasts-worldwide-public-cloud-revenue-to-grow-17-percent-in-2019
10. Balalaie A, Heydarnoori A, Jamshidi P (2016) Microservices architecture enables DevOps: migration to a cloud-native architecture. IEEE Softw 33(3):42–52
11. Bass L, Weber I, Zhu L (2015) DevOps: a software architect's perspective. Addison-Wesley Professional, New York
12. DevOps (2016) White paper. Available at: https://test.io/devops/
13. Lwakatare LE (2017) DevOps Adoption And Implementation. In: Software development practice, concept, practices, benefits and challenges. Acta Universitatis Ouluensis, 2017 ISBN 978-952-62-1710-9
14. Cito J, Leitner P, Fritz T, Gall HC (2015) The making of cloud applications: an empirical study on software development for the cloud. In: Proceedings of 10th joint meeting on foundations of software engineering. ACM Press, New York, pp 393–403

Chapter 12
Machine Learning as a Service for Software Process Improvement

Supun Dissanayake and Muthu Ramachandran

Abstract As the technology evolves, software engineering companies must follow new methodologies to implement software process improvement (SPI) to enhance their current practices. It is evident that the rise of big data processing and machine learning allows the development of new technologies to reduce manual workloads in organisations. Therefore, this chapter focuses on proposing new methods to improve SPI in organisations by developing a new maturity model and combining it with machine learning techniques. It proposes the Measurable Capability Maturity Model (MCMM), which contain measurable metrics as its key process indicators to identify maturity levels of software development companies. Compared to existing maturity models in the industry, MCMM has measurable metrics that identify maturity-level achievements through mathematical calculations. These SPI metrics were identified and developed by knowledge obtained through the literature review and qualitative research. Moreover, this chapter proposes a prototype application MLSPI (machine learning for software process improvement) by combining MCMM with machine learning functionalities. It contains functionalities to extract information from a software development organisation and identify its maturity-level achievements through mathematical calculations used in MCMM. Moreover, it provides feedback for SPI and predicts data patterns variations for each metrics using machine learning functionalities.

Keywords Software process improvement · Maturity model · Machine learning · Actionable analytics · Software engineering · Key process indicators

S. Dissanayake (✉)
University of Colombo School of Computing, Colombo, Sri Lanka
e-mail: sjd@ucsc.cmb.ac.lk

M. Ramachandran
School of Built Environment, Engineering, and Computing,
Leeds Beckett University, Leeds, UK
e-mail: m.ramachandran@leedsbeckett.ac.uk

12.1 Introduction

Software process improvement is the implementation of measures to improve business processes in a software development company [1]. According to Paulk et al. [2], software processes are a set of practices, activities, methodologies and transformations that promote the sustainability and the development of software products.

These practices benefit organisations by improving the efficiency of the work processes through development time reduction and productivity level improvement. As a result, it provides a competitive edge for the organisation.

Prior to the application of SPI, it is vital to identify the structure of a software engineering organisation [2]. This allows SPI to be separated in terms of development processes, reliability management, cost analysis, etc. Then measurable metrics could be identified to apply SPI. Figure 12.1 depicts a simple example of an SPI framework.

As shown in Fig. 12.1, a simplified SPI process consists of assessment which aims to identify improvement needs using structure key practices, capability determination which aims to determine existing strengths and weaknesses in the current process, and finally proposes a set of recommendation for improvement. Maturity models can be used to implement software process improvement in organisations [2]. They allow the identification of the quality factors of the functions within a company [3]. Both CMM and CMMI were popular maturity models that were introduced by the Carnegie Mellon University in 1989 and 2002 [4]. They consist of key process areas (KPAs) to identify the maturity level; however, one of the most significant factors of these KPA is that they lack the ability to quantify the maturity level. This is due to the fact that most of these KPA are too vague. Therefore, this chapter provides a solution by developing the MCMM, which consists of measurable metrics as key process indicators (KPIs) for each KPA. Thus, organisations can produce a measurement of achievement for each KPA.

Furthermore, this maturity model was interconnected with the MLSPI prototype, which was developed as part of this study to provide feedback for process

Fig. 12.1 SPI life cycle

improvement. Machine learning (ML) will process results from SPI metrics and generate predictions for the organisation. This implies that the user will be able to check the current maturity status of the organisation and its changes depending on ML predictions. Moreover, it has the functionality to provide feedback to move up the maturity levels.

Thus, this paper proposes a methodology where machine learning can be used to aid software process improvement. It introduces a unique maturity model with specifically developed key process indicators that are measurable for software process improvement. Furthermore, it depicts the use of machine learning for software process efficiency prediction using the above maturity model and provides valuable feedback for business organisations. Furthermore, feedback gathered for the concept and the simulations carried out with the developed prototype depict its effectiveness in applying machine learning for software process improvement.

12.2 Overview of Software Process Improvement

Software process improvement (SPI) is a very popular aspect of the software engineering industry [5]. Software processes are human-centred activities that result in unintentional actions within the company [6]. Thus, it directly affects the quality of the software developed by the organisation. IEEE depicts software quality to be an absolute necessity to enhance processes, systems and components [7]. Software processes must be validated continuously in an organisation to guarantee the efficiency of the development criteria. Hence, weak development processes will be identified and rectified to improve the end product.

Multiple software engineering companies use defect management processes as tools to apply SPI [8]. During the development process, defects can appear at any development stages such as design, engineering, testing and deployment. These defects could be software failures, faults, errors, etc. However, these small faults can cause gigantic damages in terms of the company's reputation by deteriorating the quality of software. This clearly depicts the vital importance of having an SPI within a software development organisation. Therefore, it sets a background for research ideologies to develop to improve SPI techniques used in the industry, which support the rationale of this study. One of the methodologies that could be used for SPI is a maturity model.

12.2.1 Maturity Models

Maturity models are used in organisations to identify software process quality. Following subsections denote maturity models that could be used for SPI. These maturity models are critically evaluated to identify their advantages, disadvantages and relevance to this study.

This section aims to evaluate maturity models that are currently used and proposed in the industry for software process improvement. It pinpoints areas that could be improved and depicts how MCMM will take a similar or different approach to rectify current issues.

12.2.1.1 Capability Maturity Model Integration—CMMI

Carnegie Mellon University introduced the CMMI maturity model in 2002 to measure maturity levels of software engineering companies [9]. 5000 software development organisations around the world have adopted this methodology to their projects.

These maturity levels contain several KPAs [10]. Therefore, the company must achieve each KPA at its current level, and KPAs in the level above move up to the next maturity level. This will allow the organisation to improve its current business processes and focus towards adding further functionalities to enhance business processes. Moreover, CMMI is an accepted standard in the software engineering industry; therefore, companies can use it as an indicator to depict the quality of the organisation for their business dealings [2]. However, CMMI does not have KPAs that are applicable for development methods such as Component-based Software Engineering (CBSE). This denotes that it will not be beneficial for an organisation with CBSE practices along with traditional software development [11]. Thus, this implicates that a maturity model with KPAs that are applicable to a wide variety of development methods should be developed to solve this issue. Moreover, KPIs of each maturity level cannot be quantified using CMMI, which validates the requirement of a maturity model with measurable metrics to enhance current process improvement practices.

12.2.1.2 ISO/IEC WD 15504

ISO/IEC WD is a process model that is used by software engineering companies to assess the quality of software processes [12]. It validates capability in terms of risks, strengths, weaknesses, deployment success, etc. [12].

Compared to CMMI, ISO/IEC WD 15504 shows a similar progression from its first to the final level of capability maturity. CMMI focuses on software development processes in the company; however, ISO/IEC WD 15504 focuses thoroughly on individual processes [12], which prevent processes residing in one maturity level. Moreover, ISO/IEC WD 15504 is more informative and detailed compared to CMMI; multiple base practices are connected to each process compared to CMMI [12]. Therefore, suitable aspects of ISO/IEC WD 15504 that support both Agile and CBSE were chosen meticulously when the MCMM was developed.

12.2.1.3   Integrated Component Maturity Model—ICMM

ICMM is a maturity model that is specifically developed for CBSE compared to CMMI and ISO/IEC WD 15504 [13]. Similar to CMMI, it contains five maturity levels [11]. ICMM contains KPAs that are solely developed to support CBSE [13]; hence, it is advantageous for CBSE organisations compared to the use of CMMI. Moreover, the naming criteria of each level are adapted from the CBSE Lifecycle, which further validates its connection with CBSE compared to standard software engineering practices. However, it is important to denote that ICMM is not an industry-accepted method compared to CMMI since it still requires further research to validate its effectiveness [13].

12.2.1.4   Software Component Maturity Model (SCMM)

SCMM is another maturity model that assesses CBSE; it was developed by the University of Pernambuco, Brazil [14]. Its main functionalities are to assess the quality of software through risk identification, development techniques and evaluations [14]. Furthermore, it enhances reliability, improves software quality when changes were made to the software and improves cost-benefit analysis [14]. The reusability, quality and efficiency of the component are much higher if results are closer to the highest maturity level. Therefore, it proposes five maturity levels from SCMM 1 to SCMM 5. It specifically provides a set of techniques to measure the maturity depending upon programming language, development environment and domain platform [14]. This implies its provision of versatility to support a variety of different organisations. However, SCMM is still in its development stages, and it is not used in the industry [14]. Furthermore, it requires a cost model to enhance its characteristics to improve its evaluation criteria. This implies that SCMM is very similar to ICMM in terms of functionality and its industrial usage. Hence, a combination of KPIs in CMMI, ISO/IEC WD, ICMM and SCMM was used to develop KPA and KPI for the MCMM, which supports Agile software engineering and CBSE.

12.3   Measurable Metrics for SPI

MCMM is proposed to contain KPIs that are measurable to calculate achievability of KPAs of each maturity level through results gathered from qualitative research. Therefore, it is vital to identify measurable metrics that support SPI in terms of Agile development and CBSE. One of the research studies carried out by Dissanayake [15] is used to identify measurable metrics for the development of KPI of MCMM model. Supun Dissanayake is one of the co-authors of this chapter, and the content of this chapter is the development of research carried out in the above-mentioned research study.

Dissanayake [15] carried out an array of interviews with people who are working in software engineering organisations to identify the types of data that are collected in organisations from employees and development processes. Then using this research information, Dissanayake [15] identifies a set of metrics from the literature that could be used to quantify the process effectiveness for the collectable data in originations. Table 12.1 depicts the summary of processes and metrics identified by Dissanayake [15] through qualitative research.

This research study is developing upon these findings; hence, the MCMM model and the MLSPI prototype was developed using the above metrics.

12.4 Overview of Machine Learning

Machine learning (ML) is an intelligent ability used by specially designed software to carry out meticulous tasks without being pre-programmed [16]. Hence, it is used in the software development industry to develop applications that can make their own decisions without human interactions. Alpaydin [16] denotes that due to the vast amount of research carried out in this subject area, modern ML software can make intelligent decisions from data that it obtains through software. Barnes [17] depicts that traditional software engineering solely focuses on outputting results based on gathered data and programming procedures. Conversely, ML uses data and the output from the program and produces a new program via careful assessment of data by reverse engineering the program [17]. Moreover, ML is beneficial for big data processing since it identifies data patterns meticulously and interprets big data. Therefore, machine learning techniques are used in this study to process data and apply data pattern predictions for big data gathered from software development companies.

Table 12.1 SPI metrics summary

Identified process	Measurable metric
Reusability	Reuse improvement effort
Development efficiency	Work in progress (WIP), throughput, cycle time, lead time
Cost	Cost analytics, productivity analytics, productivity
Reliability	Mean time to IPL
Software quality	Weighted method per class (WMC), depth of inheritance tree (DIT), response for a class (RFC), number of children (NOC), coupling between object classes (CBO), lack of cohesion of methods (LCOM)
Complexity	Component coupling, constraints complexity, configuration complexity

12.4.1 *Azure ML*

Azure ML is produced by Microsoft to aid software developers with predictive analytics, data mining, data visualisation, etc. [18]. Fryer [18] depicts that it provides a platform to develop evaluations and experiments to predict and validate results. Therefore, Azure ML was used as an external component to process data gathered from software development organisations and make predictions. Furthermore, Azure database allows the storage of data from the organisation on a cloud platform. This is very advantageous since Azure ML can be connected to the Azure database [19]. The use of Azure ML allows the identification of improvements in the organisation. Hence, the organisation can use that information to improve their maturity level. The use of Amazon ML, Cloud ML, etc., can be regarded as alternative approaches to carry out this exact procedure; however, these ideas were dismissed since the use of Microsoft products for all the aspects of the MLSPI prototype will dismiss any platform compatibility errors. However, prior to the use of Azure ML, it is important to identify data analysis methodologies and ML algorithms that are suitable for MLSPI application. This was achieved through regression, train models and score models that are available in Azure ML studio.

12.4.2 *Linear Regression*

It is important to use a robust ML algorithm that consists of predictive data analysis functionalities to make predictions for SPI. Therefore, linear regression is used to apply this functionality to MLSPI prototype. The main purpose of the linear regression is to identify the best-fit line for the data; thus, it makes predictions depending upon characteristics of current data. This best-fit line reduces the prediction error to the maximum amount [20]. This clearly implicates that linear regression will allow the MLSPI prototype to extract current data from SPI metrics calculations described in Sect. 12.3 and identify future trends for software process improvement. It also identifies functional dependencies between two sets of numbers [20]. Therefore, it is vital to identify both independent and dependent variables before the application of linear regression. This implies that multiple data sources from the organisation can be compared to identify data patterns to clarify software efficiency and make predictions about reliability, cost, effectiveness, etc. Alternatively, an algorithm such as logistic regression could be used to carry out the same functionality. However, it was discarded since it only provides answers in binary format (1 and 0 for yes and no) where there is only one outcome compared to continuous outcome obtained through linear regression [21]. Therefore, any unexpected values that can be obtained through linear regression would not be obtained through logistic regression. Another alternative ML algorithm that could be used is Poisson's regression. However, this was also discarded since it is used to identify counts of events and will not provide sufficient predictions for requirements like logistic regression [22]. This implies that linear regression is the most suitable ML algorithm for data pattern prediction for MLSPI.

12.4.3   Train Model

Prior to making intelligent predictions through linear regression, the ML model must first learn from its input data [23]. This process is known as "training". During this process, the input data is analysed and evaluated by the ML algorithm. It analyses data in terms of data distribution, compiling statistics, data types and data patterns [23]. Therefore, it validates the strength and effectiveness of the predicted values.

Training regression model requires a large data set to carry out the above tasks. Therefore, when the data set is sent through the train model, the data should be cleaned, skipped and purified to prevent missing values from entering the train model to prevent unprecedented errors in the result [23]. After the training process is complete, the output of the trained model will be ready to make sufficient predictions through the ML algorithm [23]. Then this will be passed into the scoring model to make predictions. The application of a train model can be achieved using Azure ML studio. Therefore, it further validates the importance of using Azure ML for the development of the MLSPI prototype.

12.4.4   Score Models

Once the input data are trained using an ML algorithm, the next step is to pass that data through a score model. The functionality of a score model is to make predictions for the required data field through a trained regression model [24]. As it depicts in Sect. 12.4.2, MLSPI prototype is using linear regression; thus, the score model was used to denote predicted numerical values for the desired data column of the imported data set. After obtaining results from the score model, the results can be evaluated through an Evaluate model, which depicts the accuracy of results developed via the score model. Moreover, these scored models can be saved as data sets [24]. This is very important for MLSPI since it stores the entirety of its data in a cloud server. Thus, results from the score model were transferred to the Azure cloud database, which was accessed by the MLSPI prototype. Thus, it further validates the importance of using Azure ML studio for the development of MLSPI application since it allows the cross-platform compatibility between the cloud data server, machine learning studio and the development environment.

This section presented various SPI methods, and the following section provides the research methodology and SPI model developed for CBSE and for Service Component-based Software Engineering which is the key to achieve quality of services in cloud computing paradigm.

12.5 Qualitative Research

Before the development of the MCMM model and MLSPI prototype, further qualitative research was carried out to identify the requirements and suitable criteria. Participants are employees of software engineering companies. These interviews were conducted to validate the information in the literature with real-life scenarios and validate the proposed ideology of this study. Following are the questions that were asked from the interviewees.

1. What types of data does your organisation collect from development processes?
2. How does your organisation collect data mentioned in Q1 and how are they stored?
3. Do you have a software process improvement method in the organisation? (if yes, then please explain its functionality)
4. What is your organisation's project management method and how it benefits the process quality?
5. Will your company consider using machine learning for process improvement in the organisation?

The first question was asked to identify data collection methodologies in organisations from development processes. This allows the development of measurable KPI for the MCMM in terms of development processes. Moreover, it could validate some of the results of Dissanayake's [15] research which were used to justify the measurability criteria of development processes in this research study. The answers to this question include complexity calculations, work in progress, velocity, deadline achievements, throughput, cycle time, reliability, complexity calculations, technical sales calculations and service-level agreement targets. Since these results are comparable with the metrics identified in Sect. 12.3, these metrics are used as measurable KPI in MCMM.

The second question was developed to identify whether the organisation has any special software that they use to collect data from each employee and business processes. Hence, answers to this question were used to identify the best methodologies to collect information from employees and business processes of the organisation. Answers to this question depicted multiple software and database platforms such as JIRA, SVN, Octopus, Time Sheets, Microsoft Project, Trello, Kanban Board, Tableau and Visual Paradigm.

This clearly indicates that software development companies use various intelligent software to collect and store data about their employees and development processes. Most of the traditional software was used by smaller companies compared to large companies. Hence, the MLSPI prototype was developed by identifying key features of each of the mentioned applications.

The third question was asked to identify the organisation's use of SPI methods (e.g. ISO, CMMI). Therefore, this question provided a clear identification of most popular SPI methodologies used in the industry, which was then compared with literature gathered for CMMI, ISO, ICMM and SCMM. Figure 12.2 shows that

50% said YES and 50% said NO to this question. This implicates that it is an equal split between two extremes. However, those who said NO mentioned that this is mainly due to their organisation being a small software development company. All the participants from large corporations suggested that they follow different types of SPI methods to improve workflow efficiency. This answer clearly validates the rationale for this research study since companies do not tend to apply SPI due to costs that it would bring to the organisation. Therefore, the organisation can apply SPI using MLSPI application without further cost by using their existing and easy to apply measurable metrics.

The fourth question was asked to identify project management methods used by different organisations (e.g. Agile, Waterfall). The answer to this question helps the identification of project management method that could help SPI. Figure 12.3 shows that 90% said Agile, 20% said Waterfall, 30% said CBSE and 10% said OTHER. This implies that Agile is the most popular project management method within software development organisations. Moreover, some participants explained that Agile project management improves SPI through faster delivery, early risk and

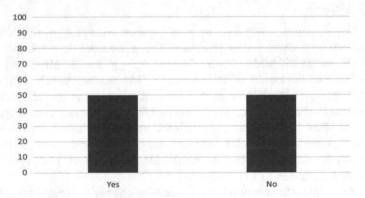

Fig. 12.2 SPI adaptations

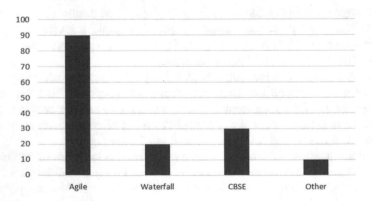

Fig. 12.3 Project management methods

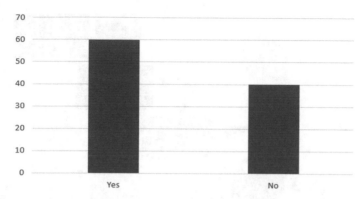

Fig. 12.4 ML usage

defects identification, task divisions, etc. Therefore, the answers to this question were used to identify metrics for SPI. Moreover, some participants have mentioned that they still use CBSE even though Agile is their main project management methodology. This implicates that CBSE can exist as a subsection of Agile for aid software process enhancement.

The fifth question was asked to identify whether current software engineering organisations are willing to use machine learning to aid software process improvement methods of the organisation. The answer to this question validates the directive of this research study. Figure 12.4 shows that 60% said YES and 40% said NO to this question. This implicates that more organisations are inclined to use ML to aid SPI. Moreover, as it depicts from interview transcripts in Appendix A, some of the participants said it will help to reduce costs by automating the software improvement process through machine learning. Moreover, those who said NO gave reasoning as their organisations being too small to apply machine learning. Moreover, they suggested that they might be more inclined to use this methodology when their company expands. This clearly denotes the use of machine learning for SPI as a valid methodology. Moreover, applications such as MLSPI can act as a cheap solution, thus attracting even smaller companies to use ML for SPI.

12.6   Development of the Maturity Model

After the identification of measurable metrics in Sect. 12.3, it is necessary to build a unique maturity model, which can identify its KPA/KPI through these measurable metrics. Therefore, these metrics were implemented through the development of MCMM. Its KPAs were developed to match answers identified in the interviewing process. Then KPIs were added for each of these KPAs by using measurable metrics by combining interview results and SPI metrics identified in Sect. 12.3. Figure 12.5 depicts the proposed MCMM.

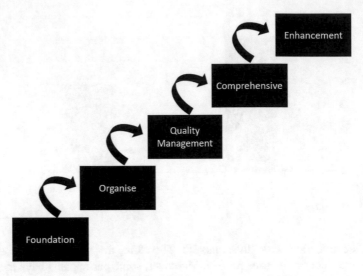

Fig. 12.5 Measurable Capability Maturity Model

12.6.1 Level 1: Foundation

The level 1 maturity (Table 12.2) includes small organisations without valid development processes. Most of these organisations are carrying out development in an ad hoc manner. Therefore, these organisations will have few employees; hence, they would follow informal software engineering methodologies. Thus, the efficiency of software development methodology can be identified and measured via the calculation of the throughput. Moreover, since there is no set development strategy, there will be no documentation; however, the progress and the efficiency of the development process could be identified via software development lifecycle (SDLC) monitoring.

Table 12.2 Foundation-level KPA and KPI

Key process areas	Key process indicators
Informal software engineering methodology	Calculate throughput
Non-documented/informal development procedures	Application of development structure analysis methodology (SDLC monitoring)

12.6.2   Level 2: Organise

Organisations at level 2 maturity (Table 12.3) should have an industry-accepted software development framework. This could be identified through SDLC monitoring to separate the development process and its efficiency. Software used for development processes should have multiple domain compatibility characteristics, which is one of the key features of the CBSE life cycle [25]. Moreover, documentation should be developed by the management to depict the development environment, functionality, compatibility, component selection, integration, acceptance, etc. Therefore, software characteristics can be recognised, and metrics for SPI can be identified. Moreover, the quality of the software should be assessed by measuring complexity levels through metrics such as constraints complexity and configuration complexity. Requirements for software should be identified to assess the cost of the development process as it was identified in the interviews. This will allow the organisation to decide upon whether to develop new components, reuse components or use COTS [26]. These costs can be measured for predictions through metrics proposed in Sect. 12.3. Moreover, the organisation should have technical support methods for the development process to improve the speed and efficiency of development processes. Software components should be tested to identify their evaluation criteria for reuse, which can be achieved by using the Depth of Inheritance Tree metric. Furthermore, the use of software repositories will benefit the organisation to store COTS and inbuilt components.

Table 12.3 Organise-level KPA and KPI

Key process areas	Key process indicators
Industry-approved software/component development method	Identification of the usage of a software development methodology
Domain compatibility	Validate domain compatibility
Component integration, acceptance and selection guidelines	Analysis of component integration, acceptance and selection
Product and methodology of quality assurance	Measure constraints complexity and configuration complexity
Requirements management for cost analysis	Cost analysis metrics
Technical support	Evaluation of technical support provided
Testing for reusability	Calculate the depth of inheritance tree (DIT)
Component storage	Application of repository evaluation methodology
Reusability analytics methods	Measurement of reuse improvement effort

12.6.3 Level 3: Quality Management

At level 3 (Table 12.4) maturity model, the managerial layer of the organisation should have methods to plan, monitor and control project documents, which should be developed at project initiation. Therefore, the organisation should follow an industry-approved project management method such as Agile, Prince2 and PMBOK [27]. Therefore, the efficiency of these measures could be identified by calculating Little's law as it was depicted by Dissanayake [15] to evaluate the efficiency of the management process. Therefore, the company must calculate the cycle time and lead time measurement to validate the efficiency of the development model. It should carry out complexity calculations via measuring component coupling, constraints complexity and configuration complexity as it was explained by Dissanayake [15] to numerically measure product quality. Moreover, the organisation should develop quality assurance audits at specific intervals to measure the quality levels of software components [28].

Table 12.4 Quality management-level KPA and KPI

Key process areas	Key process indicators
Project planning, control and monitoring	Work in progress (WIP)/throughput and cycle time calculation (Little's law) to evaluate the management criteria
Developing components and software by following a component and software specific development models	Cycle time and lead time measurement to validate the development model
Project quality assurance	Complexity calculations (component coupling, constraints complexity and configuration complexity)
Component/software documentation	Evaluation of component documentation
Defect management	Calculate weighted method per class (WMC) to make sure software complexity will not cause defects in the final product
Requirements development	Measure lead time and cycle time and evaluate the efficiency of the requirement in the development process
Testing	Measure the depth of inheritance tree (DIT), number of children (NOC) to test the robustness and efficiency of the component
Conflict analysis	Measure component coupling
Measurement of software productivity	Productivity measurement calculation
Management of repository for components	Identification of repository efficiency by calculating the number of times that it has been used for development
Training programs to improve the programming ability of employees	Evaluation of training program through a test

Every component/software developed by the organisation should be documented accordingly to access details about different development stages. Moreover, software/components should be certified through an independent third-party organisation to validate their quality. The organisation should have a defect management program where they would calculate Weighted Method per Class (WMC) to make sure software complexity will not cause defects in the final product. The organisation should have a requirements development procedure when new software is developed, and the organisation should measure lead time and cycle time to evaluate the development process efficiency. There should be a valid testing procedure for software; therefore, the organisation should specifically measure DIT and NOC to test the robustness and the efficiency of the component as it was depicted in Sect. 12.3. Furthermore, component conflicts can be analysed by measuring coupling between object classes. Furthermore, the organisation should have a software component repository management methodology with a repository manager whose task is to manage and update the component repository. Also, the organisation should have knowledge enhancing programs to train employees to provide the latest technologies to improve the efficiency of their programming abilities.

12.6.4   Level 4: Comprehensive

Level 4 (Table 12.5) maturity model quantitatively analyses the managerial aspects, the robustness of components and software similar to the level 4 of CMMI [2]. This allows the enhancement of the quality of the organisation. In order to validate the efficiency of the software, measurements should be taken to validate the quality of the software. This could be achieved through software quality metrics mentioned in Sect. 12.3 and Dissanayake [15]. Therefore, the organisation should use LOC to identify reusability percentage, DIT to measure testability and understandability, RFC to validate amenability, NOC to improve efficiency, CBO to improve modularity and LCOM to carry out a quality measurement for the class cohesiveness. Furthermore, recordings must be taken by the management level to cover legal and security aspects of the company [13]. The organisation should have close attention to the reliability of machinery and software that are being used in the organisation; hence, MTI metric that was depicted in Sect. 12.3 can be used to achieve this feature. The organisation should be using a maturity model to analyse existing defects and improve upon software components and development processes. Moreover, the organisation should be divided into subcategories with specialised job roles and departments to oversee a smooth administration process [13]. Moreover, the organisation should have a component library with manageable and searchable facilities.

Table 12.5 Comprehensive-level KPA and KPI

Key process areas	Key process indicators
Official documentation for all the key stages of the software development lifecycle	Each document should be analysed and approved by senior personnel in the organisation, and results should be added to a data log
CBSE evaluation metrics	LOC to identify reusability percentage DIT to measure testability and understandability RFC to validate amenability NOC to improve efficiency CBO to improve modularity LCOM to carry out a quality measurement for the class cohesiveness
Legal and security aspects of the components	Log of files regarding security and legal aspects of company, employee, security, etc.
Reliability assessments	Calculation of MTI
Application of SPI through a maturity model	Identification of maturity model usage
Departments and jobs are designed to meet company requirements	Calculation of positive work output from each employee and department
Component repository with searchable characteristics	The efficiency of the search criteria of the component repository

Up to the fourth level of the MCMM, it depicts different layers of evaluation criteria for current and potential practices. Thus, finally the fifth level implicates methodologies to enhance and evolve current practices.

12.6.5 Level 5: Enhancement

The final stage of the maturity model should focus on optimisation and innovation. Jalote [29] denotes that it should consist of factors that determine productivity and the quality of the organisation to improve development and product standards continuously. Therefore, the organisation should extract data gathered via software efficiency metrics such as LOC, DIT and RFC and try to implement methodologies to increase their efficiency levels. The development process can be improved through the application of measurements to enhance throughput level and WIP to improve cycle time and lead time as it was depicted by Dissanayake [15]. Also, the organisation must continuously check existing technologies to make sure that they are up to standards with current technologies in the industry [25]. Moreover, data gathered through the MTI metric and complexity metrics in Sect. 12.3 can be used to predict upcoming issues; thus, measures can be taken in advance to avoid defects. Moreover, the organisation should follow innovative approaches to improve available technology to enhance the development process and product

Table 12.6 Enhancement-level KPA and KPI

Key process areas	Key process indicators
Evaluation of available components to improve efficiency and functionality	Improve the current status of the code via results gathered through LOC, DIT, RFC, NOC, CBO and LCOM
Continuous evaluation of the development process	Improve the cycle time and lead time efficiency by making changes to throughput and WIP
Application of newest technologies	Software updates and licence renewals. Data from these processes should be added to the database
Innovation to develop new technologies	Analysis of newly developed software/technology innovations through collaborative discussions
Defect recognition and resolutions	Improvement of results gathered through complexity metrics
Identification of potential upcoming issues	Application of linear regression, train model and score models for data pattern prediction

quality by predicting upcoming issues. This could be achieved using machine learning algorithms (Table 12.6).

Overall, this section implicates the development of the MCMM. It was evident from previous sections that widely acclaimed maturity models used in the industry do not contain metrics that are measurable to quantify the maturity-level achievement of a software development company. Therefore, this maturity model acts as a solution to resolve this matter. Furthermore, the MCMM model is critically analysed and evaluated to depict its effectiveness in following sections. The next stage of this research study is the application of MCMM through the development of MLSPI prototype with Machine Learning functionalities to make predictions about organisations.

12.7   Prototype Development

After carrying out sufficient research and analysing interview findings, it was decided to develop a prototype application (MLSPI) with the ability to provide information about current maturity, SPI and predict future trends using machine learning. This prototype was integrated with the MCMM in the previous section.

It shows the use of metrics explained in Sect. 12.3 to promote SPI in a software development company that is following Agile and CBSE practices. This implementation is tested by carrying out simulations with a mock data set that is developed to depict a software development company. This data set is developed by carefully analysing results gathered for interview questions. The rest of this section depicts how the MLSPI application could be used to analyse development processes in organisations. The overall maturity representation in the home page and the analysis of development process complexity are depicted in this section.

Figure 12.6 depicts the MLSPI prototype homepage. It displays six buttons: Reusability, Development Efficiency, Complexity, Reliability, Software Quality Analytics and Cost Analytics. In Sect. 12.3, a wide variety of metrics were identified for SPI. Moreover, these metrics were validated through qualitative research as explained in Sect. 12.5. Therefore, each of these buttons will open forms where these metrics were applied to provide feedback to improve SPI. Moreover, underneath the buttons, the MCMM is displayed. It clearly depicts the achievement of each maturity level to give an understanding of the present maturity level of the company. Data depicted in each of these charts were developed depending upon the results of all the metric calculations in the MLSPI prototype.

When the Complexity button is clicked, the user will be presented with the Complexity Analytics form (Fig. 12.7).

This form is designed to carry out the calculation of metrics to identify component coupling, constraints complexity and configuration complexity as was depicted in Sect. 12.3. Figure 12.8 shows the structure of the complexity SPI process. Initially, it collects stored company data from the Azure cloud platform. Then it processes these data using component coupling, constraints complexity and configuration complexity metrics as it was implicated by Dissanayake [15]. These metrics would act as KPI of the MCMM and update MCMM feedback depiction. Moreover, these data would be passed to the Azure Machine Learning suite to

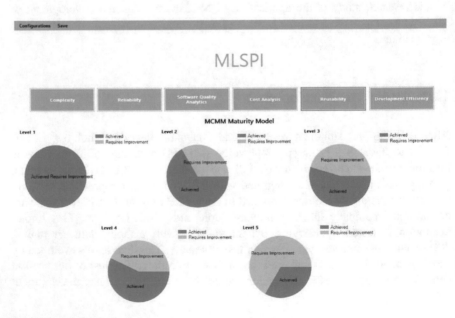

Fig. 12.6 MLSPI home page

Complexity Analytics

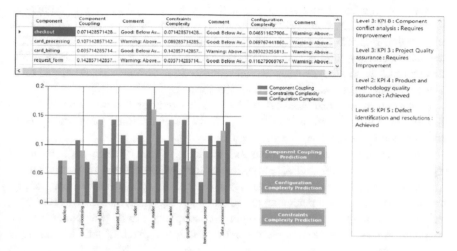

Fig. 12.7 Complexity analysis

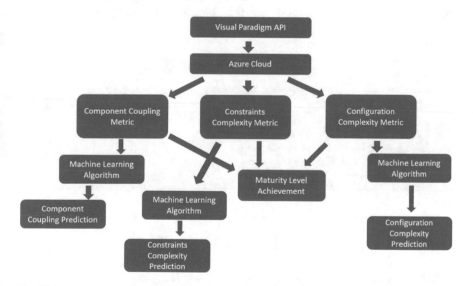

Fig. 12.8 Complexity software process improvement for MLSPI

predict the future behaviour. Hence, the predicted result would be displayed on the screen.

Data for these calculations were obtained from the cloud server and these data should be added and updated in the cloud server whenever a new component is developed or updated by the organisation. Cloud servers are widely used by organisations as it was identified through research. The chart represents the comparison between these metrics to depict their overall effect for each component. Moreover, after each complexity calculation, a comment is given to educate the organisation regarding the status of each complexity level; thus, the organisation can take necessary actions if the status is alarming. Moreover, the MCMM model uses complexity to analyse organisation maturity in its levels 2, 3 and 5, and these levels were measured and displayed on the form.

Furthermore, this form consists of three further buttons that will open forms to display component coupling prediction (Fig. 12.9), constraints complexity prediction and configuration complexity prediction.

Data for these forms were obtained through ML algorithms such as linear regression, train models and score models as they were depicted in Sect. 12.4.

Initially, it imports data for constraints complexity, configuration complexity, sharing attributes, reuse percentages, statements, development cost, external data, external functional calls and component coupling from the cloud server. A condition was added to remove the entire row if one piece of data is missing in order to prevent the occurrence of any anomalous data. The next step re-selects these cleaned data splits the data into 10% and 90% and passes into train model and score model. Thus, data were sent through both training and testing models (Sect. 12.4). Data were trained on this occasion for component coupling,

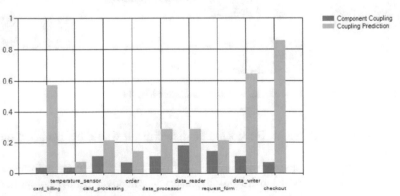

Fig. 12.9 Component coupling prediction

constraints complexity and configuration complexity. Then the model scans the data to identify correlations and make predictions for those three data columns. Finally, these predictions are exported back to the cloud server and displayed on the predicted results table and chart.

This clearly depicts the usage of organisation data to identify process maturity through a specially developed maturity model and then provide feedback using machine learning/actionable analytics to enhance current processes. Thus, a combination of these innovative methodologies allows organisations to revolutionise their current practices via actionable analytics.

12.8 Evaluation

12.8.1 MCMM Evaluation

The maturity model is developed by closely following results gathered through the interviews described in Sect. 12.5. MCMM model was developed to match development procedures of Agile and CBSE practices. 90% of participants said they follow Agile project management. 50% of that demography suggested that CBSE acts as a subdivision of their organisations. This clearly validates the decision to develop the maturity model to match the characteristics of both Agile and CBSE.

It was evident from results that most of the organisations do collect measurable metrics from their employees such as cycle time, throughput and WIP; however, they do not use these methodologies directly for SPI. Some participants suggested they use these for SPI; however, they have separate department such as "Business Excellence" to handle these criteria. This clearly validated the requirement for a maturity model like MCMM with the ability to obtain measurable metrics and provide numerical feedback and data patterns to improve SPI in an organisation without additional human input. Moreover, the development of MLSPI application combined with MCMM will further benefit the organisation by reducing cost since it will prevent them from having special departments for SPI.

Furthermore, it was identified that organisations use specialised software to collect information such as JIRA and Tableau. Therefore, data that could be collected through this software were considered when KPIs were developed for KPA in MCMM. Since 100% of interview participants work in software engineering industry, the information obtained could be deemed extremely valid.

It is important to compare MCMM with other maturity models that were recognised in Sect. 12.2.1 to depict the way it differs from these models. Table 12.7 was developed to implicate the availability of measurable KPIs in other maturity models compared to MCMM. Since the rationale of this research study was developed on lack of measurable metrics identification in modern-day maturity models, this table clearly depicts advantages of MCMM for SPI compared to

CMMI, ISO/IEC WD 15504, ICMM and SCMM. Process improvement areas defined in the first column were derived from results obtained from the interview in Sect. 12.5; thus, it can be proven that these SPI areas are valid in the software engineering industry.

As it can be seen from row 1, reusability identification is available in ICMM and SCMM due to their CBSE-oriented background; however, it is missing in CMMI and ISO/IEC WD 15504 due to their Agile-driven generic development background. It is important to denote that these four maturity models contain KPAs for reusability; however, they do not propose methodologies to measure reusability levels of software and develop assessments for improvement. Conversely, MCMM

Table 12.7 Maturity model comparison

Software process improvement area	CMMI	ISO/IEC WD 15504	ICMM	SCMM	MCMM
Reusability	Not available and no measurable KPI	Not available and no measurable KPI	Available but no measurable KPI	Available but no measurable KPI	Available with measurable KPI
Development efficiency	Available but no measurable KPI	Available but no measurable KPI	Available but no measurable KPI	Available but no measurable KPI	Available with measurable KPI
Complexity	Available but no measurable KPI	Available but no measurable KPI	Available but no measurable KPI	Available but no measurable KPI	Available with measurable KPI
Reliability	Not available and no measurable KPI	Not available and no measurable KPI	Not available and no measurable KPI	Not available and no measurable KPI	Available with measurable KPI
Software quality analysis	Available but no measurable KPI	Available but no measurable KPI	Available but no measurable KPI	Available but no measurable KPI	Available with measurable KPI
Cost analytics	Available but no measurable KPI	Available but no measurable KPI	Not available and no measurable KPI	Available but no measurable KPI	Available with measurable KPI
Quality prediction	Available but no measurable KPI	Available but no measurable KPI	Available but no measurable KPI	Not available and no measurable KPI	Available with measurable KPI

contains KPI that directly focuses upon assessing Reusability, for example level 2: KPI 9: Reusability Analytics Methods (Measurement of improvement effort), which can be directly used to obtain a measurement regarding reusability.

Row 2 implicates the availability of development efficiency KPA in all four models; however, they lack measurable KPI similar to KPI in MCMM. For example, level 1/KPI 1, level 3/KPI 2 and level 3/KPI 6 contain measurements to measure development efficiency criteria by following Little's law. This denotes that MCMM can identify issues with the development process on demand with data trends and patterns. Then it provides feedback to the organisation through MLSPI prototype much more effectively with statistics compared to other maturity models that depend on the help of a separate department to research into development efficiency-related issues.

Complexity identification methodologies are vaguely spread across all four maturity models. However, MCMM measures complexity through component coupling, constraints complexity and configuration complexity and satisfies level 3/KPI 8, level 3/KPI 3, level 2/KPI 4 and level 5/KPI 5. This implicates that MCMM identifies industry-approved complexity quality assurance metrics to identify numerical metrics for quality assurance for SPI compared to other four maturity models.

Reliability is one of the SPI methodologies that is not assessed by all four maturity models. However, this is added to MCMM since it is one of the criteria that were identified through interviews. Reliability is proposed to be measured by MTI metric identified in the literature review, and it allows the achievement of level 4/KPI 4 and level 5/KPI 5 of MCMM.

Identification of software quality is existent in all four maturity models; however, they do not specify any methodologies to carry out this process. Conversely, MCMM proposes the calculation of software quality by using metrics such as WMC, DIT, NOC, CBO and LCOM and satisfies maturity levels such as level 2/KPI 7, level 3/KPI 5, level 3/KPI 8, level 4/KPI 2 and level 5/KPI 1. Thus, the organisation will have numerical feedback to improve software quality.

Problems with cost could be analysed and improved with all the models except ICMM. Cost can be analysed via cost efficiency metrics introduced in Sect. 12.5, and this also allows the identification of productivity of the organisation. Therefore, MCMM proposes the measurement of cost through level 2/KPI 5 and level 3/KPI 9. Moreover, it proposes methods to predict data patterns for the future to take actions prior to the occurrence of any issues regarding these six SPI areas.

This clearly implicates the effectiveness of MCMM model compared to other maturity models identified in this research study. Its provision of measurable metrics allows organisations to calculate a numerical value of maturity for comparison of SPI rather than making hypothetical assumptions by following a generic maturity model. This implies the superiority of MCMM compared to existing maturity models.

12.8.2   Prototype Overview and Effectiveness

It is important to denote that MLSPI is in its prototype stage, and further methodologies such as extraction of data for assessment via API should be added in the final application. However, it is important to implicate that it consists of all the algorithms identified for SPI, and it carries out all the proposed machine learning functionalities. Figure 12.10 shows the architecture of the MLSPI prototype. As it was identified through the qualitative research, it gathers data from measurable API such as Trello, Bitbucket and Cost data. from the organisation. Then that data are passed to the Azure cloud. Then these data are displayed on the prototype to display the achievement of maturity levels using SPI metrics, and moreover, data predictions are carried out using Azure ML. Thus, these data streams could be efficiently used for software process improvement.

MLSPI application was tested to validate the effectiveness of the code. Therefore, unit tests were carried out to validate the robustness of programming techniques adapted during software development. Unit testing is a white-box testing procedure for software product evaluations [30]; thus, it was meticulously used to depict code validity. Unit tests were carried out for every single class during the development process. This allowed the validation of the error-free nature of each class during its development stage. Hence, it allowed the prevention of errors in early stages and validated its robustness.

The usability criteria of MLSPI were assessed to identify its ease of usage. This was achieved by using the System Usability Scale (SUS). SUS is a usability assessment that checks systems' capability in terms of efficiency, satisfactory level and effectiveness; it was calculated through ten questions [31]. Therefore, this test

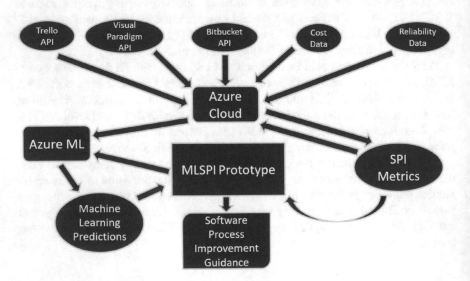

Fig. 12.10 MLSPI architecture

was carried out with ten personnel who work in the software engineering industry. These were the same people who participated in the interview. The overall score received for MLSPI was 81. Hence, according to Brooke [31], the usability of MLSPI can be regarded as "Good". This implicates that users were satisfied with the usability of the MLSPI prototype. The user satisfaction can be further improved by identifying further methodologies for data extraction. For example, quantitative research should be carried out with more than 50 participants to gather information about data extraction methodologies of organisations following both Agile and CBSE practices. This will allow further automation of data extraction philosophies followed by MLSPI.

12.9 Conclusion and Further Research

This research study proposes a software process improvement methodology with machine learning algorithms. It identifies a variety of SPI metrics that could help organisations by processing large data sets collected from employees, departments and development processes. It was identified that big data analytics in organisations has become a vital functionality due to the growth of digital data processing in the past few decades. Thus, combining big data analytics with ML provides the ultimate mechanism to automate SPI in a software development company. The effective use of ML in software development organisations provides a competitive advantage against their rivals in the industry. Usage of ML platforms such as Amazon ML and Azure ML allows the application of machine learning for software applications easily without having extremely professional knowledge about machine learning. Therefore, this research study provides a solution to combine SPI metrics for big data with machining learning algorithms through MCMM and MLSPI prototype. The development of measurable KPI for MCMM to identify measurable data in an organisation allows the meticulous calculation process of maturity levels and improvements. This model was combined with MLSPI prototype with machine learning functionalities to predict changes to the data patterns and provide feedback for necessary improvements.

There is still room for improvement for MCMM. For example, further SPI metrics such as Attributing Hiding Factor to find further class efficiencies, Method Hiding Factor to find method efficiencies, Specialisation index per class to measure inheritance efficiencies, etc. [32] can be added to enhance the effect of MCMM and increase KPI for maturity levels. Furthermore, a quantitative research procedure (Questionnaire) should be carried out with more than 50 participants to identify further measurable metrics to enhance KPI in the maturity model and make MCMM more relevant to the software development industry.

There are multiple strategies that could be implemented in the MLSPI application for further improvements. Data that were used to test the functionality of SPI algorithms and machine learning algorithms were developed to simulate a software development organisation. Hence, these were mock data and not real data from an actual company. Thus, it is recommended to test this prototype by connecting to data streams of a software development company in the industry to carry out a pilot study to further validate its effectiveness. As it was depicted in Sect. 12.7, the designing process hypothesised multiple data extraction methodologies according to data gathered through the interview process. Therefore, it is recommended to implement these processes by attaching APIs of this external software into MLSPI application. Thus, it will completely automate the data gathering processes. This could include the application of further machine learning methodologies to gather data by applying intelligent data analytics functionalities. This implies the development of a plugin component for MLSPI application that monitors development processes and feeds data into the cloud data store to be assessed by MLSPI application. Moreover, further machine learning functionalities improve data prediction. For example, data prediction could be improved via the addition of neural networks that will predict targets with multiple numerical values [33]. It could include the use of decision trees to develop regression models by enhancing results gathered through linear regression [34]. Moreover, the use of Bayes-point machine algorithm will allow the identification of Bayesian average for linear classifiers in data sets to improve predicted data patterns [35].

Overall, this chapter proposes a methodology to automate SPI in software development companies using a new maturity model with measurable metrics combined with a prototype application with machine learning functionalities. The validation of this ideology via qualitative research and validation of its efficiency with a prototype clearly implicates the effectiveness of MCMM and MLSPI prototype; therefore, it promotes a promising research area associated with actionable analytics for software process improvement.

References

1. Olson T, Humphrey W, Kitson D (1989) Conducting SEI-assisted software process assessments. Technical report, CMU/SEI-89-TR-7, Pittsburgh
2. Paulk M, Curtis B, Chrissis M, Weber C (1993) Capability maturity model for software, version 1.1. Research paper. Software Engineering Institute, Carnegie Mellon University, Pittsburgh, Pennsylvania
3. Herbsleb J, Zubrow D, Goldenson D, Hayes W, Paulk M (1997) Software quality and the capability maturity model. Commun ACM 40(6):30–40
4. Bayrasken H (2009) A report on the capability maturity model. Research paper. Computer Science Department, University of Nottingham, Nottingham
5. Sjoberg D, Dyba T, Jorgensen M (2007) The future of empirical methods in software engineering research. Future of Software Engineering (FOSE), Minneapolis, pp 358–378
6. Fuggetta A (2000) Software process: a roadmap. Proceedings conference on the future of software engineering. Limerick, Ireland, pp 25–34
7. Galin D (2004) Software quality assurance: from theory to implementation. Pearson Education Limited, UK
8. Noor R, Khan MF (2014) Defect management in agile software development. IJMECS 6(3): 55–60

9. Khraiwesh M (2014) Process and product quality assurance measures in CMMI. Intl J Comput Sci Eng Surv (IJCSES) 5(3):1–15
10. Gibson D, Goldenson D, Kost K (2006) Performance results of CMMI-based process improvement, CMU/SEI-2006-TR-004. Software Engineering Institute, Carnegie Mellon University, Pennsylvania
11. Gupta R (2009) A maturity model for CBSE. Paper presented at the proceedings of the 2nd India software engineering conference, Pune, India, pp 127–128
12. Grottke M (2001) Software process maturity model study, Research Paper, IST-1999-55017, pp 1–32
13. Tripathi AK, Ratneshwer G (2009) Some observations on a maturity model for CBSE. In: 14th IEEE international conference on engineering of complex computer systems, pp 273–281
14. Alvaro A, De Almeida ES, Meira SL (2007) A software component maturity model (SCMM), In: 33rd EUROMICRO conference on software engineering and advanced applications, SEAA 2007, Art. no. 4301068, pp 83–90
15. Dissanayake S (2018) Measurable metrics for software process improvement. Eur J Comput Sci Inf Technol 6(1):33–43
16. Alpaydin E (2010) Introduction to machine learning. MIT Press, Cambridge, pp 1–579
17. Barnes J (2015) Azure machine learning, microsoft azure essentials, 1st edn. Microsoft
18. Fryer A (2015) Microsoft Azure machine learning [PowerPoint presentation], Microsoft. Available from: http://www.ecmwf.int/sites/default/files/elibrary/2015/13314-microsoft-azure-machine-learning.pdf. Accessed 04 Mar 2016
19. Feurer M, Klein A, Eggensperger K, Springenberg J, Blum M, Hutter F (2015) Efficient and robust automated machine learning. Adv Neural Inf Process Syst 28:2944–2952
20. Chatterjee S, Price B (1991) Regression analysis by example, 2nd edn. Wiley, New York
21. Hosmer DW, Lemeshow S (2000) Applied logistic regression, 2nd edn. Wiley, New York
22. Zou G (2004) A modified poisson regression approach to prospective studies with binary data. Am J Epidemiol pp 159–706
23. Microsoft (2017a) Machine learning [Internet]. Available from: https://msdn.microsoft.com/en-us/library/azure/dn905846.aspx/. Accessed 15 June 2017
24. Microsoft (2017b) Score model [Internet]. Available from: https://msdn.microsoft.com/en-us/library/azure/dn905995.aspx/. Accessed 15 June 2017
25. Cai X, Lyu MR, Wong KF, Ko R (2000) Component-based software engineering: technologies, development frameworks, and quality assurance schemes. In: Proceedings of Asia-Pacific software engineering conference, pp 372–379
26. Khan AI, Qayyam N, Khan U (2012) An improved model for component based software development. In: Paper presented at the proceedings of the 2nd India software engineering conference, Pune, India, vol 2, no 4, pp 138–146
27. Karaman E, Kurt M (2015) Comparison of project management methodologies: prince 2 versus PMBOK for it projects. Int J Appl Sci Eng Res 4(5):572–579
28. Nautiyal L, Umesh K, Sushil C (2012) Elite: a new component-based software development model. Int J Comput Technol Appl 3(1):119–124. ISSN:2229-6093
29. Jalote P (2005) An introductory approach to software engineering, 2nd edn. Narosa Publishing House, pp 65–72
30. Parkin R (1997) Software unit testing. In: IV & V Australia: the independent software testing specialist
31. Brooke J (2013) SUS: retrospective, research paper. J Usabil Stud 8(2):29–40
32. Harrison R, Counsell S, Nithi R (2001) An overview of object-oriented design metrics. In: International conference on software technology and engineering practice, (STEP). IEEE Computer Society Press, pp 230–234

33. Microsoft (2017a) Multiclass neural network [Internet]. Available from: https://msdn.
    microsoft.com/library/azure/dn906030.aspx/. Accessed 15 June 2017
34. Microsoft (2017b) Decision forest regression [Internet]. Available from: https://msdn.
    microsoft.com/library/azure/dn905862.aspx/. Accessed 15 June 2017
35. Microsoft (2017c) Two-class Bayes point machine [Internet]. Available from: https://msdn.
    microsoft.com/library/azure/dn905930.aspx/. Accessed 15 June 2017

Chapter 13
Comparison of Data Mining Techniques in the Cloud for Software Engineering

Kokten Ulas Birant and Derya Birant

Abstract Mining software engineering data has recently become an important research topic to meet the goal of improving the software engineering processes, software productivity, and quality. On the other hand, mining software engineering data poses several challenges such as high computational cost, hardware limitations, and data management issues (i.e., the availability, reliability, and security of data). To address these problems, this chapter proposes the application of data mining techniques in cloud, the environment on software engineering data, due to cloud computing benefits such as increased computing speed, scalability, flexibility, availability, and cost efficiency. It compares the performances of five classification algorithms (decision forest, neural network, support vector machine, logistic regression, and Bayes point machine) in the cloud in terms of both accuracy and runtime efficiency. It presents experimental studies conducted on five different real-world software engineering data related to the various software engineering tasks, including software defect prediction, software quality evaluation, vulnerability analysis, issue lifetime estimation, and code readability prediction. Experimental results show that the cloud is a powerful platform to build data mining applications for software engineering.

Keywords Software engineering · Cloud computing · Data mining · Classification

13.1 Introduction

Modern software systems are inherently complex, and their development is time-consuming and subject to error-prone, since they generally involve large numbers of requirements and components. Moreover, they are becoming increasingly complicated with the advances in hardware and network technologies. This

K. U. Birant · D. Birant (✉)
Department of Computer Engineering, Dokuz Eylul University, Izmir, Turkey
e-mail: derya@cs.deu.edu.tr

© Springer Nature Switzerland AG 2020
M. Ramachandran and Z. Mahmood (eds.), *Software Engineering in the Era of Cloud Computing*, Computer Communications and Networks,
https://doi.org/10.1007/978-3-030-33624-0_13

complexity slows development activities, leads to software defects, increases the maintenance requirement, and eventually makes them well suited to data mining.

Large amount of data in various forms have been recently collected from software engineering products and processes, during the different software development phases. Software engineering (SE) data can be mainly categorized into three groups: (i) *transactions* such as test runs, static features extracted from source code, execution traces collected at runtime, and historical code changes; (ii) *graphs* such as design diagrams, static call graphs extracted from source code, dynamic call graphs collected at runtime; and (iii) *text* such as bug reports, requirement specifications, mail archives, code comments, and documentations. The increased availability of SE data allows us to apply data mining methods on the data to improve software productivity and quality.

Using well-established data mining methods, software engineers can explore valuable and meaningful patterns and relationships in data in order to better manage their projects. In addition, data mining can help them to produce higher-quality software systems that better meet business objectives and to manage projects that are delivered on time and within budget. On the other hand, mining software engineering data poses several challenges such as: (i) it is computationally expensive in the case of large volume of software engineering data; (ii) it requires data repository management; (iii) it should be easily accessible from anywhere by software team members; (iv) security issues should be considered in a comprehensive way; and (iv) it should be cost-effective. Since cloud technology provides opportunities for all these issues, this chapter proposes the utilization of cloud resources in data mining for software engineering.

The main contributions of this chapter are as follows:

- It provides a brief survey about data mining for software engineering and aids software engineers on the selection of appropriate data mining techniques for their work.
- It proposes the application of data mining methods on software engineering data in cloud platform, due to cloud computing benefits.
- It presents experimental studies conducted on five different real-world software engineering data to compare five different classification algorithms: (decision forest (DF), neural network (NN), support vector machine (SVM), logistic regression (LR), and Bayes point machine (BPM)) in terms of accuracy and runtime performance in the cloud environment.

This chapter is organized as follows. In the following section, related literature and previous works on the subject are summarized. Section 13.3 gives information about applying data mining algorithms in the cloud to various software engineering tasks. This section also explains the main steps of mining SE data in the cloud. The major advantages of the proposed approach are also listed as a whole. Section 13.4 firstly gives the description of software engineering datasets used in this study and then presents the experimental results with discussions. Finally, the last section gives some concluding remarks and future directions.

13.2   Related Works

In this section, related literature and previous works on the subject are summarized. Firstly, cloud-based data mining solutions are reviewed, and then, data mining studies conducted for software engineering are presented in detail.

13.2.1   Data Mining in the Cloud

Performing data mining (DM) in cloud computing is a recent trend in knowledge discovery field [1]. Cloud-based data mining frameworks have already begun to use in many different areas such as health [2], industry [3], environmental science [4], and production [5]. Differently, from these studies, our chapter proposes the use of data mining techniques in the cloud for software engineering (SE) domain.

Data mining in the cloud (DMCC) allows users to centralize data storage and software management, with many other benefits such as computational efficiency, scalability, availability, and cost efficiency. So, it simplifies the data mining lifecycle, including data preparation, knowledge extraction, and model evaluation. However, DMCC, from a technical point of view, requires a special infrastructure capable of supporting data storage technologies and data processing.

The main issues to be solved in DMCC are to deploy a cloud data processing platform and to prepare a visual friendly user interface to interact with the data mining engine and results. Several articles [6, 7] have recently been published on how cloud technologies can be used to implement an effective environment for building scalable DM systems. A typical DMCC framework can be built as a four-layer architecture model, including *application layer* (AL), *data mining service layer* (DMSL), *data access layer* (DAL), and *data layer* (DL). The AL provides the users interactions with an interface, the DMSL executes the data mining algorithms, the DAL provides standardized access mechanisms to data objects, and the DL is responsible for storing and managing data.

13.2.2   The Role of Data Mining in Software Engineering

The primary purpose of DM for SE is to create models which can be able to give further insight into support decision making related to software engineering [8]. It is possible to find SE tasks in all SE phases that have been facilitated by data mining integrated solutions. As shown in Fig. 13.1, software data can be mined in each phase of the development process: requirement analysis, design, development, testing, and maintenance.

- In the **requirement analysis** phase, text mining helps requirements engineers automatically extract requirements from policy documents [9]. Process mining,

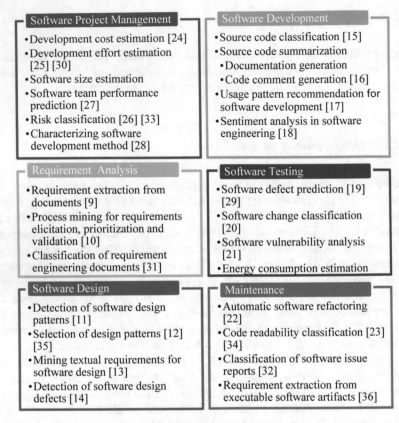

Fig. 13.1 Data mining studies in different phases of the software development process

which is a special type of data mining, has also been used for requirements elicitation, prioritization, and validation [10]. The motivation of these works is to make requirements documents more accurate and complete, and to provide time saving for analysts.

- In the *software design* phase, software design patterns can be detected by using software metrics and classification-based techniques [11]. Text mining methods can be used for the automatic selection of a fit design pattern to solve a design problem [12]. Furthermore, mining textual requirements can be useful to assist architectural software design [13]. Relational association rule mining has been used to detect software design defects in software systems [14].

- *Software development* phase is a popular stage for software data miners. Since software development is a complex combination of activities, it is important to understand the factors that influence development performance. During the development of a software project, it is possible to accumulate relevant data to provide insights by data mining that can help guide development. Source code classification has been studied to assign programs into one of the different categories according to criteria such as programming languages [15] or quality (well/badly written). Source code summarization based on data mining has received special attention, ranging from the use for code comment generation [16] to documentation generation. Data mining also helps software engineers on the other software development tasks, including usage pattern recommendation for software development [17] and sentiment analysis in software engineering [18].

- In *testing* phase, software defect prediction [19] is the top topic studied by the researchers. The main objective of software defect prediction is to classify software modules into fault-prone and non-fault-prone ones by means of software metrics. Mining software code changes is also an important problem worth studying, aiming to help to achieve the goal of testing software modules timely and effectively [20]. Software vulnerability analysis [21] is also another active research area in data mining.

- Besides software testing, *maintenance* is also one of the major phases to attempt data mining. For instance, data mining techniques can be used to discover and classify software refactoring according to their application scenarios [22]. Code readability classification, which refers to categorization of a source code as either readable or unreadable, has become a research area with increasing importance [23].

- For *software project management*, data mining may contribute to decrease risks and uncertainty in the decision-making process and to increase project success rates. Development cost [24] and effort estimation [25] based on data mining are the top topics studied by the researchers since the accurate prediction of them is one of the crucial and challenging tasks for software project management. Classification of risks in software development projects [26] is also an important issue for software project management. Data mining also helps in other SE tasks, such as project scheduling, software team performance prediction [27], characterizing software development method [28], and efficient allocation of resources.

The main techniques used for mining software engineering data are classification, clustering, and association rule mining. *Classification* is the process of building classifiers, from a set of records that contain class labels, to assign an object to one or more predefined categories. *Clustering* is the grouping together of similar objects into a number of clusters, on the basis of attributes of the objects. *Association rule mining* is the discovery of interesting relationships or correlations, in the form of "if-then" statements, among a set of items in a dataset. These are further elaborated as follows.

Classification

This has been commonly used for software engineering to predict software defects [29] and development efforts [30]. Classification has also been applied in various other software engineering tasks, including classification of software design patterns [11], classification of requirement engineering documents [31], classification of software issue reports [32], and classification of risks in software development projects [33].

Clustering

For software engineering, this has been studied by a number of researchers. Bishnu and Bhattacherjee presented a software cost estimation model based on a modified K-modes clustering algorithm [24]. Niu et al. used clustering algorithms for usage pattern recommendation for software development [17]. Scalabrino et al. used DBSCAN clustering algorithm to construct a comprehensive model for source code readability [34]. Hussain et al. proposed a framework to organize and select the correct software design pattern(s) for a given design problem by using clustering technique [35]. They compared four clustering algorithms: k-means, hierarchical (agglomerative), fuzzy c-means, and partition around medoids (PAM). Some software engineering studies [19, 25] have been introduced that used clustering as a preprocessing stage since performing clustering before classification has generally a positive effect on the progress of the classification.

Association Rule Mining (ARM)

This mechanism can provide meaningful knowledge that can be easily understood by software engineers. The effectiveness of finding association rules in software engineering data and also the benefits of ARM techniques to uncover hidden patterns in software systems architecture has been proven in several studies [14, 36]. For example; it is possible to discover association rules to deal with software defect prediction problem [14]. For instance, Ackermann et al. [36] used ARM technique to automatically extract requirements from test cases to reduce the difficulty of software maintenance. In the software engineering field, *sequential pattern mining*, a further promotion of association rules mining, has also been applied to discover rules in software datasets [37]. Sequential pattern mining is the process of extracting frequently occurring patterns in a time-related format or other ordered formats.

13.3 Materials and Methods

The study presented in this chapter combines two techniques, namely data mining and cloud, for software engineering to benefit from their capabilities together. It includes the application of classification algorithms (DF, NN, SVM, LR, BPM) in the cloud environment on five different real-world data related to the different SE tasks, including software defect prediction, software quality evaluation, vulnerability analysis, issue lifetime estimation, and code readability prediction.

This section gives information about: (i) how DM algorithms can be applied in the cloud for various SE tasks, (ii) what are the main steps that need to be taken, and (iii) the major benefits of the proposed approach.

13.3.1 Data Mining in the Cloud for Software Engineering

Mining software engineering data has recently emerged as a research topic toward the goal of improving the software productivity and quality for a given project. Its success has already been proven in both theoretical and practice studies. However, much less attention has been paid to deploy them on the cloud platform. Recently, several studies [38, 39] have been conducted that bring three concepts together: data mining, cloud, and software engineering.

However, many open issues still remain to be investigated. Which SE tasks have been facilitated by data mining in the cloud? What are the main steps that miners should follow to mine SE data in the cloud? How software engineers benefit from cloud-based data mining? This chapter answers these questions to help researchers who want to develop practical data mining models for SE tasks in the cloud environment.

Recent years, several cloud-based data mining studies have been carried out in the field of software engineering. Okumoto et al. presented a platform, named Bell Labs Software Reliability Engine (BRACE), which is a cloud-based platform for software reliability tools, including software testing and defect prediction [40]. They indicated that these tools were developed Software as a Service (SaaS) for development teams. They also mentioned that BRACE includes a fully automated software reliability growth modeling (SRGM) tool which is currently being used by a series of software projects in telecom for software defect prediction. Similarly, Ali et al. also present a parallel framework for software defect prediction and metric selection in the cloud environment [41]. Differently, from these studies, we have also run data mining algorithms in the cloud environment for other SE tasks such as prediction of software issue lifetime, code readability, and vulnerability.

How to realize fast and efficient data mining for software data analysis is one of the problems of software engineers. If data mining algorithms are implemented on the cloud, the large computational problems in the field of software engineering can be solved. For this reason, this chapter proposes the development of cloud-based data mining tasks for software engineering data.

Figure 13.2 shows a robust, powerful, scalable, and flexible cloud-based system which integrates data mining techniques such as classification, clustering, and ARM to provide deeply insights into how to improve SE processes. The system can be simultaneously accessed by multiple software engineers from anywhere at any time. It can be able to deal with the increasing production of SE data and will create the conditions for the efficient mining of massive amounts of SE data. Data mining services are taking a significant position in the cloud-based system. The complexity of data mining services is not only limited to technical or conceptual considerations, but also requires a complete specification. The definition of services is provided on three different dimensions. The *structure dimension* determines the ability of a provider to deliver the service for software engineering. The *process dimension*

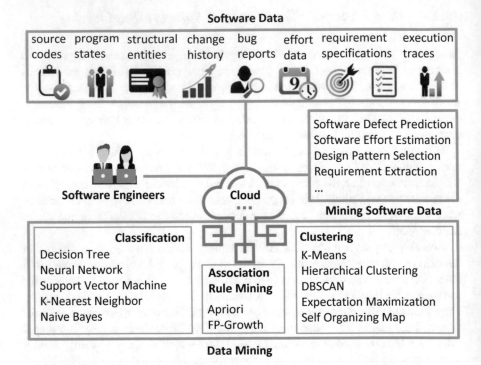

Fig. 13.2 Data mining for software engineering in the cloud

explains services as tasks which are performed on software engineering data. The *outcome dimension* refers to the outputs as a result of rendering a data mining service.

Three major cloud models are available for software engineers to implement data mining solutions in the cloud: Software as a Service (SaaS), Platform as a Service (PaaS), and Infrastructure as a Service (IaaS). The main principles of the three models are given as follows:

- **Data Mining Software as a Service**: Software engineers can use ready-to-use data mining tools that can be directly accessed through a Web browser or a mobile application. By this way, they can focus on data mining applications instead of on infrastructure.
- **Data Mining Platform as a Service**: The cloud provides supporting platforms that software engineers can use to build their own data mining applications within a short time without concern about the technical infrastructure details.
- **Data Mining Infrastructure as a Service**: The cloud provides a set of virtualized resources that software engineers can use as an infrastructure to store large software engineering datasets, execute data mining algorithms, or implement data mining systems to obtain the required knowledge.

Popular examples of machine learning SaaS and PaaS frameworks in the cloud are Azure Machine Learning Studio, Watson Studio, SageMaker and Cloud Machine Learning Engine.

13.3.2 Main Steps of Mining SE Data in the Cloud

Figure 13.3 shows an overview of the five main steps in mining software engineering data in the cloud. These steps and the main roles of software engineering experts in these phases are explained as follows:

- **Determine SE Task**: This phase focuses on identifying the objectives from a software engineering perspective, understanding application domain involved and the knowledge that's required. After that, a preliminary plan is designed to achieve these objectives by converting them into a DM problem definition. It is also necessary to define the available resources, restrictions about the process, cost, benefits and success criteria of the study, and the cloud environment where the knowledge discovery process will take place. Software engineers can use either a *problem-driven approach* (start with a SE problem of interest) or a *data-driven approach* (start with a SE data to mine); however, in practice, they generally use a hybrid of these two approaches. The software engineering predictive problems differ from other areas in that they are project oriented rather than transactional, and the subjects require advanced SE knowledge.
- **Collect SE Data**: This phase focuses on the collection of SE data related to an SE task from various sources and storing in the cloud. As shown in Fig. 13.3, software engineering data includes source codes (i.e., code comments or code

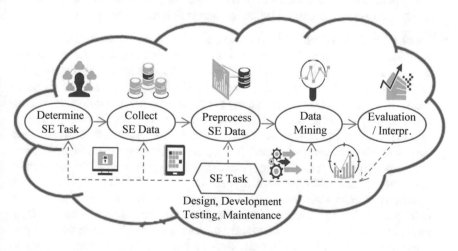

Fig. 13.3 Basic steps of mining software engineering data

images), software metrics (i.e., lines of code), program states (i.e., data or data structures), execution traces (i.e., execution time of a function), structural entities (i.e., patterns, layers, packages, classes, and methods), code changes (i.e., versions, refactoring), bug reports (i.e., the execution error results), and documentations (i.e., requirement specifications). It may be required to integrate all the SE data obtained from different sources into one dataset. This process is very important because the DM discovers from the available data. Researchers frequently obtain data from open-source projects like Apache Ant, Eclipse, JEdit, and Mozilla. Software engineers also play an important role in achieving SE data that can be used for data mining.

- *Preprocess SE Data*: This phase includes all activities to construct the final dataset from the initial raw SE data, including data selection, data cleaning, and data transformation.

  - *Data selection*: It involves extracting essential data from the raw software engineering data. Data mining professionals can decide which certain parts of the data will be included in the final dataset based on software engineers' knowledge. Feature selection techniques can also be used to reduce dimensionality while improving data mining performance.

  - *Data cleaning*: Data cleaning is the step where noise, inconsistent and duplicate data are removed from the raw SE data, and missing and incorrect values are filled with correct values. Software engineering experts have helpful knowledge regarding data cleaning, which should be provided by data miners before applying data mining methods.

  - *Data transformation*: In this phase, SE data is transformed into forms appropriate for mining. For example, call sequences or call graphs can be extracted from source code or keywords can be detected from software documents. Discretization, aggregation, or normalization operations can also be performed on SE data to improve data quality. This step can be crucial for the success of the entire DM study, and it is usually very study-specific.

- *Mining SE Data*: Based on the objective of SE task, an appropriate data mining technique is selected and then applied on SE data in the cloud. Some common data mining techniques are classification, regression, clustering, association rule mining, sequential pattern mining, and anomaly (outlier) detection. Two major models that can be built in data mining are predictive and descriptive. *Predictive* models, i.e., classifiers, present the complex and nonlinear relationship between input and output data observations and are used to make predictions about future events or trends. *Descriptive* models, i.e., clusters, present understandable and useful patterns and associations describing a complex dataset in summative, informative, and discriminative forms. In order to build a predictive or descriptive model, data miners need to try different data mining algorithms to determine which algorithm is most appropriate for the SE data. Since the input parameters of the data mining algorithms can considerably influence the performance of SE predictive models, data miners should use some methods to automatically find the optimal parameter values for the available dataset.

- *Evaluation/Interpretation*: In this stage, the constructed models or mined patterns are evaluated and interpreted in terms of their usefulness with respect to the goals defined in the first step. If the models or patterns are not useful, then the process might again start from any of the previous steps until a satisfied result is obtained. Software engineers may have the key knowledge to state the reasons for poor-performed models. Software engineers may use their domain knowledge to identify why certain models do not satisfy the objectives. If the models or patterns are useful, the knowledge discovered is represented to the software engineers in easy-to-understand format. Mostly, visualization techniques are being used to represent knowledge in a structured way to assist the SE task. Hereafter, software engineers can use the constructed model to support decision making in different ways. For instance, they may use to gain information on significant relationships between input features and the variable of interest. For example, they can use to make predictions for previously unseen input features. Software engineers may also use various sets of input features to find which of them would be capable of producing more desirable outcome. High understandability and high prediction accuracy increase the acceptance of the constructed models by software engineers. When new software engineering data is available, the model should be updated incrementally in the cloud environment.

13.3.3 Advantages

The major advantages of mining software engineering data in the cloud can be listed as follows:

More Processing Power

Data mining is time-consuming and computationally costly; therefore, it needs high-performance devices, especially when huge volume of software engineering data is available. Cloud computing maximizes the effectiveness of shared resources by the utilization of available infrastructures; thus, it can provide more processing power and effectively reduces the execution time of DM algorithms on software engineering data. The cloud-based data mining system can improve software data processing ability several times, and so, it eliminates time limitations.

Information Sharing

The software engineering data is ordinarily stored in a local storage unit on a computing node, and so, it can be locally available for data processing. Because of the sharing feature of the cloud, the software engineers can access the information from anywhere at any time. Therefore, cloud computing-based data mining system can better support data sharing. In addition, the cloud environment makes data storage and its management easier. Massive software engineering data can be processed in the cloud without needing to move it.

High Reliability

In the cloud architecture, the underlying data is divided into a collection of parts stored in different nodes. By this way, the cloud provides a fault tolerance mechanism for data. This mechanism ensures the integrity and reliability of software engineering data, even if any storage node is damaged. The data mining system can continue to provide service to software engineers without interruption through virtual machines and mirrors.

Lower Cost

Cloud computing-based data mining systems offer on-demand pricing options with affordable prices. The software engineers do not need to purchase servers and other related hardware equipment to apply data mining techniques on software data, and so, cloud computing eliminates the investment cost needed for hardware and software architecture. They only need to pay for the resources they used. In addition, the infrastructure maintenance cost is significantly reduced as well, which is also provides saving in the time and cost of human resources.

Availability

The cloud environment enables software engineers to access data mining applications using a Web browser, regardless of their location or which device they use (i.e., personal computer, laptop, or mobile phone). Thanks to cloud environment that DM applications can be accessed through the World Wide Web, and software engineers can connect from various locations at any time.

Good Scalability

With the rapid development of DM technology, new DM algorithms have emerged constantly. Cloud computing systems can be able to cover these through their scalability.

Improved Security

In the cloud computing architecture, providers are subjected to devote resources to solve security issues.

Providing Services in Real-Time

The software engineers have requirements for the effectiveness of data mining to make the correct management decisions. However, the standard DM systems usually do not provide real-time services. Through cloud computing, data mining result can be acquired in a shorter time, and so, real-time information feedbacks can be provided for critical software engineering tasks.

Although benefits of mining software engineering data in the cloud have been proven, there are also some concerns and drawbacks that can be summarized as follows:

- Requires high-speed network and connectivity constantly to provide an attractive service for both data transfer and running applications.
- External dependency on the cloud service provider.
- Non-interoperability between cloud-based systems.
  (If the user wants to move from one service provider to another, then the transfer of large volume of data from the old to new provider can often be painful and time-consuming).
- The regulations and laws specified by government and other organizations may not be based on cloud type of procedures.

13.4 Experimental Studies

The experimental studies in this chapter are concerned with the use of classification algorithms in the cloud to provide useful knowledge about how to improve software engineering processes and products. We focus on classification task, instead of other DM techniques such as clustering, because many review articles reported that classification is the most commonly applied DM technique in various fields [42], as well as in software engineering [43].

Estimation, risk, and uncertainty are key terms in a software project management environment on which success of a project is dependent. The aim of the presented studies in this chapter is to provide decision support related to these key terms. We built and compared several classification models in the cloud environment which are able to provide further decision support related to software engineering.

The experimental studies were performed on Azure Machine Learning Studio, which is a cloud-based computing platform that is accessible through a Web-based interface. The effectiveness of classification algorithms varies from dataset to dataset. Therefore, data miners need to try different algorithms to determine which algorithm is most appropriate for given SE data. In this study, five classification algorithms were compared in the cloud environment in terms of both classification performance and runtime efficiency, including decision forest (DF), neural network (NN), support vector machine (SVM), logistic regression (LR), and Bayes point machine (BPM). These algorithms were chosen because they are among the most popular classification algorithms used in data mining studies [44]. They have gained a lot of attention in the past decade due to their good predictive performances. They have a wide range of applications in many different fields [44], as well as in software engineering [45, 46].

In each experiment, 10-fold cross-validation was performed on the datasets. SMOTE technique was used to balance some datasets.

All algorithms were applied using their default parameters in Azure ML studio. These default parameters can be summarized as follows:

- *Decision forest*: Resampling method is bagging, the number of decision trees is set to 8, maximum depth of the decision trees is initialized to be 32, and the number of random splits per node is set to 128.
- *Neural network*: Both the number of hidden nodes and iterations are set to 100, and learning rate is initialized to be 0.1.
- *Support vector machine*: Lambda parameter, which is the weight for L1 regularization, is assigned as 10-3.
- *Logistic regression*: Both L1 and L2 regularization weights are set to 1.
- *Bayes point machine*: The number of training iterations is assigned as 30.

13.4.1 Classification Algorithms

Although many classification algorithms are available in literature, in this study, only most popular ones were compared in the cloud environment. They are briefly explained below.

Decision Forest

The decision tree algorithm predicts unknown class labels using a tree structure derived from training data. The structure of decision tree consists of nodes, branches, and leaves that represent attributes, attribute values, and class labels, respectively. Decision forest is an ensemble learning method that builds a number of decision trees for classification task. Each tree in the decision forest produces an output and the final prediction is determined based on the aggregation process.

Neural Network

A neural network is an interconnected group of processing units (neurons) that are constructed to learn the relations between input and output data for classification. This network consists of a set of interconnected layers: input layer, hidden layer(s), and output layer. All nodes in a layer are connected by the weighted edges to nodes in the next layer. The network is trained over a set of examples by adjusting the weights of the interconnections.

Support Vector Machine (SVM)

SVM finds an optimal hyperplane that maximizes the margin between classes. It can effectively perform linear and nonlinear classification. For nonlinear classification, SVM maps the provided data into a higher dimensional space by using a kernel trick.

Logistic Regression (LR)

LR is a well-known statistical technique used to predict the probability of event occurrence by fitting data to a logistic function.

Bayes Point Machine (BPM)

This is a binary classifier that uses a Bayesian approach to linear classification. It approximates the Bayes optimal decision by estimating the mean of the posterior distribution of classifier parameters. Since the BPM is a Bayesian-based technique, it is not prone to overfitting to the training data.

13.4.2 Dataset Description

This chapter presents experimental studies conducted on five different real world related to the different SE tasks, including software defect prediction, software quality evaluation, vulnerability analysis, issue lifetime estimation, and code readability prediction. Table 13.1 presents the main characteristics of the datasets employed in the experiments, in terms of the number of records (instances), the number of attributes (features), and the types of SE problems they deal with. All these datasets include a target attribute having two different class values, so they are proper for binary classification. Brief descriptions about the datasets are given as follows:

- **PC5**: This dataset consists of defect information (buggy or bug free) and various software metrics such as lines of codes, decision count, parameter count, number of operators, and so on. These software metrics were obtained from the functions developed in the NASA flight software for earth orbiting satellite.
- **Mozilla**: This dataset stores the history of modification made to C++ classes in the open-source Mozilla project [47]. The observation period was between May 29, 2002 and Feb 22, 2006. There are 6 attributes in the dataset: class id, start

Table 13.1 Main characteristics of software engineering datasets

Datasets	Type of problem	Number of records	Number of attributes	Description
PC5[1]	Defect detection	17,186	39	Software defect prediction (fault-prone or non-fault-prone)
Mozilla4[2] [47]	Software quality	15,545	6	Recurrent event modeling (1 for the class with an event, else 0)
Moodle[3] [48]	Vulnerability analysis	2942	14	Predicting vulnerable software components
Combined[4] [49]	Issue lifetime	47,516	18	Issue close time prediction (<365 or $\geq$ 365)
Readability[5] [50]	Code readability	360	105	Automatic source code readability estimation

[1] https://zenodo.org/record/268439
[2] https://zenodo.org/record/268450
[3] https://seam.cs.umd.edu/webvuldata/
[4] https://zenodo.org/record/197111
[5] https://dibt.unimol.it/report/readability/

and end time interval, event (1 if a defect fix takes place, 0 otherwise), size (lines of code), and state (1 for the class with an event, 0 otherwise).

- **Moodle**: This dataset contains vulnerability information and computed software metrics for Moodle 2.2.0 web application [48]. Some of the software metrics in the dataset are lines of code, number of methods in a file, total external calls, maximum nesting complexity, Halstead's volume, internal and external functions called. All vulnerabilities in this dataset were verified by the dataset's authors and localized to individual files by hand.
- **Combined**: This dataset contains records for predicting the amount of time required to close issue reports in software repositories [49]. Namely, it contains issue lifetime values collected from 10 large software projects.
- **Readability**: This dataset contains a large set of features (104) for code readability, which are organized into four categories: visual, spatial, alignment, and linguistic [50]. Some of *visual features* are the number of loops, assignments, and parentheses; *spatial features* are the number of comments, keywords, numbers, strings, and literals; *alignment features* are the number of operators and expressions; *linguistic features* are the relative number of identifiers composed by words present in an English dictionary. The dataset was collected for code readability prediction. To collect the dataset, a survey was conducted with 5K+ human annotators judging the readability of 360 code snippets written in three different programming languages (i.e., Java, Python, and CUDA).

13.4.3   Experimental Results

To evaluate the performances of the different classification models, we reported on various well-known metrics such as accuracy, precision, recall, F1-score, receiver operating characteristic (ROC), and area under the curve (AUC). These metrics are explained with their formulas and definitions in Table 13.2. *Accuracy* is the percentage of data which are correctly classified and calculated as (TP + TN)/ (TP + TN + FP + FN), where TP, TN, FP, and FN represent the number of true positives, true negatives, false positives ,and false negatives, respectively. In addition to accuracy, precision and recall are valuable measures to evaluate classifiers. The higher *precision* indicates the capability of avoiding false matches, while the higher *recall* indicates the capability of detecting correct events. *F1-score* is the harmonic mean of precision and recall metrics, and therefore presents a single measure that incorporates both aspects of classifier effectiveness.

Table 13.3 presents the empirical results obtained from experimental studies conducted on five different real-world software engineering datasets. When the average classification accuracies for all used datasets are considered (the last column in Table 13.3), it is possible to say that, DF method outperforms the other algorithms, with an accuracy of 90.25%. This is probably because of its ensemble structure. The DF algorithm works by building a series of decision trees to improve

Table 13.2 Detailed information about classifier performance measures

Measure	Formula	Description
Accuracy	$Acc = \dfrac{TP+TN}{TP+TN+FP+FN}$	The ratio of the number of correctly classified samples to the total number of samples
Precision	$Prec = \dfrac{TP}{TP+FP}$	The ratio of correct assignments of a class to the total number of assignments to that class
Recall	$Rec = \dfrac{TP}{TP+FN}$	The ratio of correct assignments of a class to the total number of test samples from this class
F1-score	$F = \dfrac{2*precision*recall}{precision+recall}$	The harmonic mean of precision and recall
AUC	AUC is the area under the receiver operating characteristic (ROC) curve	
ROC	The curve which is generated to compare correctly and incorrectly classified samples	

Table 13.3 Comparison of classification algorithms in the cloud on software engineering datasets

Algorithms		Datasets					Avg. Acc.
		PC5	Mozilla4	Moodle	Combined	Readability	
Decision forest	Acc.	97.21	94.81	96.79	84.95	77.50	90.25
	Prec.	0.887	0.947	0.954	0.848	0.794	
	Rec.	0.857	0.977	0.825	0.787	0.794	
	F1	0.871	0.962	0.883	0.816	0.789	
	AUC	0.987	0.970	0.982	0.923	0.829	
Neural network	Acc.	95.22	90.21	93.22	79.74	77.50	87.18
	Prec.	0.750	0.920	0.782	0.716	0.735	
	Rec.	0.852	0.935	0.748	0.875	0.910	
	F1	0.797	0.928	0.764	0.785	0.811	
	AUC	0.977	0.948	0.969	0.900	0.819	
Support vector machine	Acc.	93.29	77.03	85.68	71.35	80.00	81.47
	Prec.	0.747	0.799	0.533	0.682	0.791	
	Rec.	0.595	0.879	0.172	0.610	0.848	
	F1	0.661	0.837	0.259	0.644	0.817	
	AUC	0.944	0.859	0.875	0.808	0.869	
Logistic regression	Acc.	93.40	85.34	84.63	75.96	79.72	83.81
	Prec.	0.738	0.891	0.426	0.721	0.800	
	Rec.	0.623	0.891	0.114	0.708	0.827	
	F1	0.675	0.891	0.178	0.714	0.811	
	AUC	0.957	0.888	0.878	0.827	0.869	
Bayes point machine	Acc.	93.45	84.28	85.51	74.90	76.11	82.85
	Prec.	0.742	0.878	0.529	0.711	0.791	
	Rec.	0.623	0.890	0.170	0.689	0.754	
	F1	0.676	0.884	0.256	0.700	0.767	
	AUC	0.959	0.882	0.866	0.823	0.866	

the generalization ability and then voting on the most popular output class. NN has also good performance (87.18%) on the software engineering domain.

According to the results, DF gives the best performance among all of the classifiers, with an accuracy of 97.21%. This result indicates that this classifier can be reliably used to enable the prediction of the presence of defects in software being developed. Similarly, the classification accuracy 96.79% of the DF algorithm indicates that the application can enable software engineers make informed decisions regarding vulnerability so as to meet set their needs. Both DF and NN methods performed well, where their F1 scores all close to 1.

On the code readability dataset, the SVM algorithm obtained the best result (80%). The accuracy rate of LR method is also very close to this value, since it reaches 79.72% on this dataset. On the Mozilla4 dataset, the DF algorithm represented a significant achievement with an accuracy of 94.81%, where the ensemble approach achieved a gain of 17.78% according to the individual algorithm. In general, the BPM algorithm seems not suitable for the software engineering datasets due to its lower accuracies in classifying.

The area under the *ROC curve* is also widely used measure of performance of classifiers. Figure 13.4 shows the ROC curves of the algorithms (DF, NN, SVM and LR from left to right, respectively) for each dataset separately. According the graphs, it is possible to say that all algorithms performed well on the PC5 dataset, while they tended to be wrong on the readability dataset. The DF algorithm is generally better on the datasets compared with the others, with an AUC of 0.94 on average.

Figure 13.5 shows the cumulative ranks of the methods for all SE datasets. In the ranking method, each method was rated according to its classification accuracy performance on the datasets. This process was performed by assigning rank 5 to the most accurate method, rank 4 to the second best, and so on. Thus, the method with high rank has better performance than others. In the case of tie, the average ranking was assigned to each method. The DF algorithm is most often ranked highest. The NN algorithm has also generally good performances since its rank values are generally 4. Compared with other methods, 3 out of the 5 cases the SVM algorithm is ranked as the lowest.

The execution times of the classification algorithms taken for 10-fold cross-validation in the cloud are given in Table 13.4 to compare their effectiveness. This comparison is critical to realize fast and efficient data mining for software data analysis. According to the results, the LR and SVM algorithms are significantly faster than the others. This means that these algorithms can be efficiently used to process big software engineering data and to provide real-time estimations for critical software engineering tasks. In terms of both classification accuracy and runtime efficiency, the DF algorithm can be preferred for mining software engineering data. The results also indicate that the data mining application benefits from flexible computing resources of the cloud environment required for data analysis.

The main findings from these experiments can be summarized as follows:

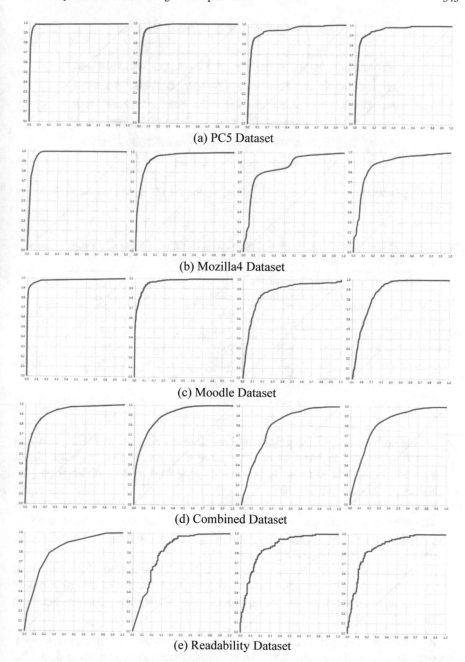

(a) PC5 Dataset

(b) Mozilla4 Dataset

(c) Moodle Dataset

(d) Combined Dataset

(e) Readability Dataset

Fig. 13.4 Comparison of algorithms, including DF, NN, SVM, and LR (from left to right), by ROC Curves for each dataset separately

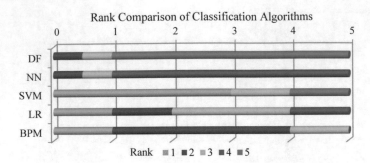

Fig. 13.5 Cumulative ranks of the classification algorithms (5: highest accuracy, 1: lowest accuracy)

Table 13.4 Comparison of classification algorithms on software engineering datasets in terms of runtime efficiency in the cloud

Algorithms	Execution time (sec)					Avg.
	PC5	Mozilla4	Moodle	Combined	Readability	
Decision forest	5.84	6.13	4.06	24.74	4.07	8.97
Neural network	7.92	41.31	13.41	127.88	6.35	39.37
Support vector M.	4.97	4.50	6.96	6.99	4.05	5.49
Logistic regression	5.72	4.57	4.12	6.85	4.17	5.09
Bayes point machine	34.67	9.38	5.54	41.02	5.78	19.28

- On the software engineering domain, ensemble learning methods (i.e., decision forest) often enable users to achieve more accurate predictions with higher generalization abilities than the predictions generated by an individual (single) model in the classification task.
- Only accuracy is not enough to assess the effectiveness of the data mining model constructed for software engineering. Therefore, various measures such as precision, recall, F1-score, AUC, and ROC curve should also be utilized to examine the predictive quality of classifiers.
- When big software engineering data is available, the selection of the algorithm is critical in terms of execution time needed to get result. The cloud environment allows us to generate a data mining model faster, saving considerable execution time.

As shown in the experimental studies, cloud environment provides a centralized, robust, scalable, and powerful platform for mining software engineering data. The data mining application can be provided Software as a Service to software engineers, and do not only include data mining algorithms, but also a unified user interface. Furthermore, the system is scalable enough to allow real-time estimations and the necessary computation power. By being cloud-based, the data mining application is not only well suited for implementation, but also can be efficiently used for fast computing. Finally, cloud platform allows for quick deployment and for easily integration with software engineering data.

In big companies, the main motivation of cloud-based data mining is that project team members are distributed geographically and perform different SE tasks. Cloud platform also makes the application easy to be shared by remote software engineers. Cloud-based data mining application provides centralized and unified access to remote team members for different SE tasks. So, data mining application in the cloud can be used by a number of software engineers with considerable prediction capability.

13.5 Conclusion

In software engineering, it has been proven that data mining improves software productivity and quality. However, mining software engineering data poses several challenges: computationally expensive, requiring large data storage, security mechanism, accessibility from anywhere from any time, and cost efficiency. To overcome these problems, this chapter proposes the utilization of cloud resources in data mining for software engineering since cloud technology provides opportunities for all these issues.

The purpose of this chapter is to provide information about how data mining can be applied on the cloud environment for the software engineering domain through presenting different examples of studies. Mining for software engineering falls into three main categories: *classification*—predicting classes based on already labeled SE data, *clustering*—grouping SE data into clusters, and *association rule mining*— finding correlations between items. This chapter presents the usefulness of data mining in all stages of a software development life cycle—from planning, analysis, design, implementation, and testing to maintenance. It explains the main steps that miners should follow to mine SE data in the cloud, including determining SE task, collecting, preprocessing, mining SE Data, and evaluating/interpreting data mining results.

This chapter presents experimental studies conducted on five different SE datasets related to the various SE tasks, including software defect prediction, software quality evaluation, vulnerability analysis, issue lifetime estimation, and code readability prediction. In the experimental studies, the performances of five classification algorithms (decision forest, neural network, support vector machine, logistic regression, and Bayes point machine) are compared in the cloud in terms of both classification accuracy and runtime efficiency. Experimental results show that the cloud is a powerful platform to build data mining applications for software engineering.

As a future work, instead of using default parameters, the optimal input parameters can be identified for each algorithm and for each dataset. In this way, classification accuracies of the algorithms can be improved.

References

1. Sarkar A, Bhattacharya A, Dutta S, Parikh KK (2019) Recent trends of data mining in cloud computing, proc emerging technologies in data mining and information security (IEMIS 2018). Adv Intell Syst Comput 813:565–578
2. Chen J, Li K, Rong H, Bilal K, Yang N, Li K (2018) A disease diagnosis and treatment recommendation system based on big data mining and cloud computing. Inf Sci 435:124–149
3. Dahmani D, Rahal SA, Belalem G (2016) Improving the performance of data mining by using big data in cloud environment. J Inf Knowl Manage 15(4):2016
4. Rajarajeswari P, Pradeep Kumar J, Vasumathi D (2018) Design and implementation of weather fore casting system based on cloud computing and data mining techniques. Int J Eng Technol 7:219–224
5. Xu H, Fan G (2019) Application of big data mining technology in intelligent safe production on cloud computing platform. Adv Intell Syst Comput 842:1255–1262
6. Marozzo F, Talia D, Trunfio P (2018) A workflow management system for scalable data mining on clouds. IEEE Trans Serv Comput 11(3)
7. Zhou G (2015) Cloud platform based on mobile internet service opportunistic drive and application aware data mining. J Electr Comput Eng, Article no 357378, 21 Jan 2015
8. Minku LL, Mendes E, Turhan B (2016) Data mining for software engineering and humans in the loop. Progr Artif Intell 5(4):307–314
9. Massey AK, Eisenstein J, Anton AI, Swire PP (2013) Automated text mining for requirements analysis of policy documents. In: Proceedings of 21st IEEE international requirements engineering conference (RE 2013), Rio de Janeiro, Brazil, pp 4–13, 15–19 July 2013
10. Ghasemi M (2018), What requirements engineering can learn from process mining. In: Proceedings of 1st international workshop on learning from other disciplines for requirements engineering (D4RE 2018), Banff, Canada, Article number 8595126, pp 8–11, 20 Aug 2018
11. Dwivedi AK, Tirkey A, Rath SK (2018) Software design pattern mining using classification-based techniques. Front Comput Sci 12(5):908–922
12. Hamdy A, Elsayed M (2018) Towards more accurate automatic recommendation of software design patterns. J Theor Appl Inf Technol 96(15):5069–5079, 15 Aug 2018
13. Casamayor A, Godoy D, Campo M (2012) Mining textual requirements to assist architectural software design: a state of the art review. Artif Intell Rev 38(3):173–191
14. Czibula G, Marian Z, Czibula IG (2015) Detecting software design defects using relational association rule mining. Knowl Inf Syst 42(3): 545–577
15. Gilda S (2017) Source code classification using neural networks. In: Proceedings of the 14th international joint conference on computer science and software engineering (JCSSE), Thailand, 12–14 July 2017
16. Zheng W, Zhou H, Li M, Wu J (2019) CodeAttention: translating source code to comments by exploiting the code constructs. Front Comput Sci 1–14
17. Niu H, Keivanloo I, Zou Y (2017) API usage pattern recommendation for software development. J Syst Softw 129:127–139
18. Calefato F, Lanubile F, Maiorano F, Novielli N (2018) Sentiment polarity detection for software development. Empir Softw Eng 23(3):1352–1382
19. Siers MJ, Islam MZ (2018) Novel algorithms for cost-sensitive classification and knowledge discovery in class imbalanced datasets with an application to NASA software defects. Inf Sci 459:53–70
20. Zhu X, Niu B, Whitehead EJ, Sun Z (2018) An empirical study of software change classification with imbalance data-handling methods. Softw Pract Exp 48(11):1968–1999
21. Ghaffarian SM, Shahriari HR (2017) Software vulnerability analysis and discovery using machine-learning and data-mining techniques: a survey. ACM Comput Surv 50(4):1–36, , Article number 56
22. Liu W, Liu H (2016) Major motivations for extract method refactorings: analysis based on interviews and change histories. Front Comput Sci 10(4):644–656

23. Mi Q, Keung J, Xiao Y, Mensah S, Gao Y (2018) Improving code readability classification using convolutional neural networks. Inf Softw Technol 104:60–71

24. Bishnu PS, Bhattacherjee V (2016) Software cost estimation based on modified K-Modes clustering Algorithm. Nat Comput 15(3):415–422

25. Dejaeger K, Verbeke W, Martens D, Baesens B (2012) Data mining techniques for software effort estimation: a comparative study. IEEE Trans Softw Eng 38(2):375–397

26. Zavvar M, Yavari A, Mirhassannia SM, Nehi MR, Yanpi A, Zavvar MH (2017) Classification of risk in software development projects using support vector machine. J Telecommun Electron Comput Eng 9(1):1–5

27. Gilal AR, Jaafar J, Capretz LF, Omar M, Basri S, Aziz IA (2018) Finding an effective classification technique to develop a software team composition model. J Softw Evolut Process 30(1):1–12

28. Shawky DM, Abd-El-Hafiz SK (2016) Characterizing software development method using metrics. J Softw Evolut Process 28(2):82–96

29. Mauša G, Galinac Grbac T (2017) Co-evolutionary multi-population genetic programming for classification in software defect prediction: an empirical case study. Appl Soft Comput 55:331–351

30. Iwata K, Nakashima T, Anan Y, Ishii N (2017) Machine learning classification to effort estimation for embedded software development projects. Int J Softw Innov 5(4):19–32

31. Werner CM, Berry DM (2017) An empirical study of the software development process, including its requirements engineering, at very large organization: how to use data mining in such a study. In: Proceedings of 4th symposium on Asia-Pacific requirements engineering symposium (APRES 2017), Melaka, Malaysia, 9–10 Nov 2017. Communications in Computer and Information Science, vol 809, pp 15–25

32. Pandey N, Sanyal DK, Hudait A, Sen A (2017) Automated classification of software issue reports using machine learning techniques: an empirical study. Innov Syst Softw Eng 13 (4):279–297

33. Chaudhary P, Singh D, Sharma A (2016) Classification of software project risk factors using machine learning approach. In: Intelligent systems technologies and applications, pp 297–309

34. Scalabrino S, Linares-Vásquez M, Oliveto R, Poshyvanyk D (2018) A comprehensive model for code readability. J Softw Evolut Process 30(6):1–23

35. Hussain S, Keung J, Sohail MK, Khan AA, Ilahi M (2019) Automated framework for classification and selection of software design patterns. Appl Soft Comput 75:1–20

36. Ackermann C, Cleaveland R, Huang S, Ray A, Shelton C, Latronico E (2010) Automatic requirement extraction from test cases. In: Barringer H et al (eds) Runtime verification (RV 2010), vol 6418. Lecture notes in computer science. Springer, Berlin, pp 1–15

37. Sartipi K, Safyallah H (2010) Dynamic knowledge extraction from software systems using sequential pattern mining. Int J Softw Eng Knowl Eng 20(6):761–782

38. Lu H, Wang L, Ye M, Yan K, Jin Q (2018) DNN-based image classification for software GUI testing. In: Proceedings of IEEE SmartWorld, ubiquitous intelligence & computing, advanced & trusted computing, scalable computing & communications, cloud & big data computing, internet of people and smart city innovation, 8–12 Oct 2018, Guangzhou, China, pp 1818–1823

39. Jiang Y, Huang J, Ding J, Liu Y (2014) Method of fault detection in cloud computing systems. Int J Grid Distrib Comput 7(3):205–212

40. Okumoto K, Asthana A, Mijumbi R (2017) BRACE: cloud-based software reliability assurance. In: Proceedings of IEEE 28th international symposium on software reliability engineering workshops, Toulouse, France, 23–26 Oct 2017

41. Ali MM, Huda S, Abawajy J, Alyahya S, Al-Dossari H, Yearwood J (2017) A parallel framework for software defect detection and metric selection on cloud computing. Cluster Comput 20(3):2267–2281

42. Baitharu TR, Pani SK (2013) A survey on application of machine learning algorithms on data mining. Int J Innov Technol Explor Eng 3(7), December 2013

43. Halkidi M, Spinellis D, Tsatsaronis G, Vazirgiannis M (2011) Data mining in software engineering. Intell Data Anal 15:413–441
44. Alzubi J, Nayyar A, Kumar A (2018) Machine learning from theory to algorithms: an overview. J Phys: Conf Ser 1142
45. Malhotra R (2015) A systematic review of machine learning techniques for software fault prediction. Appl Soft Comput 27:504–518
46. Azeem MI, Palomba F, Shi L, Wang Q (2019) Machine learning techniques for code smell detection: a systematic literature review and meta-analysis. Inf Softw Technol 108:115–138
47. Koru G, Zhang D, Liu H (2007) Modeling the effect of size on defect proneness for open-source software. In: Proceedings of 3rd international workshop on predictor models in software engineering (PROMISE'07: ICSE Workshops 2007), Minneapolis, MN, USA, 20–26 May 2007
48. Walden J, Stuckman J, Scandariato R (2014) Predicting vulnerable components: software metrics vs text mining. In: Proceedings of IEEE 25th international symposium on software reliability engineering (ISSRE), Naples, Italy, pp 23–33, 3–6 Nov 2014
49. Rees-Jones M, Martin M, Menzies T (2017) Better predictors for issue lifetime CoRR abs/170207735
50. Dorn J (2012) A general software readability model Master's Thesis. University of Virginia, Department of Computer Science. Accessed 12 Apr 2019. http://www.cs.virginia.edu/%jad5ju/publications/dorn-mcs-paper.pdf

Index

© Springer Nature Switzerland AG 2020
M. Ramachandran and Z. Mahmood (eds.), *Software Engineering in the Era of Cloud Computing*, Computer Communications and Networks,
https://doi.org/10.1007/978-3-030-33624-0

351

Printed in the United States
By Bookmasters